The Longman
Literary Companion to Science

The Longman Literary Companion to Science

EDITED BY

WALTER GRATZER

Longman Group UK Limited
Longman House, Burnt Mill, Harlow
Essex CM20 2JE, England
and Associated Companies throughout the world

First Published in 1989

British Library Cataloguing in Publication Data

The Longman Literary companion to science.
1. English literature. Special subjects: Science – Anthologies
I. Gratzer, Walter
820, 8′0356

ISBN 0-582-03372-1

Set in Linotron 202 10/12pt Bembo

Printed and Bound in Great Britain
at the Bath Press, Avon

Acknowledgements

We are grateful to the following for permission to reproduce copyright material:

the author's agent on behalf of the Estate of Nigel Balchin for an extract from *The Small Back Room* & an extract from *A Sort of Traitors* by Nigel Balchin; the author's agents for an extract from *Kepler* by John Banville, © John Banville 1981; the author's agents for extracts from *The Weeping Wood* & an extract from *Helene* by Vicki Baum; The Bodley Head Ltd & Harper & Row, Inc for the story 'The Gold-Makers' in *The Inequality of Man* by J. B. S. Haldane; The Bodley Head Ltd & Summit Books, Inc, a Division of Simon & Schuster, Inc for an extract from *If this is a Man* by Primo Levi, US title *Survival in Auschwitz* by Primo Levi, translated by Ruth Feldman, copyright © 1985 Summit Books, Inc; The Boxwood Press & the author, Ralph A. Lewin, for the poem 'Elks, Welks & Their Ilk' in *Biology of the Algae & Diverse Other Verses*, copyright 1987 by The Boxwood Press; Cambridge University Press for extracts from *What Little I Remember* by O. R. Frisch; Jonathan Cape Ltd & the author, Lionel Davidson, for an extract from *The Sun Chemist*; Jonathan Cape Ltd & Dell Publishing Co for 'Address to the American Physical Society' in *Wampeters, Foma & Granfalloons* by Kurt Vonnegut; Jonathan Cape Ltd & Harcourt Brace Jovanovich, Inc for extracts from *Arrowsmith* by Sinclair Lewis, US copyright 1925 by Harcourt Brace Jovanovich, Inc, renewed 1953 by Michael Lewis; Jonathan Cape Ltd on behalf of the Estate of Robert Frost & Henry Holt & Co, Inc for the poems 'Why Wait for Science' & 'A Wish to Comply' by Robert Frost in *The Poetry of Robert Frost* ed by Edward Connery Latham, US copyright 1947 by Holt, Rinehart & Winston, Inc & renewed 1975 by Lesley Frost Ballantine; Jonathan Cape Ltd & Henry Holt & Co, Inc for an extract from *Asleep in the Afternoon* by E. C. Large; Jonathan Cape Ltd & Charles Scribner's Sons, an imprint of Macmillan Publishing Co, for extracts from *Sugar in the Air* by

E.C. Large, copyright 1937 Ernest Charles Large, renewed © 1965 Ernest Charles Large; Carcanet Press Ltd for the poem 'Note on θ, φ and ψ' by Michael Roberts from *Selected Poems & Prose* edited by Frederick Grubb; Century Hutchinson Publishing Group Ltd & E. P. Dutton & Co, Inc for an extract from *Not By Bread Alone* by Dudinstev; Chatto & Windus Ltd for an extract from *Alexander Fleming* by Gwyn MacFarlane; Chatto & Windus Ltd on behalf of Mrs Laura Huxley & Harper & Row, Inc for extracts from *Antic Hay* by Aldous Huxley, US copyright 1923 by Aldous Huxley, renewed 1951 by Aldous Huxley, an extract from *Point Counter Point* by Aldous Huxley, US copyright 1928 by Aldous Huxley, & an extract from *After Many a Summer* by Aldous Huxley, US copyright 1939 by Aldous Huxley; Chelsea Publishing Co, Inc for extracts from *Quest: An Autobiography* by Leopold Infield, copyright © 1941 & 1980 by Helena Infield; the author's agents for extracts from *Einstein* by Ronald Clark & an extract from *JBS* by Ronald Clark; William Collins Sons & Co Ltd & the author's agents for extracts from *The Tin Men* by Michael Frayn, ©Michael Frayn 1965; Columbia University Press for an extract from *The Rise & Fall of T. D. Lysenko* by Zhores Medvedev, copyright © 1969 Columbia University Press; the author's agents for extracts from *Come Out To Play* by Alex Comfort; Constable & Co Ltd for extracts from *Life of Elie Metchnikoff* by Olga Metchnikoff; the author, William Cooper, for extracts from his books *The Struggles of Albert Woods & Memoirs of a New Man*; J.M. Dent & Sons Ltd for an extract from *Pasquier Chronicles* by Georges Duhamel, trans by Beatrice de Holthoir; André Deutsch Ltd for extracts from *Experiencing Science* by Jeremy Bernstein; André Deutsch Ltd & Alfred A Knopf, Inc for an extract from *Roger's Version* by John Updike, copyright © 1986 by John Updike, an extract from *Couples* by John Updike, copyright © 1968 by John Updike, the poems 'To Crystallization' & 'Ode to Growth' in *Facing Nature* by John Updike, copyright © 1985 by John Updike & 'Cosmic Gall' in *Telephone Poles & Other Poems* by John Updike, copyright © 1960 by John Updike, first published in *The New Yorker*; Diogenes Verlag AG for a translation of an extract from *The Physicists* by Fredrich Dürrenmatt; Gerald Duckworth & Co Ltd & the author's agents for the poem 'The Microbe' by Hillaire Belloc in *More Beasts for Worse Children*; E.P. Dutton & Co, Inc and the author's agents for an extract from *The Scientists* by Eleazar Lipsky; Faber & Faber Ltd for Section VI from the poem 'The Kingdom' in *The Collected Poems of Louis MacNeice* by Louis

MacNeice & an extract from *Experiment in Autobiography* by H. G. Wells; Faber & Faber Ltd & Macmillan Publishing Co for an extract from the poem 'Four Quartz Crystal Clocks' in *The Complete Poems of Marianne Moore* by Marianne Moore, US title *Collected Poems* by Marianne Moore, US Copyright 1941, renewed 1969 by Marianne Moore; the author, Martin Gardner, for extracts from his book *Fads & Fallacies*; Victor Gollancz Ltd for an extract from *Brighter Than a Thousand Suns* by Robert Jungk, trans by J. Cleugh; Victor Gollancz Ltd & Pocket Books, a Division of Simon & Schuster, Inc, for extracts from *Timescape* by Gregory Benford, copyright © by Gregory Benford & Hilary Benford; the author's agents for an extract from *Courier to Peking* by June Goodfield, copyright © 1973 by June Goodfield Toulmin; Mrs Irene Goudsmit for an extract from *Alsos* by Samuel A. Goudsmit, copyright 1947 by Henry Schuman, Inc; the author's agents on behalf of the Executors of the Estate of Robert Graves for Stanza 33 of *The Marmosite's Miscellany* by Robert Graves; Hamish Hamilton Ltd for an extract from *Trial & Error* by Chaim Weizmann; Hamish Hamilton Ltd & The Putnam Publishing Group for extracts from *Most Secret War* by R. V. Jones, copyright © 1978 by R. V. Jones; Harcourt Brace Jovanovich, Inc for extracts from *The Sweeping Wind: A Memoir* by Paul de Kruif, copyright © 1962 by Paul de Kruif; Harper & Row, Inc for extracts from *The Enigmas of Chance* by Marc Kac; Harper & Row, Inc & the author's agents for the poem 'The Miraculous Countdown' in *Everyone But Thee, Me and Thee* by Ogden Nash (Lippencott, 1962), copyright © by Ogden Nash, 1962; Harvard University Press for extracts from *Science & Government* by C. P. Snow, © 1960, 1961 by The Presidents & Fellows of Harvard College, © 1988 by Philip C. H. Snow; William Heinemann Ltd & Harper & Row, Inc for an extract from *The Black Cloud* by F. Hoyle (1957); the author's agents for an extract from 'The Tissue-Culture King' by Julian Huxley & the poem 'Cosmic Death' in *The Captive Shrew & Other Poems* by Julian Huxley; the author's agents for an extract from *The Sleepwalkers* by Arthur Koestler; Macmillan Ltd, London & Basingstoke, for an extract from *The Prof* by Roy Harrod; Macmillan Ltd, London & Basingstoke, & Charles Scribner's Sons, an imprint of Macmillan Publishing Co, for an extract from *The Affair* by C. P. Snow, copyright © 1960 C. P. Snow; Macmillan Ltd, London & Basingstoke, & the author's agents on behalf of the Estate of C. P. Snow for an extract from *A Variety of Men* by C. P. Snow, copyright © 1967 by C. P. Snow

& an extract from *The Search* by C. P. Snow, copyright C. P. Snow; McClelland & Stewart Ltd & the author, Michael Bliss for an extract from *The Discovery of Insulin*; Oberlin College Press for the poem 'Evening in the Lab' by Miroslav Holub, trans by Stuart Friebert & Dana Habova, first published in *Sagittal Section*; Oxford University Press for an extract from *Otto Warburg: Cell Physiologist, Biochemist & Eccentric* by Hans Krebs & an extract from *Memoir of a Thinking Radish* by Peter Medawar © Peter Medawar 1986; Oxford University Press & Prometheus Books for extracts from *Science: Good, Bad & Bogus* by Martin Gardner; Penguin Books Ltd for an extract from *Bouvard & Pecuchet* by Gustave Flaubert, trans by A. J. Krailsheimer, copyright © Alban Krailsheimer, 1976 & the poem 'Suffering' in *Selected Poems* by Miroslav Holub, trans by Ian Milner & George Theiner, copyright © Miroslav Holub, 1967, trans copyright © Penguin Books 1967; Penguin Books Ltd & Schocken Books, Inc, for an extract from *The Periodic Table* by Primo Levi, US trans by Raymond Rosenthal, trans copyright © 1984 by Schocken Books; the author, Paul Preuss, for an extract from his book *Broken Symmetries*, copyright © 1983 Paul Preuss; Random House, Inc & International Creative Management, Inc for extracts from *Nobel Dreams* by Gary Taubes, copyright © 1986 by Gary Taubes;The Royal Society & the author, Sir Mark Oliphant, for an extract from 'Working with Rutherford' in *Royal Society's Notes & Records* Vol 27 (1972); Charles Scribner's Sons, an imprint of Macmillan Publishing Co, for extracts from *Adventures of a Mathematician* by Stanislaw M. Ulam, copyright © 1976, 1983 S. M. Ulam; Sidgwick & Jackson Ltd & Atlantic Monthly Press for extracts from *Invisible Frontiers, The Race to Synthesize a Human Gene* by Stephen S. Hall, copyright © 1987 by Stephen S. Hall;Simon & Schuster, Inc & the author's agents for extracts from *Betrayers of the Truth* by William Broad & Nicholas Wade, copyright © 1982 by W. Broad & N. Wade; The Society of Authors on behalf of the Bernard Shaw Estate for extracts from *In Good King Charles's Golden Days* and *The Doctor's Dilema* by George Bernard Shaw;The Society of Authors as the literary representative of the Estate of John Masefield & Macmillan Publishing Co for extracts from *Multitude & Solitude* by John Masefield; The Souvenir Press Ltd for extracts from *Strong Medicine* by Arthur Hailey; Ticknor & Fields, a Houghton Mifflin Co, for extracts from *The Life it Brings: One Physicist's Beginnings* by Jeremy Bernstein, copyright © 1987 by Jeremy Bernstein; University of Central Florida Press for the poem 'Next Slide, Please' by Roald Hoffmann in The Metamict State; Unwin Hyman Ltd for extracts from *La Statue Intérieure* by François Jacob; Unwin Hyman Ltd & W W Norton & Co, Inc for extracts from *Surely You're Joking Mr Feynman!*, Adventures of a Curious Character, by Richard P. Feynman, as told to Ralph Leighton, US copyright © 1985 by Richard P. Feynman & Ralph Leighton & an extract from *What Do* You *Care What Other People Think*?, Further Adventures of a Curious Character, by Richard P. Feynman, as told to Ralph Leighton, US copyright © 1988 Gweneth Feynman & Ralph Leighton; VAAP, Moscow, for a translated extract from *The Geneticist* by Daniil Granin; the author, Kurt Vonnegut, for an extract from his book *Cat's Cradle*; George Weidenfield & Nicolson Ltd for extracts from *The Double Helix* by J. D. Watson; Egon A. Weiss as the literary Executor of Leo Szilard for extracts from *Voice of the Dolphin* by Leo Szilard; the author's agents on behalf of the literary Executors of the Estate of H. G. Wells for 'Chapter 19' in *The Invisible Man*, 'Chapter 1' in *Food of the Gods*, an extract from *Ann Veronica*, an extract from *Love & Mr Lewisham* & the story 'The Moth' in *The Complete Short Stories of H. G. Wells,* all by H. G. Wells.

We have been unable to trace the copyright holders in the following and would appreciate any information that would enable us to do so:

'Pâté de Foie Gras' in *The Edge of Tomorrow* by Isaac Asimov; *Refusnik* by Mark Azbel; *Haphazard Reality* by H. G. B. Casimir; *Lawrence & Oppenheimer* by Nuel Pharr Davis; *My World Line* by George Gamow; *The Walnut Trees of Altenburg* by Andre Malraux; 'Bell & Langley' by Thomas McMahon in *Granta* Issue 16; the poem 'Atomic Architecture' by A. M. Sullivan; a riposte to Alexander Pope by J. C. Squire in *Poems in One Volume; My Life & Hard Times* by James Thurber; *The Nobel Prize Duel* by Nicholas Wade; *Live With Lightning* by Mitchell Wilson.

Contents

Acknowledgements vii
Introduction xv

The Laboratory Scene 1
The Illuminati 85
Science Collides with the World 175
The Thrill of the Chase: Discovery 315
Science and War 393
The Pathology of Science 437

References 497
Index of Names and Institutions 509
Index of Titles 515

Introduction

A while ago there appeared in the personal column of the *New York Review of Books* an ad, which ran something like this:

> Petite, attractive, intelligent WSF, 30, fond of music, theatre, books, travel, seeks warm, affectionate, fun-loving man to share life's pleasures with view to lasting relationship. Send photograph. Please no biochemists.

The imagination may reel but the message is clear: even those who seek their romantic attachments among the readers of the *New York Review of Books* view one of the great intellectual endeavours of our time as no more than the gristle in life's hamburger. For it is surely beyond dispute that science is the cultural activity for which the twentieth century will be pre-eminently remembered. Its Golden Age is in full flower. It is free – the tribulations of technology always excepted – from the crises of purpose and identity that have afflicted Western music, literature, architecture and painting.

It is estimated that nine-tenths of the scientists who have ever lived are alive and working. Scientists, or those with a qualification in science or in some subject in which science has a major part, are much the most abundant products of our educational system. Science then forms part of the experience of a sizeable proportion of the reading public; yet it seems barely to have ruffled the surface of our literature – the mirror of society. Perhaps after all we are not conditioned by our education; perhaps Lady Bracknell was right when she opined that 'in England, at any rate, education produces no effect whatsoever. If it did, it would prove a serious danger to the upper classes, and probably lead to acts of violence in Grosvenor Square.'

The cultivated middle-class (from which most English literature has sprung) has of course always wrinkled its nose at science. The traditional view is encapsulated by Rudyard Kipling:

> There are whose study is of smells,
> And to attentive schools rehearse

How something mixed with something else
Makes something worse.

If this seems merely a simplistic view of organic chemistry, consider Hilaire Belloc on microbiology (in *More Beasts for Worse Children*):

The Microbe is so very small
You cannot make him out at all,
But many sanguine people hope
To see him down a microscope.
His jointed tongue that lies beneath
A hundred curious rows of teeth;
His seven tufted tails with lots
Of lovely pink and purple spots,
On each of which a pattern stands,
Composed of forty separate bands;
His eyebrows of a tender green;
All these have never yet been seen –
But Scientists, who ought to know,
Assure us they must be so. . . .
Oh! let us never, never doubt
What nobody is sure about!

Such amiable disdain has probably now yielded, at least in part, to unease. W. H. Auden said that when he came among scientists he felt like a shabby curate entering a drawing-room full of dukes. More soberly, Lionel Trilling has lamented that physics at least is too intellectually demanding for all but the few, and to be excluded from the aesthetic rapture that it plainly brings to the best physicists is an affront, or as he calls it, 'a wound to our intellectual self-esteem.'

Yet it has to be said that the Victorians, confronted with what must have seemed at the time to be equally challenging concepts, did a whole lot better, and from Prince Albert down followed advances in physics and biology with earnest determination. To scholars of earlier ages, learning was likewise indivisible. Dr Johnson himself, whose name one does not generally associate with science, wrote a life of Boerhaave towards the beginning of his career and there, according to Boswell, 'discovered that love of chymistry which never forsook him.' He had 'an apparatus for chymical experiments' in his garrett in Fleet Street, and during a journey with Boswell was 'chiefly occupied in reading Dr Watson's second volume of *Chemical Essays*'. And here is how the lawyer, politician and man of letters, Lord Brougham, writing at the end of the eighteenth century, responded to a tour around the frontiers of scientific discovery with a celebrated guide, the Scottish chemist, Joseph Black:

> 'The gratification of attending one of Black's last lecture courses exceeded all I have ever enjoyed. I . . . have heard the commanding periods of Pitt's majestic oratory – the vehemence of Fox's burning declamations – but I would without hesitation prefer, for mere intellectual gratification . . . . to be once more allowed the privilege . . . of being present, while the first philosopher of his age was the historian of his own discoveries, and be an eyewitness of those experiments by which he had formerly made them, once more performed by his own hands.'

Even in the very dawn of English literature, science makes its appearance:

> I wol yow telle, as was me taught also,
> The foure spirites and the bodies seven,
> By ordre, as ofte I herde my lord hem nevene.
> The firste spirit quik-silver called is,
> The second orpiment, the thridde, y-wis,
> Sal armoniak, and the firthe brimstoon.
> The bodies sevene eek, lo! hem heer anoon:
> Sol gold is, and Luna silver we threpe,
> Mars yron, Mercurie quik-silver we clepe,
> Saturnus leed, and Jupiter is tin,
> And Venus coper, by my fader kin!

Thus Chaucer in *The Canon's Yeoman's Tale* (anno 1387), giving a fair run-down of inorganic chemistry as it then stood.

The divide between C. P. Snow's two cultures, then, is a twentieth-century manifestation, and it is with us yet (though surely narrowing slowly). Writers of fiction, according to their tastes and experience, have explored practically every human pursuit and walk of life: you will find novels by the score set in coalmines, in lawyers' chambers, in ducal palaces, on the battlefield, in the brothel and in the drawing-room. All these would confront the anthologist with an impossible embarrassment of choice. But what of the laboratory? As I have argued, the average reader is rather more likely to be familiar with laboratories than with solidering, for example, and correspondingly more likely to experience in such a setting that delicious 'shock of recognition', which is one of the prime pleasures that fiction can afford.

There have, to be sure, always been a few writers from whose work science has been inseparable. First among these is H. G. Wells, who not only wrote novels and especially short stories about science, but also saw the popularisation of science (for instance in the form of the two-volume compendium of biology, that he brought out in association with his son, G.P. and Julian Huxley) as his most important mission. Many of the literati among Wells's contemporaries were scornful of his *oeuvre*: Max Beerbohm described his prose as 'cold rice-pudding spilt on the pavements of Gower Street', and this was probably part of the snobbery that science habitually attracted. A critic said of D. H. Lawrence that he led his readers to the latrine and locked them in. Perhaps they liked no better to be locked into a smelly laboratory by Wells.

A notable literary event, which seemed to reviewers in the journals *Science* and *Nature*, in 1925, to mark a new dawn in fiction, was the publication of Sinclair Lewis's magisterial novel, *Arrowsmith*, which owed much to his acquaintance with the Rockefeller Institute. Aldous Huxley's novels of the same period were much influenced by his association with some of the leading scientific thinkers in Britain, such as his brother, Julian, and J. B. S. Haldane. C. P. Snow had been a physical chemist, and the combination of his novels, in which many of the characters were scientists (often readily identifiable), and his utterances about the two cultures (a term that he coined, and invested with the implication that he alone among contemporary writers was a Colossus bestriding both) made him for a while the prophet of a new technological age. William Cooper wrote one admirable scientific novel (better, it seems to me, than anything of Snow's). But the finest literature overtly rooted in science is, I believe, the work of Primo Levi. The impact made by his incomparable collection of linked stories, *The Periodic Table*, surely demonstrated that there is appreciation enough of science as a literary theme. John Updike is another major novelist with an interest in science and a grasp of what it is all about.

Perhaps surprisingly, science appears more widely in poetry than in fiction, and much nineteenth-century verse especially is rich in scientific imagery. Shelley and Coleridge, among many others, were interested in the discoveries that were changing man's view of his world. Coleridge conducted an animated correspondence with Sir Humphry Davy and indeed proposed at one stage to assist him in his laboratory. There is thus an abundance of material from which to select; I have also included some examples of light or even frivolous verse.

My aim in this collection, then, has been to explore the incursions of science into literature. I have included much that is not of the first quality, but is (I hope) redeemed by historical or social interest. Some of the writers on whom I have drawn are now largely (perhaps to some extent deservedly) forgotten, and there are a few others whose sales in paperback have not secured them respectful consideration in our ancient seats of learning. Other works, widely admired in their day, have seemed to me too ponderous and stylized for modern tastes, and also sometimes to resist filleting. (One such is *The Redemption of Tycho Brahe*, by the German novelist Max Brod, in which the character of Kepler is known to represent Einstein.)

I have selected from novels, stories and poetry inspired or influenced by science or scientists. Except where it is touched by science (which happens surprisingly rarely), there is no science fiction. The exploration of imagined worlds – from Jules Verne, through the great futurologist, Tsiolkovsky, to E. H. Abbott and many admirable contemporary writers, such as Olaf Stapledon – was not my concern. They have been abundantly anthologized elsewhere.

I have tried wherever possible to link fictitious to factual descriptions of people or events, to the extent that a correspondence has been established or surmised. I have therefore drawn profusely on biography and memoirs and equally on journalism. I do not feel that the inclusion of the last calls for any apology. Cyril Connolly has defined literature as the art of writing something that will be read twice, and journalism that which will be grasped at once. This strikes me as a false dichotomy. The best journalism has always had literary lustre. It is perhaps worth recalling that the greatest journalist of them all, H. L. Mencken, observed science with a penetrating and baleful eye; it was a typical Mencken shaft to write of the Catholic Church's attitude to birth-control that a recourse to mathematics was permitted, but to resort to physics or chemistry was forbidden.

In recent years a new breed of science writers has emerged, highly informed and close enough to the process of discovery to count as insiders. They may be said to belong to the movement termed by Tom Wolfe The New Journalism. This he defines as history recorded from within, by someone who has lived, and preferably suffered, through the events that he chronicles. It is Wolfe's thesis that much of the best literature now on offer, at least in America, is of this kind – that journalism can now be judged as an art form.

It has been no part of my purpose to present fine writing about science. There are no essays on the nature or the substance of science – no Bacon, no Gilbert White, T. H. Huxley or Lankester, nor yet any of the latter-day masters, such as Medawar, Stephen J. Gould or Lewis Thomas. I could not improve on the many excellent collections that already exist. Needless to say, my selection is personal, and I have no doubt that I shall be found to have omitted many a reader's favourite works. This will assuredly be most often due to sheer ignorance, and I am eager to be educated. I have pestered a large number of friends in the course of my search for material. Many made valuable suggestions, on which I have acted, and I thank them all. The two to whom I must record my special debt are Gilbert Beaven and Julius Marmur. Without them, many of the choicest nuggets would have lain undiscovered – at least by me. My thanks go also to Robyn Marsack, who edited and improved the text and eliminated numerous inaccuracies, false attributions and infelicities.

The Laboratory Scene

1

No better description of life in the laboratory exists than in William Cooper's splendid novel, The Struggles of Albert Woods. *Cooper was a scientist and student of C.P. Snow. He achieves in this story a verisimilitude that brings the smell of the organic chemistry laboratory to one's nostrils. The setting is the Dyson Perrins Laboratory of Oxford University. Cooper has not revealed who the inspirations were for Albert Woods and Dibdin. I have heard it suggested that Albert Woods was E.K. (Sir Eric) Rideal, Professor of Colloid Science in Cambridge, but he seems to me possibly an amalgam of A.R. Todd (Lord Todd of Trumpington, and in Cambridge known as Lord Todd Almighty) and Sir Robert Robinson, for many years Waynflete Professor of Organic Chemistry in Oxford.*

'We must put you one or two irons in the fire', was one of Dibdin's favourite promises and nobody could say he did not fulfil it.

A combination of two things makes a great experimental scientist – an outstanding ability for doing experiments and a sure instinct for knowing which experiments to do.

Dibdin did not possess an outstanding ability for doing experiments. He lacked the quickness of eye, the delicacy of touch, the physical tirelessness that Albert Woods, for instance, was already showing signs of. His pupils tried to keep him away from their benches because ineptitude did not damp his enthusiasm for participating in their experiments.

As for a sure instinct for knowing which experiments to do – how can you judge a man who at one time or another suggests all experiments imaginable, from the inspired to the impossible?

Even Dibdin's enemies, who said he had no capacity at all for doing experiments, had to admit that he was most ingenious in inventing experiments for other people to do. Dibdin's school had more irons in the fire per man than any other school in the country. One of the things with which the department of experimental chemistry hummed was fertility. (To say it hummed a little with dizziness as well is neither here nor there.)

Dibdin had his enemies among the senior chemists of the day. They read with interest the stream of papers that came from his school: often they were impressed by the experiments that had been done. But obstinately they went on repeating that the experiments did not do very much towards elucidating scientific truth. This opinion they would give – justly, had they been correct – as a reason for not electing him to the Royal Society.

Election to a fellowship of the Royal Society is what every research scientist looks on as the most desirable of all signs of recognition. The number of fellows of the society is about five hundred. The number of professional scientists in the country doing original research is of the order of ten thousand, possibly rather less. You may think these figures make the society sound less august than I made out, and the competition for getting into it less fearsome. But it is just this degree of competition that is the most fearsome. If the society were say one-tenth its present size, the odds against election would be so great that all but a few would feel they had not got a chance. But at twenty to one everyone feels he has got a chance: he enters the competition. And so at the age of thirty the thought of being elected is to men like Dibdin, Woods and Smith inspiring: at the age of forty-five the thought of not having been elected is agonizing.

Dibdin's research pupils were aware of his peculiar gifts. They were alternately illumined, hypnotized and maddened by them. They found it difficult to believe that a man they knew so well, whose weaknesses were so innocently paraded before them, was a great scientist. Yet at the bottom of their hearts, whether they liked to admit it or not, they all felt loyalty and a curious confidence in him.

Albert Woods, after a year in the laboratory, had not lost his enthusiastic respect for Dibdin. Refusing to have any irons put in the fire for him, he had obstinately gone on studying non-typical Wurmer–Klaus reactions: it gave a pleasant inflation to his pride to think that Dibdin had neither initiated his experiments nor taken any serious part in them. He felt freer than the others to recognize Dibdin's considerable if peculiar gifts.

'He brings things out of the hat,' Albert was wont to say loyally if he heard Dibdin being criticized. It was a curiously apt expression.

When in the following March Dibdin was not elected to the Royal Society in that year's list, Albert was indignant.

'He brings things out of the hat, and that's more than a lot of the old gentlemen do!'

His stable-mate took the irreverence of the remark ill.

On the whole Albert and Clinton shared their room amicably. Albert had now got an elaborate experiment going which involved working at low pressures: he spent hours pumping out his apparatus and sealing up leaks in the taps and joints with extraordinary care and patience. He was proud of his work, and Dibdin had several times complimented him on it.

One day Albert came back from luncheon rather later than usual. He went straight into his room because it was time to begin a new set of readings. The pump was whirring harmoniously. He was feeling excited and pleased with himself.

To his surprise one of the pressure gauges was giving a completely unexpected indication. He exclaimed in dismay, looked at another one. Also quite wrong. With his eyes popping he ran his fingers over the whole system.

Two of the taps were open and it was obvious that air had been let into the apparatus.

'Sabotage!' he cried.

At that moment Clinton came in.

'Somebody's sabotaged my experiment.'

'Don't look at me, Woods, I haven't been near it.'

Albert was making feverish adjustments. His hair fell over his spectacles as he bent down. 'Give me a hand.'

'What do you want me to do?'

Albert told him. Clinton did it.

'Five weeks' work gone up the spout,' Albert kept saying. His face was given up to expressing chagrin.

The immediate remedies had been administered. They faced each other.

'Who's done it?'

Silence for a moment. Then simultaneously they lifted their noses, Albert his stubby one, Clinton his beaky. They smelt the same thing – pipe-tobacco, fresh on the air, stale in the pipe, immediately identifiable.

'It's him,' Albert cried.

'Agreed,' said Clinton. His eye gleamed. 'He likes taking a hand in experiments.'

'But with my apparatus! Why in Heaven's name with my apparatus?'

'He does it to all of us.'

'Five weeks' work up the spout!'

'He likes to come in and take readings.'

Albert was speechless.

In this silence they heard footsteps outside the door. Albert ran across the room. He saw Dibdin quietly padding up the corridor.

'Dr Dibdin!'

Dibdin turned round. 'Yes.' He took his pipe out of his mouth.

'Someone's interfered with my apparatus.'

'What apparatus?'

'My apparatus – the apparatus I'm working with.'

'When?'

'While I was out at lunch, of course.'

'I'm sorry to hear it, Mr Kelly.'

'My name's not Kelly, it's Woods!'

'I beg your pardon, Mr Woods.'

Albert was beside himself with fury.

Dibdin was recovering from his alarm. 'I know what we'll do,' he said.

'What?'

'I'll come back with you myself and help you to put it right again.'

'No!' burst out Albert. 'Please no! I mean, don't bother, Dr Dibdin.'

Dibdin had put his pipe back again in his mouth. He looked hurt when Albert turned his back on him and went down the corridor to his own room.

It was a bright summer morning. Dibdin was sitting in his room reading a small sheaf of typewritten pages. Sunlight was reflected into his window from the building opposite. Blue pipe-smoke rose into the air. As he read, Dibdin's big sad wily eyes were lit with mild pleasure. The pages were an account of Albert Woods's recent work.

There was a tap on the door and Albert came in.

'Thank you for coming, Mr Woods. Please sit down.'

Dibdin studied the papers for a moment longer. He blew out a cloud of smoke.

At first Albert glanced nervously at Dibdin's face. Dibdin's expression reassured him. Albert settled himself comfortably in his chair and watched, his lips pursed in a smile.

'You know what I wanted to talk to you about, Mr Woods.'

'My paper.'

Dibdin quietly turned over the last page again.

He looked up. 'I think it's a very satisfactory piece of work.'

Albert blushed faintly.

'We've got some interesting irons in the fire here.' Dibdin tapped the page with a blunt not very clean finger. 'Very interesting.'

Albert did not speak but pride welled up in him.

'There are one or two points I want to discuss with you before we go on to publication.'

'Yes,' said Albert, confident they could not be serious.

'We mustn't rush into publication.' Dibdin smiled mischievously.

Albert nodded without much conviction.

'If you'll accept the guiding hand,' said Dibdin, 'of an old hand at the game.'

'Of course,' Albert said.

Dibdin put the papers down on the desk, took his pipe out of his mouth and examined the stem. The window was wide open and the sound of a lab-boy whistling down below came in. The fresh air was warm and summery. Albert was wearing an open-necked shirt with the collar pulled outside his tweed jacket. The gold rims of his spectacles gleamed.

'It's always my practice to go through any paper that one of my pupils is preparing, point by point first of all,' Dibdin said easily. 'It clears the air.' He glanced at Albert. 'You may have noticed that about our papers in this department. The air of them's always clear.' Before Albert could reply he went on: 'Now. Let's start with your preamble.' He smiled friendlily, almost affectionately. 'I think we can make the air of your preamble a little clearer.'

Dibdin was right. Albert's literary style was not sharp, in fact it was often rambling. But that it could be made sharper by Dibdin was a different matter entirely – a matter, one would have thought, for speculation.

Dibdin waved his pipe. 'Just a little, perhaps. But every little helps.'

Albert took Dibdin's first modest emendations quite humbly.

They passed on to the opening section, in which Albert outlined what the experiments were going to be and what his aim was in doing them.

Dibdin proposed to remodel one of the key sentences. At first Albert nodded as humbly as he had done earlier. Then, as Dibdin's voice moved on, it dawned on him that the sense had been changed.

'I beg your pardon, Dr Dibdin. I think we ought to think about the last change you were making. It makes it mean something really different.'

'Does it?' Dibdin looked up with innocent surprise.

Albert looked back with equal, if less innocent surprise.

'Of course it does.'

'You surprise me by saying that, Mr Woods.'

'Really?' Albert could scarcely believe him.

Dibdin looked first slightly taken aback and then shrewd.

Albert began: 'Don't you see that would imply that non-typical reactions don't all involve the same type of energy changes –'

'Don't they?'

'Don't they?' Albert's voice began to thicken: 'Of course they don't. That's the whole point –' He broke off. It suddenly looked to him as if Dibdin could not have understood his work properly at all, in fact had possibly got the work on the whole subject muddled.

They embarked on a long discussion. It was soon heated on Albert's side, evasive on Dibdin's: on both sides, after twenty minutes, it was getting very involved as well.

Albert simply could not tell if Dibdin understood what he was talking about or not. At one moment Dibdin would be saying something penetrating – Albert felt he had learnt something: at the next moment Dibdin would be making a mistake that was childish.

However in the end they came back to Albert's original sentence. Albert sat back trying to conceal his triumph. And Dibdin said with satisfaction:

'You see, Mr Woods, what I mean about clearing the air.' He looked triumphant also. 'Don't be discouraged.' He began to puff his pipe. 'I think we shall make something out of you. That's one of our jobs here.'

Albert's mixture of emotions was almost too much for him to contain. All the same something in Dibdin's tone made Albert glance at him sharply. Dibdin now had a mysterious expression.

'We don't forget the labourer is worthy of his hire,' he said.

'Yes,' said Albert, feeling he was not a labourer and wondering what could be coming next.

'I've been thinking the time will soon be ripe for us to fix you up with a fellowship of a college.'

Albert's heart jumped. 'There's nothing I'd like more.'

'It's time you had a bit of home and beauty.'

Albert was momentarily incoherent with delight. Dibdin caught the word 'grateful'. Then Albert said: 'What do I have to do? Do I have to take any steps?'

'The prize,' said Dibdin.

Albert stopped.

'We'll enter you for the prize fellowship.' Dibdin explained that their college, the one of which he was a fellow and Albert nominally a B.A., awarded each year a prize and research fellowship. Albert had known about it for nearly two years already. The only thing Albert did not know was which year he should enter. The prize depended on the candidate's record of research, plus an essay, plus an interview. The former was the most important.

'Next year,' Dibdin said, cheerfully.

'Not this?' said Albert.

'We mustn't rush it, Mr Woods.'

Albert said nothing. He was compelled to take Dibdin's advice since Dibdin would be one of the judges.

Albert was not cast down. 'By next year my record of research will be more impressive.'

'That's the spirit,' said Dibdin. 'Every little helps.'

The interview had apparently come to an end. Dibdin's pipe had gone out and he was lighting it again. Albert stood up to go. He looked at the papers on Dibdin's desk. Dibdin put down his match in one of the tobacco-tin lids and picked up Albert's typewritten sheets.

'You'll be wanting to polish them, I expect,' he said. 'We're never too proud to polish.'

'Yes,' said Albert respectfully.

There was a hiatus while the papers passed from hand to hand. Albert glanced down – there was a nervous, faintly apprehensive look on Dibdin's face.

'There's one other little matter,' Dibdin said.

'Yes.'

'Just to clear things up.' The look of apprehensiveness vanished.

'Yes.'

Dibdin said: 'I see you've put your own name at the top of your paper, Mr Woods.' His eyes looked sad and thoughtful. 'I always make it a matter of principle to put my name as well on every paper that comes out of the department.'

'Yours?' Albert said incredulously.

'Yes,' said Dibdin, still sad and thoughtful. 'I make it a matter of principle, Mr Woods. And I like my name to come first – it makes it easier for purposes of identification.' He rounded it off. 'First come, first served.'

For a moment Albert could not speak. His face reddened, his eyes flickered behind his spectacles.

'I see.'

Albert took the papers from Dibdin's hand, and lest he should say anything more rushed out of the room.

Albert went straight to his own room. Clinton Smith looked up from his work. Albert's face was as red as a turkey-cock's. Clinton said:

'What's the matter with you?'

'The ——!' The word that made his mother scream burst out. The flat 'midland u' and the soggy 'g's' echoed round the Oxford laboratory. 'The ——! He's going to put his name on my paper. On my paper, I tell you.' Albert waved the paper in Clinton's face.

'It's perfectly in order if he wants to,' Clinton said. 'Most heads of departments do it.'

'That's the way they make their reputations, I suppose.'

'I don't propose to give you my opinion on that.'

Albert sat down. 'But on my paper, Smith. Don't you understand? I'd started this work before I ever came here.'

'What's that got to do with it?'

'What that's got to do with it is that I doubt if he even understands it. He's going to put his name on it when he doesn't understand what the theory's about and when the only time he touches the apparatus it goes up the spout for a month!' He was breathless.

Clinton maintained his fixed look.

There was a pause.

Clinton said curtly: 'What else did he say?'

'He advised me to go in for the college's ruddy prize fellowship.'

'Did he?'

Albert started at the tone. He looked at Clinton. 'Did he advise you to go in for it, too?'

'Yes.'

'Which year?'

Clinton did not answer. He stared at Albert with bright eyes under hooded lids.

Albert said: 'He advised me to wait till next year.'

'That's the year he advised me to enter.'

2

Here, from Invisible Frontiers, *Stephen Hall's account of the race to produce the first genetically cloned human protein, insulin, is a notion of how Woods felt when Dibdin scooped the honours.*

Axel Ullrich and Goodman were standing together in the hallway when who should round the corner with Herb Heyneker but Wally Gilbert, the rival whose progress on insulin—real or rumored—had tormented their lives for months. Gilbert walked directly up to Goodman and, from one man of few words to another, said, "Tell me about insulin."

Goodman said nothing. Caught off guard by Gilbert's sudden appearance, he just stood there. "Howard was a terrible communicator in the best of times," John Shine recalls, "and Wally just put him on the spot." "It was really awkward," Ullrich recalls, "because nobody really said anything. Nobody wanted to confirm it." Ullrich and Shine made futile attempts at small talk. A look of concern flickered across Gilbert's face. He could later dismiss the strange interlude as "just a general side of the competition," but undoubtedly Goodman's agitated reaction registered, in its own way, as silent confirmation of the UCSF success. Goodman stared back, saying nothing, looking nervous and discomfited. Finally he "just took off," as one observer recalls, without even responding to Gilbert's appeal to tell him about insulin.

Goodman had the reputation for being exceedingly reticent. In this case it was as if there was too much to say.

Around one in the afternoon on May 23, with full-throated ease, Howard Goodman and William Rutter and other members of the UCSF rat insulin team met the press for the first time. Because insulin continued to be a magic word, the response was almost overwhelming.

There can be no doubt that there was considerable political advantage to publicizing the work. Although the actual paper had not yet been published, it was due to appear within weeks in *Science*. The speed with which it was rushed into print (with some unusually significant typographical errors) suggested that there were political overtones to the editorial handling of the work at the magazine—that the experiment was an important stepping-stone to gaining public acceptance of recombinant DNA research and should be publicized without delay. When Rutter and Goodman reported to the assembled press on May 23 that they had succeeded in cloning the rat insulin gene, it sent the right kind of ripples through the rest of the country.

Accounts of the work appeared on the front pages of the *New York Times, Washington Post*, and *Los Angeles Times* the next day. The *San Francisco Chronicle* said the work "offers proof that the much-debated field of combining genes from unrelated organisms can in fact provide crucial insights into the ways by which chemistry governs heredity." The British magazine *New Scientist* observed, "The California team's experiment will stand as a landmark in genetic engineering research." The scientists were asked if the work followed the NIH guidelines; no one volunteered an account of the pBR322 interlude. Given the controversy that would erupt within months, it would have been simple to acknowledge the mistake, perhaps—as Chirgwin suggests—in a footnote to the paper. Instead, as the *San Francisco Examiner* reported, "The UC group emphasized that the gene-isolation work was carried out according to the safety guidelines put forth last year by the National Institutes of Health."

One of the more optimistic comments to emerge from the press conference was Rutter's assessment of the next step. Rat insulin could be *expressed*—that is, produced by bacteria—within six months, he contended, and expression of human insulin was only a year or two farther down the road. Inserting the human gene into bacterial cells was P4 work, requiring the equivalent of a biological warfare laboratory; Rutter said that if necessary, his team would go to the Army's Fort Detrick facility. The next stages of the work were laid out.

For all its superficial success, the press conference created very real problems at UCSF. "There was a feeling of unreality about the whole thing," recalls Chirgwin; other workers in the Rutter group were, in his opinion, doing equal or better work, and the burst of publicity "made me feel a little awkward." He was not alone. No other phenomenon has come to so sharply define tensions between scientists, particularly that intimate enmity between lab directors and the postdoctoral fellows who work with them, as the press conference. Public recognition of scientific work inevitably involves simplification, and with it, an untidy compression of credit; the process becomes positively slapdash, however, in the setting of the press conference, where reporters represent a broad spectrum of expertise and ignorance, where requests to explain the medical implications of basic research often elicit over-simplified or exaggerated claims, and where journalists, by habit and convention, reflexively cite the head of the participating laboratory for an overview of the work, the summational comment, the key quote. The contributions of younger scientists are invariably compressed, simplified, or sometimes completely ignored when chronicled for popular consumption.

That was the rude lesson on May 23 for the San Francisco postdocs. Axel Ullrich, first author on the paper, was by consensus the key mover in the insulin project. His role, along with the rest of the postdocs, was noted in the lower paragraphs of the ensuing stories—if noted at all. Meanwhile, Goodman, who had been on sabbatical during most of the critical work and, in the view of others in the lab, had hardly expressed enthusiastic support for the project, flew in from out of town for the press conference.

The attention heaped on Rutter and Goodman caused great consternation among the postdocs, and indeed throughout the whole department. "Of course we were not very happy about that," admits one, expressing a widely shared sentiment. "Especially Goodman, who really had *nothing* to do with it. And he took all the credit. It was just incredible." Another member of the insulin team observed that what he unhappily refers to as "the-boss-did-it" syndrome began with the May 23 press conference. Even the lab directors, he says, believed their press clippings, and were swept up in the unexpected fame that public attention conferred. These appear not to be simply postdoctoral plaints; as Brian McCarthy puts it, "They did the work when Howard was gone, and after he had discouraged them from doing it. When he got back, he rushed in and took credit. . . . With respect to Bill, he hadn't given much advice because he wasn't in that area [of research]. . . They have a legitimate grievance."

The person who stood to lose the most credit was of course the person who had done most of the actual cloning work: Ullrich. "At that time," Ullrich concedes, "I was really . . . I was clearly not happy about it, but I thought, 'That's the way it works around here.' But then this continued and became even worse later on." The ill will that developed had a decidedly sour effect on the subsequent collaboration between the Rutter and Goodman laboratories. Indeed, bad feelings began to perfuse the whole department. The rush to clone medically important genes had a "boomtown" feel to it, as one postdoc put it, and some researchers began to complain that increased secrecy reflected attempts at personal financial gain on the part of the researchers. "After the initial successes," Chirgwin recalls, "people were sniffing Nobel Prizes and multimillion-dollar stock options further down the road. The lab chiefs could sort of see that they were going to be in the National Academy, and at least had a shot at Stockholm further down the line. That was when the enticements of power and fame, as a major corrupting force, were just appearing." Rutter concedes that "we hadn't planned on a lot of publicity, and I don't think we handled it particularly well," but insists that the lab chiefs deserve credit for setting up the programs in the first place.

The family atmosphere that had once marked relations was supplanted by factionalism, suspicion, and a certain degree of secrecy. "In 1977 following the cloning of the rat insulin DNA," wrote Keith Yamamoto, a faculty member at UCSF, "a fellow faculty member confided that the ensuing patent application and press coverage seemed to be affecting the motivation of some of his post-doctoral colleagues, and that the potential for financial gain was an apparent consideration in the planning of their experiments." "Most people who have a historical perspective will tell you that there were really two phases to the department," says Brian McCarthy. "Before recombinant DNA and after recombinant DNA." The watershed event, McCarthy says, was probably the rat insulin experiment.

3

And again:

Gilbert instructed Efstratiadis to call *Boston Globe* science writer Robert Cooke and give him the story. It was on page one of the *Globe* the following day.

Once again, the word "insulin" worked its magic. Press reaction was swift and overwhelming. The Harvard experiment received widespread coverage in newspapers, on the radio, and in the scientific press. *The Wall Street Journal* acknowledged that it was "a major step toward producing human insulin for diabetics by bacteria"; *Chemical and Engineering News*, an authoritative trade magazine, reported the result as if handicapping a race. "Recombinant DNA research advocates have been promising insulin production as one of the practical achievements that the new field might bring," began the account. "And several research groups in the U.S. and Canada are speeding toward that end of making insulin in the test tube with a little help from some microbes. A team of scientists led by Dr. Walter Gilbert of Harvard University has just edged into the lead." The Harvard researchers, however, learned the same hard lessons as the UCSF postdocs had a year earlier. "Everyone was very disappointed by the *Globe* article," Efstratiadis says, "because although I had organized the research, Wally's name was all over the place and our names were at the bottom of an inside page."

4

The view of the research supervisor as a monstrous incubus on his protégés is prevalent in the literature and journalism of science. Gregory Benford sets his novel, Timescape, *partly in the Caltech of 1962, partly in the decaying Cambridge (England) of some years hence (1998). It is pleasantly packed with circumstantial detail.*

California trades bucks for brains. They'd gotten Herb York, who used to be Deputy Director of the Defense Department, to come in as the first Chancellor of the campus. Harold Urey came, and the Mayers, then Keith Brueckner in nuclear theory, a trickle of talent that was now turning into a steady stream. In such waters a fresh Assistant Professor had all the job security of live bait.

Gordon walked down the third floor hallways, looking at the names on the doors. Rosenbluth, the plasma theorist some thought was the best in the world. Matthias, the artist of low temperatures, the man who held the record for the superconductor with the highest operating temperature. Kroll and Suhl and Piccioni and Feher, each name summoning up at least one incisive insight, or brilliant calculation, or remarkable experiment. And here, at the end of the fluorescent and tiled sameness of the corridor: Lakin.

"Ah, you received my note," Lakin said when he answered Gordon's knock. "Good. We have decisions to make."

Gordon said, "Oh? Why?" and sat down across the desk from Lakin, next to the window. Outside, bulldozers were knocking over some of the eucalyptus trees in preparation for the chemistry building, grunting mechanically.

"My NSF grant is coming up for renewal," Lakin said significantly.

Gordon noticed that Lakin did not say "our" NSF grant, even though he and Shelly and Gordon were all investigators on the grant. Lakin was the man who okayed the checks, the P.I. as the secretaries always put it – Principal Investigator. It made a difference. "The renewal proposal isn't due in until around Christmas," Gordon said. "Should we start writing it this early?"

"It's not writing I'm talking about. What are we to write *about?*"

"Your localized spin experiments –"

Lakin shook his head, a scowl flickering across his face. "They are still at an exploratory stage. I cannot use them as the staple item."

"Shelly's results –"

"Yes, they are promising. Good work. But they are still conventional, just linear projections of earlier work."

"That leaves me."

"Yes. You." Lakin steepled his hands before him on the desk. His desk top was conspicuously neat, every sheet of paper aligned with the edge, pencils laid out in parallel.

"I haven't got anything clear yet."

"I gave you the nuclear resonance problem, plus an excellent student – Cooper – to speed things up. I expected a full set of data by now."

"You know the trouble we're having with noise."

"Gordon, I didn't give you that problem by accident," Lakin said, smiling slightly. His high forehead wrinkled in an expression of concerned friendliness. "I thought it would be a valuable boost to your career. I admit, it is not precisely the sort of apparatus you are accustomed to. Your thesis problem was more straightforward. But a clean result would clearly be publishable in *Phys Rev Letters*, and that could not fail to help us with our renewal. And you, with your position in the department."

Gordon looked out the window at the machines chewing up the landscape, and then back at Lakin. *Physical Review Letters* was the prestige journal of physics now, the place where the hottest results were published in a matter of weeks, rather than having to wait at *Physical Review* or, worse, some other physics journal, for month after month. The flood of information was forcing the working scientist to narrow his reading to a few journals, since each one was getting thicker and thicker. It was like trying to drink from a fire hose. To save time you began to rely on quick summaries in *Physical Review Letters* and promised yourself you would get around to reading the longer journals when there was more time.

"That's all true," Gordon said mildly. "But I don't have a result to publish."

"Ah, but you do," Lakin murmured warmly. "This noise effect. It is most interesting."

Gordon frowned. "A few days ago you were saying it was just bad technique."

"I was a bit temperamental that day. I did not fully appreciate your difficulties." He combed long fingers through his thinning hair, sweeping it back to reveal white scalp that contrasted strongly with his deep tan. "The noise you have found, Gordon, is not a simple aggravation. I believe, after some thought, that it must be a new physical effect."

Gordon gazed at him in disbelief. "What kind of effect?" he said slowly.

"I do not know. Certainly something is disturbing the usual nuclear resonance process. I suggest we call it 'spontaneous resonance' just to have a working name." He smiled. "Later, if it proves as important as I suspect, the effect may be named for you, Gordon – who knows?"

"But Isaac, we don't understand it! How can we call it a name like that? 'Spontaneous resonance' means something inside the crystal is causing the magnetic spins to flip back and forth."

"Yes, it does."

"But we don't *know* that's what's happening!"

"It is the only possible mechanism," Lakin said coolly.

"Maybe."

"You do not still treasure that signal business of yours, do you?" Lakin said sarcastically.

"We're studying it. Cooper is taking more data right now."

"That is nonsense. You are wasting that student's time."

"Not in my judgment."

"I fear your 'judgment' is not the only factor at work here," Lakin said, giving him a stony look.

"What does that mean?"

"You are inexperienced at these matters. We are working under a deadline. The NSF renewal is more important than your objections. I dislike putting it so bluntly, but –"

"Yes, yes, you have the best interests of the entire group in mind."

"I do not believe I need my sentences finished for me."

Gordon blinked and looked out the window. "Sorry."

There was a silence into which the grating of the bulldozers intruded, breaking Gordon's concentration. He glanced into the stand of jacaranda trees further away and saw a mechanical claw rip apart a rotten wooden fence. It looked like a corral, an aged artifact of a western past now fading. On the other hand it was more probably a remnant of the Marine land the University had acquired. Camp Matthews, where foot soldiers were pounded into shape for Korea. So one training center was knocked down and another reared up in its place. Gordon wondered what he was being trained to fight for here. Science? Or funding?

"Gordon," Lakin began, his voice reduced to a calming murmur, "I don't think you fully appreciate the significance of this 'noise problem' you're having. Remember, you do not have to understand everything about a new effect to discover it. Goodyear found how to make tough rubber accidentally by dropping India rubber mixed with sulfur on a hot stove. Roentgen found *x*-rays while he was fumbling around with a gas-filled electrical discharge experiment."

Gordon grimaced. "That doesn't mean everything we don't understand is important, though."

"Of course not. But trust my judgment in this case. This is exactly the sort of mystery that *Phys Rev Letters* will publish. And it will bolster our NSF profile."

Gordon shook his head. "I think it's a signal."

"Gordon, you will come up for review of your position this year. We can advance you to a higher grade of Assistant Professor. We could even conceivably promote you to tenure."

"So?" Lakin hadn't mentioned that they could also, as the bureaucratese went, give him a "terminal appointment."

"A solid paper in *Phys Rev Letters* carries much weight."

"Uh huh."

"And if your experiment continues to yield nothing, I am afraid I will, regretfully, not have very much evidence to present in support of you."

5

The agony of the graduate student, struggling for survival in an unfriendly environment, emerges in Invisible Frontiers. *The Professor here appears as no 'guide, philosopher and friend' to his student. Leon Lederman, the 1988 Nobel Laureate in Physics, says of the research students in physics: 'Night is the time that graduate students are on shift, and they don't lie; they haven't learned yet.'*

In order to escape Gilbert's hectoring, Fuller took to working on his experiments at night, spending as little of the day as possible in the Bio Labs. He felt under enormous pressure to have his blunt-end cloning procedures in working order for the insulin project, for Gilbert had arranged to use the laboratory of Susumu Tonegawa in Basel, Switzerland, that summer to attempt the insulin cloning (Harvard still didn't have its P3 facility). Fuller remembers, "Wally keeps asking me, 'Are you ready yet?' And I keep saying, 'Another four weeks . . .' or 'Another two weeks. . . .' " In truth, Fuller's experiments had not achieved any consistency. Sometimes the blunt-end cloning technique appeared to work, sometimes it didn't.

All of Fuller's bad luck boiled to a head in June 1977 as he made final preparations for the trip to Basel. Efstratiadis had slaved months to prepare one microgram of rat insulin cDNA—one *millionth* of a gram—and this he gave to Fuller for further workup prior to the trip. To this minuscule amount, Fuller planned to attach linkers. This would give the rat DNA a set of couplers on either end. It would be this rat insulin "gene"—a complementary DNA with linkers on either end—that he would plug into bacteria while he was in Basel.

At one stage in these preparations, the material had to be placed in a vacuum desiccator, an instrument that dries materials under a vacuum. The DNA rested in an Eppendorf tube—an inch-long vessel. Ordinarily, the tube would have a cap of Parafilm (a plastic similar to Saran Wrap) with holes poked in it, but Fuller admits he had neglected this step, and when another Harvard researcher went to retrieve material from the same drying chamber, both the vacuum and Fuller's hopes suffered calamitous rupture.

"He popped open the drying chamber too quickly," Fuller recalls, "and the air came in and threw out two-thirds of this cDNA into this alkali. It was on the bottom as a drying agent, okay? So this is how things go: here was this *gold* sitting there, and I hadn't covered it." Horrified, Fuller realized that more than half of the precious genetic payload for their experiment in Basel had vanished in a gust of man-made wind. "It was all I could do to keep myself calm." he recalls.

Questions raced through his mind. "What am I doing? What have I just done? It's my fault, you know. I can't tell Wally. Wally can't know this." Gilbert, Fuller admits now, never learned about the mishap. Fuller later regarded the incident as "probably the worst day of my life at that point, I think. Not losing out to these guys in California. It was losing that two-thirds of the material."

As it was, relations were tense. They were due to leave for Basel in a matter of days. Now some 0.6 micrograms of Efstratiadis's DNA was gone. Quietly, without telling anyone, Forrest Fuller put linkers on what remained of the DNA and then girded himself for what promised to be a difficult month in Switzerland.

6

Robert Frost does not lie either. Here, in 'A Wish to Comply', is his reaction to viewing the moving oil drops in R.A. Millikan's apparatus for measuring the charge on the electron. Millikan got the Nobel Prize (see no. 87), but a recent examination of his notebooks suggests that he reported what he saw a good deal less truthfully than did Robert Frost.

A WISH TO COMPLY

Did I see it go by,
That Millikan mote?
Well, I said that I did.
I made a good try.
But I'm no one to quote.
If I have a defect
It's a wish to comply
And see as I'm bid.
I rather suspect
All I saw was the lid
Going over my eye,
I honestly think
All I saw was a wink.

7

James Thurber had rather similar difficulties, memorably set down in My Life and Hard Times, *from which the following extract is taken. (Thurber's problem of course was bad eyesight. In a letter to Dr Bruce, his oculist, he once wrote: When it comes to a comparison with the lower animals, I have found that I can see just about as well as the water buffalo, one of the few animals that can lick a tiger. Up to now the zoologists have believed that the tiger's stripes form protective coloring for him. As a matter of fact, he is striped so that the water buffalo can see him coming. This is my own contribution to the science of zoology and none of my colleagues know about it yet.)*

I passed all the other courses that I took at my University, but I could never pass botany. This was because all botany students had to spend several hours a week in a laboratory looking through a microscope at plant cells, and I could never see through a microscope. I never once saw a cell through a microscope. This used to enrage my instructor. He would wander around the laboratory pleased with the progress all the students were making in drawing the involved and, so I am told, interesting structure of flower cells, until he came to me. I would just be standing there. "I can't see anything," I would say. He would begin patiently enough, explaining how anybody can see through a microscope, but he would always end up in a fury, claiming that I could *too* see through a microscope but just pretended that I couldn't. "It takes away from the beauty of flowers anyway," I used to tell him. "We are not concerned with beauty in this course," he would say. "We are concerned solely with what I may call the *mechanics* of flars." "Well," I'd say, "I can't see anything." "Try it just once again," he'd say, and I would put my eye to the microscope and see nothing at all, except now and again a nebulous milky substance—a phenomenon of maladjustment. You were supposed to see a vivid, restless clockwork of sharply defined plant cells. "I see what looks like a lot of milk," I would tell him. This, he claimed, was the result of my not having adjusted the microscope properly, so he would readjust it for me, or rather, for himself. And I would look again and see milk.

I finally took a deferred pass, as they called it, and waited a year and tried again. (You had to pass one of the biological sciences or you couldn't graduate.) The professor had come back from vacation brown as a berry, bright-eyed, and eager to explain cell-structure again to his classes. "Well," he said to me, cheerily, when we met in the first laboratory hour of the semester, "we're going to see cells this time, aren't we?" "Yes, sir," I said. Students to right of me and to left of me and in front of me were seeing cells; what's more, they were quietly drawing pictures of them in their notebooks. Of course, I didn't see anything.

"We'll try it," the professor said to me, grimly, "with every adjustment of the microscope known to man. As God is my witness, I'll arrange this glass so that you see cells through it or I'll give up teaching. In twenty-two years of botany, I—" He cut off abruptly for he was beginning to quiver all over, like Lionel Barrymore, and he genuinely wished to hold onto his temper; his scenes with me had taken a great deal out of him.

So we tried it with every adjustment of the microscope known to man. With only one of them did I see anything but blackness or the familiar lacteal opacity, and that time I saw, to my pleasure and amazement, a variegated constellation of flecks, specks, and dots. These I hastily drew. The instructor, noting my activity, came back from an adjoining desk, a smile on his lips and his eyebrows high in hope. He looked at my cell drawing. "What's that?" he demanded, with a hint of a squeal in his voice. "That's what I saw," I said. "You didn't, you didn't, you *did*n't!" he screamed, losing control of his temper instantly, and he bent over and squinted into the microscope. His head snapped up. "That's your eye!" he shouted. "You've fixed the lens so that it reflects! You've drawn your eye!"

Another course that I didn't like, but somehow managed to pass, was economics. I went to that class straight from the botany class, which didn't help me any in understanding either subject. I used to get them mixed up. But not as mixed up as another student in my economics class who came there direct from a physics laboratory. He was a tackle on the football team, named Bolenciecwcz. At that time Ohio State University had one of the best football teams in the country, and Bolenciecwcz was one of its outstanding stars. In order to be eligible to play it was necessary for him to keep up in his studies, a very difficult matter, for while he was not dumber than an ox he was not any smarter. Most of his professors were lenient and helped him along. None gave him more hints, in answering questions, or asked him simpler ones than the economics professor, a thin, timid man named Bassum. One day when we were on the subject of transportation and distribution, it came Bolenciecwcz's turn to answer a question. "Name one means of transportation," the professor said to him. No light came into the big tackle's eyes. "Just any means of transportation," said the professor. Bolenciecwcz sat staring at him. "That is," pursued the professor, "any medium, agency, or method of going from one place to another." Bolenciecwcz had the look of a man who is being led into a trap. "You may choose among steam, horse-drawn, or electrically propelled vehicles," said the instructor. "I might suggest the one which we commonly take in making long journeys across land." There was a profound silence in which everybody stirred uneasily, including Bolenciecwcz and Mr. Bassum. Mr. Bassum abruptly broke this silence in an amazing manner. "Choo-choo-choo," he said, in a low voice, and turned instantly scarlet. He glanced appealingly around the room. All of us, of course, shared Mr. Bassum's desire that Bolenciecwcz should stay abreast of the class in economics, for the Illinois game, one of the hardest and most important of the season, was only a week off. "Toot, toot, too-tooooooot!" some student with a deep voice moaned, and we all looked encouragingly at Bolenciecwcz. Somebody else gave a fine imitation of a locomotive letting off steam. Mr. Bassum himself rounded off the little show. "Ding, dong, ding, dong," he said, hopefully. Bolenciecwcz was staring at the floor now, trying to think, his great brow furrowed, his huge hands rubbing together, his face red.

"How did you come to college this year, Mr. Bolenciecwcz?" asked the professor. "*Chuffa* chuffa, *chuffa* chuffa."

"M'father sent me," said the football player.

"What on?" asked Bassum.

"I git an 'lowance," said the tackle, in a low, husky voice, obviously embarrassed.

"No, no," said Bassum. "Name a means of transportation. What did you *ride* here on?"

"Train," said Bolenciecwcz.

"Quite right," said the professor. "Now, Mr. Nugent, will you tell us——"

8

The most candid description of the relations (to which after that digression we now return) between the research grandee, out for his Nobel Prize, and the armies (at least in particle physics) of his vassals is to be found in Gary Taubes's book, Nobel Dreams, *based on months spent as a fly on the walls of the CERN laboratories in Geneva.*

The meeting room quickly filled to overflowing for the two-thirty meeting. Rubbia uncharacteristically arrived precisely on time. He was in a gracious mood, smiling, accepting the commotion with unusual good humor. The UA1 members joked and babbled and waited for the physics, which as it turned out they would never get to hear.

When the room had settled, Rubbia explained that he had reviewed the topics to be covered and that the list was too long. It could only be covered, he said, if they stayed until midnight, and even if then, they would never do the physics justice. He suggested that they save half the topics for a second meeting at a later date. Then he told them he would make an impromptu agenda, because the agenda was properly the responsibility of his co-spokesman, Antoine Leveque, and he did not know what Leveque had in mind. Leveque, Rubbia told them, had been called away because his mother-in-law had died. She's dead, Rubbia said, deceased. And he reiterated it four or five times.

Acting the parliamentarian, Rubbia asked for motions from the floor: Should they try to discuss all the topics, or should they put half off for another day?

Many of the physicists began to wonder what was brewing. They had seen Rubbia use such excuses before. It was generally understood that Rubbia made every important decision pertaining to the experiment, and never went into a physics meeting unprepared.

Daniel Denegri suspected what was going on. That morning, Rubbia had argued with a young Saclay physicist, Elizabeth Locci, about whether she would present her physics work in the meeting, or simply her work on the calibration of the calorimeters. Several UA1 physicists had seen Locci come out of his office, apparently infuriated, her graduate student in tears. Whatever had happened, Rubbia had made his point.

It was the rule of the collaboration. You build the apparatus, you maintain it, you recalibrate it each year. The calorimeters were French, so the French had to do the calibration. It was as critical as it was difficult and painstaking. Without accurate calibration, for instance, the energy of a W particle that the detector recorded at 80 GeV might actually be 70 or 90 or anywhere in between. More importantly, a monojet that appeared to have too much missing energy to be the product of any phenomenon of the standard model might simply be mismeasured. There was no way to tell until the machine was carefully and thoroughly calibrated. Rubbia needed the French calibration work, and that's what he wanted to hear reported.

Denegri knew that Rubbia's disciples had taken the missing-energy analysis back to Boston. If previous history was anything to go by, this would be the analysis that would be discussed in conferences and written into papers. So he assumed that Rubbia had resolved to review only half the topics in the meeting so as to push the Saclay physics work back to the next meeting, which was as yet unscheduled.

Denegri told Rubbia that the French physicists had the right to present physics data, not just calibration data. And Rubbia accused him of trying to turn a physics meeting into a political arena. Later, many of the UA1 physicists would agree with this. Denegri, they said, was out of line. But at the same time, those who knew how deeply the clashes between Rubbia and the Paris contingent had cut, knew that Denegri had to force the issue if he were ever to get a fair hearing. Denegri no longer had Rubbia's ear. If he tried once again to talk to him quietly, he would fail, or so he believed. And he was getting angry.

"It's not fair," Denegri said to Rubbia, his voice rising, "that Felicitas Pauss, Steve Geer, and Jim Rohlf are the only people who are allowed to do physics in UA1."

Rubbia whirled around to the blackboard, and on one side scrawled in large letters the names of Felicitas Pauss, Jim Rohlf, and Steve Geer. On the other side, he wrote "Fair" and circled it. Then he turned toward Denegri and, jabbing his finger at Denegri's face, said; "*You* are accusing me of being unfair. That's a serious accusation."

The show was on. Rubbia asked again how Denegri could claim that Rubbia was attempting to prevent them from speaking, when the agenda had yet to be set. Denegri replied that Rubbia had told Locci that morning that she would be allowed to present only calibration data. Rubbia pivoted, searching for Locci, and when he found her, he asked her to say whether this was true. She calmly replied that he had said to her, no physics . . .

Rubbia interrupted. "*No*," he said, refusing to submit, "don't tell me what I told you."

Locci, under attack, defended herself. She was confused, and maybe a bit scared, but she refused to back down. What she did next pitched the meeting against Rubbia. He began to lose control.

"Now you have insulted me," she told him, "so I have the right to say what I want to say." In the years that she had worked in UA1, she said, Rubbia had kept her from doing physics that she was capable of doing. She was no longer willing to do only what he told her to do. She wanted to present the physics, but she had been kept from doing so because Rubbia refused to let her discuss it in group meetings, saying she had not first briefed him in private. Yet he never seemed to find time to discuss physics with her in private. "That's all I wanted to say," she concluded, "and if you have any answer to that, go ahead."

Rubbia asked if others in the group felt that they were prevented from doing physics. He turned to a Collège de France physicist standing in the back, who had walked in just a few minutes earlier. Rubbia asked him if he agreed with Locci, and the Frenchman said yes. Rubbia was flabbergasted. "How could you agree?" he protested. "You don't even know what's being discussed."

"Basically eighty percent of the audience was on our side," Denegri said later. "This had never happened before. Usually when Rubbia accuses somebody, nobody is on your side. In this case, it was an absolutely unique situation."

Rubbia had lost control. This is ridiculous, he told them, attempting one last rally. I will not subject my group to such a display. He threatened to postpone the meeting until such time as the situation could be clarified. Pointing again at Denegri, Rubbia said that the man had made a serious charge against him and against the three people whose names he had written on the board. He wrote his own name on the board under the list of three people. "Rubbia." We have been insulted as a group, he said, and until these sorts of problems can be solved as a group, we can't discuss physics on any realistic basis.

He walked over to Felicitas Pauss, who sat with her elbows on her knees and her head in her hands. He asked her if she thought she had been insulted. Pauss smiled sadly, shook her head, and said nothing. Rubbia walked back to the board. A serious charge, he repeated. It would have to be discussed at the level of an executive committee meeting, and until then they could not talk about physics.

The meeting was over. Giorgi Salvini, the elder statesman from Italy, tried to salvage it and failed. Horst Wahl, the leader of the Austrian contingent, tried to salvage it, was insulted by Rubbia, and left. Rubbia had decreed that there would be no physics. But his was not the victory.

9

James Clerk Maxwell, one of the cloud-capped geniuses of science, had no need of teams of workers. He was besides a versifier of facility and charm: this is a parody of Burns's 'Coming through the Rye'.

RIGID BODY SINGS

Gin a body meet a body
  Flyin' through the air,
Gin a body hit a body,
  Will it fly? and where?
Ilka impact has its measure,
  Ne'er a' ane hae I,
Yet a' the lads they measure me,
  Or, at least, they try.

Gin a body meet a body
  Altogether free,
How they travel afterwards
  We do not always see.
Ilka problem has its method
  By analytics high;
For me, I ken na ane o' them,
  But what the waur am I?

10

From Nicholas Wade's inside view of the struggle between two intensely ambitious and antipathetic scientists to beat the other to the isolation of the pituitary releasing factors (thyrotropin releasing factor, or TRF, was one of them), The Nobel Prize Duel, *here is a scientific emperor, for once on the receiving end of some unscrupulous dealings. (The race ended more or less in a dead-heat.)*

Guillemin had gravely misinterpreted the silence of the Schally-Bowers-Folkers alliance at the New York meeting. Their reticence stemmed from policy, not ignorance. A lifetime in the pharmaceutical industry had given Folkers a keen appreciation of the importance of patents and of not publishing too much too soon.

Guillemin's strategy was to publicize every half step forward so as to establish priority. Folkers believed in controlling information tightly so as neither to help the opposition unnecessarily nor to make premature assertions that might later prove false. "It is time to be strategic in the release of structural information, if we wish to have a reasonable chance of being first with the complete structure [of TRF]," Folkers had written to Schally in the first days of the alliance. Guillemin's assumption from the New York conference that he had a considerable lead did not make him relax his efforts. But it aggravated the shock of his later discovery that he had hardly any lead at all. That rude realization was to come very shortly.

The delicate situation in which the New Orleans–Austin team was placed, of being near to discovery but uncertain how near, was one for which Folkers' scientific and political talents were well suited. He had the strategic realities of the situation at his fingertips. When Schally had called Enzmann on May 16 to announce the activity of the doubly amidated glu-his-pro, Folkers was traveling in Europe. Enzmann immediately sent him a cable with the news. Back came a wire from Folkers in Stockholm to Schally and Bowers: "Wunderbar – recommend prove hormone structure and assay pure synthetics before publication. Secrecy vital. Patents possible. Expedite next hormone. Karl." None of these recommendations had been carried out by the time of the Endocrine Society meeting in New York. But by holding his counsel there, Folkers acquired a critical advantage over Guillemin: he gained sight of the cards in his rival's hand without showing his own. The advantage had to be exploited, but how? Neither Folkers nor Schally was happy with Enzmann's assertions that pyro-glu-his-pro-amide was the real TRF. It all seemed too easy. Besides, it was ridiculous to suppose that Enzmann should have accidentally synthesized TRF by preparing the intermediate for a quite different compound. Schally had other reasons for doubt. Not only had he known since 1966 that TRF must be a larger molecule, but Guillemin had specifically stated at New York that pyro-glu-his-pro-amide was not TRF. Trouncing Guillemin was great sport, but Schally didn't like to get on the other side to him in a scientific argument.

Folkers understood clearly that there was not the time to indulge in doubts. Something had to be done before Guillemin swept the board. The solution Folkers arrived at was one of fine judgment and subtlety. It was to compose an article which claimed neither too much nor too little. The principal contents of the article were the chemical recipes for synthesizing pyro-glu-his-pro-amide and acetyl-pyro-glu-his-pro-amide. In it Folkers noted that both substances showed TRF-like activity in bioassay. It did not claim that either substance was or was not TRF. But the wording of the title of the paper implied that an important discovery had been made.

Folkers listed himself as first author on the paper, followed by his students Enzmann and Jan Bøler, and by Bowers, Redding, and Schally. On July 24, less than a month after the New York meeting, he sent copies of the manuscript to Bowers and Schally. "I have tried to give Andrew the great credit which I believe he deserves on this problem and at the same time bring into fair comparison his work and that of R.G. [Guillemin]," Folkers wrote in a covering letter. But nothing could assuage Schally's pain that the problem to which he had devoted his life's work should be brought so near to solution in a paper that gave Folkers prime credit – Folkers, who had devoted as many months to TRF as Schally had years, who had been in Stockholm when Enzmann's compounds were assayed, whose main contribution to the problem had been the loan of Enzmann's services. Mixed with the chagrin was a paralyzing doubt: Schally was still not completely convinced that pyro-glu-his-pro-amide was the true TRF. His anguish blotted out the joy of discovery. The victory over TRF, supposing that he should be the winner, was already ashes before he had tasted it. There was nothing he could do. A fight with Folkers over the authorship of the paper would lose precious time, and would in any case be futile. Folkers' laboratory had provided the compounds; all that had been done in New Orleans was to assay them. TRF was Schally's problem, but Folkers would insist on the credit for this development. With his past honors – president of the Stanford Research Institute, vice-president of Merck, president of the American Chemical Society – Folkers would be the winner in any dispute with a scientist as little known as Schally. Schally swallowed his chagrin.

The authorship of the article posed another problem for Schally, one that forced him into a corner of his own making. Folkers had listed Tommie Redding as an author because Redding had done bioassays of Enzmann's compounds. The problem was that Schally had promised his colleague Bowers in February 1969 that Redding would do no work on TRF; Bowers would do all the TRF assays while Schally's laboratory concentrated on the chemistry. Despite the promise, Schally couldn't resist having Redding duplicate Bowers' work. Folkers, rather than get in the middle of a quarrel between his two partners in New Orleans, had been sending samples of his materials to both laboratories. The authorship of the alliance's first paper brought the issue to a head. Bowers objected to Redding's name being on the author list, on the perfectly reasonable grounds that Redding, according to Schally's prior assurances, had done no work on the TRF assays. Redding was not around to defend his interests when the dispute arose. Having seen Folkers' manuscript, he left for a vacation, happy that his seven years of working on TRF with Schally had culminated in his sharing of the credit for an advance of such importance. He returned to find that his name had been taken off the paper. Schally, having promised Bowers that Redding had not been working on TRF, had little choice but to remove Redding's name.

11

A.M. Sullivan captures in 'Atomic Architecture' the seductions of organic chemistry and of its transformation into life.

ATOMIC ARCHITECTURE

Take Carbon for example then
What shapely towers it constructs
To house the hopes of men!
What symbols it creates
For power and beauty in the world
Of patterned ring and hexagon –
Building ten thousand things
Of earth and air and water!
Pride searches in the flues of earth
For the diamond and its furious sun,
Love holds its palms before the glow
Of anthracite and purrs.
Five senses take their fill
Of raiment, rainbows and perfumes,
Of sweetness and of monstrous pain.

If life begins in carbon's dancing atoms
Moving in quadrilles of light

To the music of pure numbers,
Death is the stately measure
Of Time made plausible
By carbon's slow procession
Out of the shifting structure
Of crumbling flesh and bone.

12

*One of the most engaging of scientific autobiographies (in general an ineffably boring genre) is that of François Jacob (*La Statue intérieure*). A much admired figure, he succeeded, after an adventurous war, in finding his way to the laboratory of André Lwoff, one of the rare bright blooms amongst the weed-choked wastes of French science in the 1940s. Jacob and Jacques Monod eventually shared the Nobel Prize with their mentor.*

On the appointed day I arrived early, as is my custom. The laboratory was located on the second floor of the chemistry building, under the roof beams. To reach André Lwoff's office one had to follow a long winding corridor, crammed with all manner of apparatus: when I arrived Lwoff was taking lunch in the company of two charming persons, his technician and his secretary I later learned. In his office I laid bare my ignorance, my good intentions, my desires. His blue eyes fixed me with a long gaze; a prolonged inclination of the head. His laboratory was full. There was no room for me.

What vain hope! But I was stubborn. In the course of the winter I returned several times to see André Lwoff. And every time there was the same look, blue and mild, the same shake of the head, the same refusal. By the spring my optimism began to weaken. I set out to look for a second choice, in the event that . . . In June I hazarded a final approach to André Lwoff. Arriving, I found his eyes even more blue than usual, the inclination of his head deeper, the reception warmer. Without even waiting for me to profess again my ignorance, my good intentions, my desires, he announced: 'You know, we have just found induction of prophage.' I greeted this news with an 'Oh', into which I put all the astonishment, all the wonder, all the admiration of which I was capable, while thinking 'What is that supposed to mean? What can a prophage be? What language is he talking?' He gave me another long look, chin sunk on chest, and asked: 'Would it interest you to work on phage?' Stunned, I could only splutter 'That is exactly what I should like.' 'Then go off on your holidays and come on the first of September.'

I slowly descended the stairs, at once thrilled and disoriented, like a child that has just acquired a long-coveted toy. At the same time the story of the prophage made me uneasy. I rushed into the first library I saw to consult dictionaries. Induction? Prophage? I departed no wiser. André Lwoff had admitted me to his laboratory. Perhaps because of my persistence. Or my less than classic features. Or perhaps more simply on account of his good humour, brought on by the induction of prophage. To this day I do not know the answer. What I do know, on the other hand, is that in his place I should certainly not have taken into my department an individual like myself.

13

Jeremy Bernstein, a theoretical physicist turned writer, gives in The Life it Brings *a young American visitor's impressions of French physics only a few years later. He had come to Paris to work with a leading theoretician of the younger generation, whom he had come to know in Princeton.*

By now, Michel had returned to Paris, and we set up shop in the basement of one of the buildings in the Polytechnique, and it gradually became evident to me just how peculiar Michel's situation was. To put the matter in context, I think it is fair to say that he was at that time the best-known French theoretical physicist, along with Maurice Lévy, who was at the École Normale Superieure. The two men had a correct but not especially cordial relationship. It was typical that once I learned to speak French, I could *tutoyer* both men, though they formally '*vousvoyered*' each other. Despite Michel's international reputation, his actual position at the École Polytechnique was *maître de conférences* – a sort of associate professor – in the laboratory of one Jean Vignal, who was the professor. The unfortunate fact was that Vignal, a marine engineer, did not know any modern physics. He was an elderly gentleman whose education in physics seems to have stopped somewhere in the late 1920s. He taught students entirely incorrect things in the very field in which Michel was an acknowledged master. (Vignal, for example, had the idea that the electrons in beta radioactivity emerged with a single energy. It was the fact that they didn't that led Pauli, in the early 1930s, to invent the neutrino.) On top of that he often repeated lectures, which caused the students to chant, in chorus, '*Déjà vu!*'

Michel had an incendiary temperament and less than even the minimum tolerance for fools, which made his situation potentially explosive. Matters might have been smoothed over if the experimental group, led by the noted physicist Louis le Prince Ringuet, had had a serious interest in theory. Le Prince Ringuet was a colorful cosmic ray experimenter who distinguished himself at Rochester Conferences by giving bilingual lectures that sounded like performances by Maurice Chevalier. As far as Le Prince Ringuet was concerned, theoretical physics was an idle luxury. He was a true Baconian, who believed that all you had to do is to put your photographic plate somewhere and the Lord would provide. To do an experiment with something in mind was an anathema to him. Michel and Le Prince Ringuet hardly spoke to each other. Despite these tensions, Michel had managed to gather around him a small group of devoted and brilliant students, some of whom are now among the leaders of French physics.

For some time before I arrived, pressure had been building to move much of the scientific faculty out of Paris to a new, American-style campus that was being constructed at Orsay, a relatively short commute from town. The high-energy-theory group was being created by Lévy. Michel was to join it after the first of the year, and I was to move with him. But before that took place things developed along three fronts: French, tennis, and physics. The French was the key to everything else.

My plan had been to absorb French osmotically from taxi drivers, newspapers, my car radio, and waiters. This worked as long as all I wanted to do was to order simple meals in restaurants or take a taxi to the Champs-Elysées. But it soon became clearer and clearer to me that getting beyond that stage was going to be impossible without help – a lot of help. It also became clear that unless I learned the language, my time in Paris was going to be lonely and frustrating. Two things convinced me that I had better do something. The first was a little reception given by the commandant of the École Polytechnique for the faculty and visitors. I was invited to accompany Michel. The guests were evidently a distinguished group. The tradition of mathematics at the Polytechnique had always been very strong. When I was there, the great mathematician Jacques Hadamard, who was in his nineties, gave a retirement address, and Laurent Schwartz, considered to be one of the greatest mathematicians in the world, was then on the faculty. It would have been nice to talk to him and others at the reception. They were very polite and initially interested in my having come from America, but as soon as they found that I could not carry on a sensible conversation in French, they lost all interest. I left the reception feeling like a moron.

14

In Mitchell Wilson's novel about physics, Live with Lightning, *the hero (Erik) is known to be the great experimental physicist and eventual elder of the American physics establishment, I.I. Rabi (of whom more later). His research supervisor, Dr Haviland, may have been the East Coast patrician, George B. Pegram. The identity of the evil Professor Regan is not revealed.*

The university sat at the very crest of its own hill; clean brick and white plaster buildings in the style of New England churches. The campus was a huge spacious octagon crisscrossed by graveled tree-lined paths, which met in the center at a fountain surrounded by a wide circle of elms. The campus lawns were flecked with students, and summer droned secretly in the air. Far across the green, a group of boys played softball and the tiny figures went through the furious pantomime of "He's out . . . He's safe." On the steps of the Physics Building a boy, in corduroy slacks and sweater, and a girl sat with a pile of books between them. She had her plaid skirt pulled high to allow her thighs to be tanned by the sun. She and the boy were silently trying on each other's glasses.

Upstairs in a large, old man's room, Regan sat at a roll-top desk. His swivel chair had a high carved back and the arms were lion paws. When he rose, they saw that he had been sitting on a circular inflated rubber cushion, like a small automobile tire. He was taller than either of his visitors and one eye was so bleached that the iris was no more than a pale blue blur.

"How do you do, gentlemen," he said softly. "I have just been talking to your young associate from New York."

Erik turned and saw Fabermacher sitting very quietly by the window. He seemed very pale and suppressed, as if fighting off fear. He had been sitting with his feet close together, his hands in his lap and he gave Erik the impression that he had been waiting silently for a long time. Regan gave them a moment for greetings and then he motioned them all to be seated. He walked to the window and looked out for a while at the river far below. He wore a dark suit of a cut as old-fashioned as the round stained-glass shades of the table lamps. He held the cord of the window shade as if he were actually hanging on to the slender string for support.

"Well," he said, turning back to them with a crooked smile that had a quality of sweetness in it. The sweetness persisted and became so fixed that it was ugly. "Sooner or later, someone is going to tell you about that damned river, so I might as well be the first. If you throw a chip of wood into it, it's supposed to end up at New Orleans. But if the good people of New Orleans are still waiting for wood chips from the town of Argyle to flow down the Sanecanock, through Eustis Lake, four more rivers and the Mississippi, why then I sadly fear that the good people of New Orleans can't have a hell of a lot to do with their time. . . . Well, so much for the river," he drawled, walking stiffly back to his chair.

The spring groaned, and Regan sighed. He ran his hand back over his bald head and then with a lot of squinting put on a pair of silver-rimmed spectacles.

"Well, gentlemen, we all know the department needs new blood. Or at least an old vampire like me needs a change of diet." He paused at his own joke, smiled at them again, and then continued. Erik saw that Regan was deliberately acting the part of the crotchety old gentleman. He was playing it so broadly and so obviously that to be a spectator was to be ridiculed.

"It may take a couple of years for you to get the hang of my ways, but we'll get along," said Regan. "Now, Dr. Trasker, your papers seem pretty sound to me. I like the way you build apparatus, no gimcracks, no nonsense. Good solid engineering." His voice fell into droll afterthought. "It's true I hadn't heard of you until old Leach mentioned your name, but I suppose that's my fault, not yours." He looked mockingly at Trasker for a moment, challenging a reply, and then grunted and turned to Erik. "As for you, my dear young Dr. Gorin, I looked up your paper with Haviland at Columbia. Now that was a very nice job of work, although I can't say that I think very highly of the good Dr. Haviland." The *good* Dr. Haviland almost was drowned in the treacly voice which suddenly grew coarse and sharp. "In my day we had a name for handsome young men like him, but that day is done and gone I fear. I sadly fear," he repeated, mocking himself. The swivel chair turned once more, and for a moment more he was silent. Then, as if he were chiding a child whom he had just frightened, he said, "And now we come to you, Herr Fabermacher, sitting there so quiet and polite. You have nice manners, Herr Fabermacher, and I'm a man who appreciates what I haven't got. Now, *mein Herr*, about your work. It's all gibberish to me, but if that kind of gibberish is the style today, why I say let 'em have it. . . . If any of you take offense at anything I've said, you're just plain damn fools. I don't mean any harm. And when I do mean harm, you'll know. You most certainly will."

15

Nigel Balchin was a science graduate, who wrote several highly readable science-based novels. A Sort of Traitors *deals with the conflicts of conscience that develop when the potential for ill of an apparently innocent discovery is revealed. The story opens with a description of long, placid and monotonous labours in the bacteriological laboratory.*

About an inch and a half in the middle of the glass tube was now bright orange in the blow-pipe flame. Arthur blew gently into the end of the tube and a perfectly symmetrical bulb expanded. It looked easy and inevitable. Arthur had been blowing glass in the Haughton Laboratory for forty-five years. It was his favourite little joke when demonstrating to students. 'Professor Sewell and me came to the Haughton about the same time, him as an undergraduate and me as a lab. boy. Professor Sewell went on and studied physiology and zoology and the bio-chemistry and I went on and studied glass. We both learnt a lot since then, gentlemen.' Arthur could remember Sir Phillip Lowes and the early work on synthetic amino-acids, and he had made a gas analysis line for Chubb in 1909. He was always rather short with Professor Sewell. In private, he thought him a great man, though not, perhaps, as great a man as Lowes or Chubb.

Pearce came in, carrying a broom, and said, 'That monkey's took my cap.' Arthur cut off the air so that the blow-pipe flame became long and yellow and smoky. He flicked his dark glasses up on his forehead and slowly twirled his bulb in the cooler flame.

'Well, don't let him eat it,' he said, without looking up. 'He's on a diet.'

'Took it clean off as I was sweeping,' said Pearce indignantly.

Arthur carefully put down his bulb, sighed, and slid off his stool. He said, 'I don't see what you want to wear a cap for in here. Afraid you'll catch cold?'

They went into the long, rather narrow animal room. There was only one big cage. The rest of the place was occupied by hutches containing rabbits, rats and guinea pigs. Arthur went up to the big cage and said, 'Phillips, what you doing with that? Give it back.' The young chimpanzee was sitting at the back of the cage holding the cap in its hand and biting tentatively at its peak. It looked up for a moment and then turned its back, hunched its shoulders and clutched the cap to its chest. Pearce said, 'Took it clean off my head. Reached out see, Arthur.'

Arthur said, 'I'd better get Miss Byrne to him.'

Lucy Byrne was a dark girl of about twenty-five who looked as though she hadn't slept for a week. She went up to the cage and said, 'Oh, *Phillips*!' very reproachfully.

The chimpanzee turned its head and then, closing its eyes firmly, huddled over the cap again. Lucy said, 'All right. It's bribery but. . . ' She put her hand in the pocket of the white overall, produced half a biscuit and clicked her tongue. Phillips ambled eagerly over to the bars and put out a hand. Lucy said, 'Oh, no. Fair do's. Give Pearce his cap first.' Phillips looked at the cap and then dropped it disinterestedly on the floor of the cage near the bars. Pearce made a quick grab and seized it. Lucy put the biscuit in the chimpanzee's hand. She said, 'He hasn't hurt it, has he? He never does hurt things.'

Marriott looked up as Lucy came in and said, 'Get it?' Lucy said, 'Oh, yes. I think he was glad to get rid of it. He always has a conscience when he steals things.' Marriott said, 'As chimpanzees go, Phillips is a white elephant. What's the point of keeping him if he's too valuable to use? This is a lab., not a zoo.'

'Phillips is all right,' said Lucy. 'The silly part is keeping him in a cage and having old Pearce free. Phillips is much more intelligent than Pearce. And he's got a nicer nature.' She climbed on to her stool. 'We'd got to eighty-nine.'

Marriott bent over the Petrie dishes.

'Ninety,' he said. 'Slight growth. Ninety-one, slight growth.

Ninety-two, slight growth. Ninety-three, slight . . . no, half a mo'.' He studied it for a few moments. 'Ninety-three,' he said carefully, 'slight growth.'

Lucy gave a deep sigh. 'I wish you wouldn't do that,' she said. 'It's so disappointing.'

Marriott said, 'Ninety-four, slight growth. What's disappointing?'

'When you hesitate like that. It makes me think something exciting's coming like "no growth" or "strong growth".'

'Ninety-five, slight growth. Ninety-six, slight growth. Ninety-seven, slight growth. The trouble with you is that you've got a craving for excitement. Ninety-eight, slight growth. Anything for a change, that's your motto. You had a "no growth" only two days ago. You want to live in a whirl. Ninety-nine, slight growth. One hundred, slight growth.' He straightened up. 'And that's the end of batch two.'

16

One of the kindlier perceptions of science by the artist, is by Louis MacNeice – from The Kingdom*:*

A little dapper man but with shiny elbows
And short keen sight, he lived by measuring things
And died like a recurring decimal
Run off the page, refusing to be curtailed;
Died as they say in harness, still believing
In science, reason, progress. Left his work
Unfinished *ipso facto* which, continued,
Will supersede his name in the next text-book
And relegate him to the anonymous crowd
Of small discoverers in lab or cloister
Who link us with the Ice Age. Obstinately
He canalised his fervour, it was slow
The task he set himself but plotting points
On graph paper he felt the emerging curve
Like the first flutterings of an embryo
In somebody's first pregnancy; resembled
A pregnant woman too in that his logic
Yet made that hidden child the centre of the world
And almost a messiah; so that here

Even here, over the shining test-tubes
The spirit of the alchemist still hovered
Hungry for magic, for the philosopher's stone.
And Progress – is that magic too? He never
Would have conceded it, not even in these last
Years of endemic doubt; in his perspective
Our present tyrants shrank into parochial
Lords of Misrule, cross eddies in a river
That has to reach the sea. But has it? Who
Told him the sea was there?
Maybe he told himself and the mere name
Of Progress was a shell to hold to the ear
And hear the breakers burgeon. Rules were rules
And all induction checked but in the end
His reasoning hinged on faith and the first axiom
Was oracle or instinct. He was simple
This man who flogged his brain, he was a child;
And so, whatever progress means in general,
He in his work meant progress. Patiently
As Stone Age man he flaked himself away
By blocked-out patterns on a core of flint
So that the core which was himself diminished
Until his friends complained that he had lost
Something in charm or interest. But conversely
His mind developed like an ancient church
By the accretion of side-aisles and the enlarging of lights
Till all the walls are windows and the sky
Comes in, if coloured; such a mind . . . a man . . .
Deserves a consecration; such a church
Bears in its lines the trademark of the Kingdom.

17

This picture of laboratory life in the sixteenth century from Arthur Koestler's The Sleepwalkers *suggests that science has lost something in piquancy with the passage of the centuries.*

Tycho's observatory, the Uraniburg, built by a German architect under Tycho's supervision, was a symbol of his character, in which meticulous precision combined with fantastic extravagance. It was a fortress-like monster which is said to have been "epoch-making in the history of Scandinavian architecture", but on the surviving woodcuts looks rather like a cross between the Palazzo Vecchio and the Kremlin, its Renaissance façade surmounted by an onion-shaped dome, flanked by cylindrical towers, each with a removable top housing Tycho's instruments, and surrounded by galleries with clocks, sundials, globes, and allegorical figures. In the basement were Tycho's private printing press, fed by his own paper mill, his alchemist's furnace and private prison for unruly tenants. He also had his own pharmacy, his game preserves and artificial fishponds; the only thing he was missing was his tame elk. It had been dispatched to him from his estate but never reached the island. While spending a night in transit at Landskroner Castle, the elk wandered up the stairs to an empty apartment where it drank so much strong beer that on its way downstairs it stumbled, broke its leg and died.

In the library stood his largest celestial globe, five feet in diameter, made of brass, on which, in the course of twenty-five years, the fixed stars were engraved one by one, after their correct positions had been newly determined by Tycho and his assistants in the process of re-mapping the sky; it had cost five thousand dalers, the equivalent of eighty years of Kepler's salary. In the south-west study, the brass arc of Tycho's largest quadrant—fourteen feet in diameter—was fastened to the wall; the space inside the arc was filled with a mural depicting Tycho himself surrounded by his instruments. Later on, Tycho added to the Uraniburg a second observatory, the "Starburg", which was built entirely underground to protect the instruments from vibration and wind, only the dome-shaped roofs rising above ground level; so that "even from the bowels of the earth he could show the way to the stars and the glory of God". Both buildings were full of gadgets and automata, including statues turning on hidden mechanisms, and a communication system that enabled him to ring a bell in the room of any of his assistants—which made his guests believe that he was convoking them by magic. The guests came in an unceasing procession, savants, courtiers, princes and royalty, including King James VI of Scotland.

Life at Uraniburg was not exactly what one would expect to be the routine of a scholar's family, but rather that of a Renaissance court. There was a steady succession of banquets for distinguished visitors, presided over by the indefatigable, hard-drinking, gargantuan host, holding forth on the variations in the eccentricity of Mars; rubbing ointment on his silver nose, and throwing casual titbits to his fool Jepp, who sat at the master's feet under the table, chattering incessantly amidst the general noise. This Jepp was a dwarf, reputed to have second sight, of which he seemed to give spectacular proof on several occasions.

Tycho is really a refreshing exception among the sombre, tortured, neurotic geniuses of science. He was, it is true, not a creative genius, only a giant of methodical observation. Still, he displayed all the vanity of genius in his interminable poetic outpourings. His poetry is even more dreadful than Canon Koppernigk's, and more abundant in quantity—Tycho was never in want of a publisher, since he had his own paper mill and printing press. Even so, his verses and epigrams overflowed onto the murals and ornaments of Uraniburg and Stjoerneburg, which abounded in mottoes, inscriptions and allegorical figures. The most impressive of these, adorning the wall of his chief study, represented the eight greatest astronomers in history, from Timocharis to Tycho himself, followed by "Tychonides", a yet unborn descendant—with a caption expressing the hope that he would be worthy of his great ancestor.

18

Julian Huxley, whom we shall meet again later, occasionally expressed his polymathic vision in verse of no little skill. 'Cosmic Death' is an example.

COSMIC DEATH

By death the moon was gathered in
  Long ago, ah long ago;
Yet still the silver corpse must spin
  And with another's light must glow.
Her frozen mountains must forget
  Their primal hot volcanic breath,
Doomed to revolve for ages yet,
  Void amphitheatres of death.

And all about the cosmic sky,
  The black that lies beyond our blue,
Dead stars innumerable lie,
  And stars of red and angry hue
  Not dead but doomed to die.

19

What follows (from Michael Frayn's admirable novel, The Tin Men*) exposes, in perhaps only slightly extrapolated form, what goes on at the frontiers of computer science.*

"Yes," said Rowe, but on reflection that sounded a shade facile, and he rephrased it. "Y-e-e-e-s," he said.

"I mean," said Macintosh, "there are plenty of other people in the Institute who've got their eye on the new wing. Oh, yes. I'm told there's a group—no names, no pack drill—who want to hold some sort of huge sexual orgy in there."

"Yes, so I hear."

"They'll never get away with it, of course. I'm told Nunn's furious about it. It makes you think, though. I mean, the things I could use that wing for if only I had the time! If only I could get away from my damned Samaritans for a week or two! Did I ever tell you about my idea for programming a computer to write pornographic novels? Well, I sometimes wonder if you couldn't programme machines to perform a great deal of human sexual behaviour. It would save a lot of labour."

"Yes," said Rowe. "Yes."

"At any rate in the early stages. On the same principle you might also programme machines to go through the initial conversational moves two people make when they first meet. They're always standardised, like chess openings. You could select your gambit, then go away and make the tea while the machine played it, and come back and pick up the conversation when it started to become interesting."

"Yes," said Rowe.

"It breaks my heart to see that wing going empty when there's so much to be done. I mean, let's accept—and I owe this suggestion to my good friend Goldwasser—that all ethical systems are ossified, in which case all operations within an ethical system can be performed by computer. I should be designing circuits to demonstrate what happens when one ossified system, say a Christian one, comes into contact with another ossified system, say a liberal agnostic one. And what happens when two computers with incompatible systems try to programme a third between them.

"Ah, Rowe, Rowe, Rowe! Doesn't it move you to contemplate the great areas of life which have ossified, where activity has been reduced to the manipulation of a finite range of variables? The pity and the terror of it, Rowe! These vast petrified forests are our rightful domain. They are waiting helplessly to be brought under the efficient, benevolent rule of the kindly computer.

"Take the field of religious devotions. What computer man can survey devotional practice without lifting up his heart and thanking God for sending him such a prize? When we are called in to write a programme for automated devotion—as we shall be in five or ten years' time—we shall of course recommend performing all the religious services in the country from one central computer, which would also write its own sermons, logically developing any given topic along the lines laid down by the Thirty-Nine Articles or the Holy Office without fear of heterodoxy. But in fifteen or twenty years' time we shall be writing programmes for praying. The subjects and sentiments tend to come in a fairly limited range."

"Ah," said Rowe, "there's a difference between a man and a machine when it comes to praying."

"Aye. The machine would do it better. It wouldn't pray for things it oughtn't to pray for, and its thoughts wouldn't wander."

"Y-e-e-s. But the computer saying the words wouldn't be the same . . ."

"Oh, I don't know. If the words 'O Lord, bless the Queen and her Ministers' are going to produce any tangible effects on the Government it can't matter who or what says them, can it?"

"Y-e-e-s, I see that. But if a man says the words he *means* them."

"So does the computer. Or at any rate, it would take a damned complicated computer to say the words *without* meaning them. I mean, what do we mean by 'mean'? If we want to know whether a man or a computer *means* 'O Lord, bless the Queen and her Ministers,' we look to see whether it's grinning insincerely or ironically as it says the words. We try to find out whether it belongs to the Communist Party. We observe whether it simultaneously passes notes about lunch or fornication. If it passes all the tests of this sort, what other tests *are* there for telling if it means what it says? All the computers in my department, at any rate, would pray with great sincerity and single-mindedness. They're devout wee things, computers."

"Y-e-e-e-s. But I take it you don't believe in a God who hears and answers prayers?"

"That's not my end of the business. I'm just concerned with getting the praying done with maximum efficiency and minimum labour output."

"But, Macintosh, if you're as cynical as that, what do you think the difference between man and computer is?"

"I'm not absolutely certain, Rowe. I'm inclined to rule out what you might call 'soul'. I should think that in time we could teach computer to be overwhelmed by the sound of Bach or the sight of another computer, to distinguish between good sonnets and bad, and to utter uplifting sentiments at the sight of the Matterhorn or the sunset. Obviously it's not the faculty of choice; a computer can be programmed to choose, just as man does, rationally, anti-rationally, at random, or by any combination of the three. Some people, of course, have set a certain store by the human capacity for faith. But I don't think it helps us much here. If you tell computer that the sky is green, computer will believe you. Or you could programme computer to perceive empirically that the sky is blue, but to act on a profound, unspoken faith that it is green. Think of that, Rowe! And if you objected that this faith was forced on him willy-nilly, then you might imagine a computer programmed to choose whether to believe the evidence of his senses that the sky is blue, or whether in spite of what he saw to accept that it was green because its operator had told him it was."

"I suppose so," said Rowe.

20

Frayn may have been inspired by Dean Swift's vision of academic activity in Laputa.

This academy is not an entire single building, but a continuation of several houses on both sides of a street, which growing waste, was purchased and applied to that use.

I was received very kindly by the warden, and went for many days in the academy. Every room has in it one or more projectors; and I believe I could not be in fewer than five hundred rooms.

The first man I saw was of a very meagre aspect, with sooty hands and face, his hair and beard long, ragged, and singed in several places. His clothes, shirt, and skin were all of the same colour. He had been eight years upon a project for extracting sunbeams out of cucumbers, which were to be put in phials hermetically sealed, and let out to warm the air in raw, inclement summers.

He told me he did not doubt that in eight years more he should be able to supply the governor's gardens with sunshine at a reasonable rate; but he complained that his stock was low, and entreated me to give him something as an encouragement to ingenuity, especially since this had been a very dear year for cucumbers. I made him a small present, for my lord had furnished me with money on purpose, because he knew their practice of begging from all who go to see them.

I saw another at work to calcine ice into gunpowder; who likewise showed me a treatise he had written concerning the malleability of fire, which he intended to publish.

There was a most ingenious architect, who had contrived a new method for building houses, by beginning at the roof and working downwards to the foundation, which he justified to me by the like practice of those two prudent insects, the bee and the spider.

There was a man born blind, who had several apprentices in his own condition. Their employment was to mix colours for painters, which their master taught them to distinguish by feeling and smelling. It was indeed my misfortune to find them at that time not very perfect in their lessons, and the professor himself happened to be generally mistaken. This artist is much encouraged and esteemed by the whole fraternity.

In another apartment I was highly pleased with a projector who had found a device of ploughing the ground with hogs, to save the charges of ploughs, cattle and labour. The method is this: in an acre of ground you bury, at six inches' distance and eight deep, a quantity of acorns, dates, chestnuts, and other mast or vegetables whereof these animals are fondest; then you drive six hundred of them into the field, where, in a few days, they will root up the whole ground in search of their feed, and make it fit for sowing, at the same time manuring it with their dung. It is true, upon experiment, they found the charge and trouble very great, and they had little or no crop.

However, it is not doubted that his invention may be capable of great improvement.

I went into another room, where the walls and ceiling were all hung round with cobwebs, except a narrow passage for the artist to go in and out. At my entrance, he called aloud to me not to disturb his webs. He lamented the fatal mistake the world had been so long in of using silkworms while we had such plenty of domestic insects who infinitely excelled the former, because they understood how to weave as well as spin. And he proposed further that, by employing spiders, the charge of dyeing silks would be wholly saved, whereof I was fully convinced when he showed me a vast number of flies most beautifully coloured, wherewith he fed his spiders, assuring us that the webs would take a tincture from them; and as he had them of all hues, he hoped to fit everybody's fancy, as soon as he could find proper food for the flies, of certain gums, oils, and other glutinous matter, to give a strength and consistence to the threads.

There was an astronomer who had undertaken to place a sun-dial upon the great weathercock on the town-house, by adjusting the annual and diurnal motions of the earth and sun, so as to answer and coincide with all accidental turnings of the wind.

I visited many other apartments, but shall not trouble my reader with all the curiosities I observed, being studious of brevity.

I had hitherto seen only one side of the academy, the other being appropriated to the advancers of speculative learning, of whom I shall say something, when I have mentioned one illustrious person more, who is called among them the "Universal Artist."

He told us he had been thirty years employing his thoughts for the improvement of human life. He had two large rooms full of wonderful curiosities, and fifty men at work. Some were condensing air into a dry, tangible substance, by extracting the nitre and letting the aqueous or fluid particles percolate; others softening marble for pillows and pin-cushions; others petrifying the hoofs of a living horse, to preserve them from foundering. The artist himself was at that time busy upon two great designs: the first, to sow land with chaff, wherein he affirmed the true seminal virtue to be contained, as he demonstrated by several experiments, which I was not skilful enough to comprehend. The other was, by a certain composition of gums, minerals, and vegetables, outwards applied, to prevent the growth of wool upon two young lambs; and he hoped, in a reasonable time, to propagate the breed of naked sheep all over the kingdom.

21

Another celebrated reductio ad absurdum *occurs in Samuel Butler's dissertation on the intellectual life of* Erewhon.

It was indeed true that much was now known that had not been suspected formerly, for the people had had no foreign enemies, and, being both quick-witted and inquisitive into the mysteries of nature, had made extraordinary progress in all the many branches of art and science. In the chief Erewhonian museum I was shown a microscope of considerable power, that was ascribed by the authorities to a date much about that of the philosopher of whom I am now speaking, and was even supposed by some to have been the instrument with which he had actually worked.

This philosopher was Professor of botany in the chief seat of learning then in Erewhon, and whether with the help of the microscope still preserved, or with another, had arrived at a conclusion now universally accepted among ourselves – I mean, that all, both animals and plants, have had a common ancestry, and that hence the second should be deemed as much alive as the first. He contended, therefore, that animals and plants were cousins, and would have been seen to be so, all along, if people had not made an arbitrary and unreasonable division between what they chose to call the animal and vegetable kingdoms.

He declared, and demonstrated to the satisfaction of all those who were able to form an opinion upon the subject, that there is no difference appreciable either by the eye, or by any other test, between a germ that will develop into an oak, a vine, a rose, and one that (given its accustomed surroundings) will become a mouse, an elephant, or a man.

He contended that the course of any germ's development was dictated by the habits of the germs from which it was descended, and of whose identity it had once formed part. If a germ found itself placed as the germs in the line of its ancestry were placed, it would do as its ancestors had done, and grow up into the same kind of organism as theirs. If it found the circumstances only a little different, it would make shift (successfully or unsuccessfully) to modify its development accordingly; if the circumstances were widely different, it would die, probably without an effort at self-adaptation. This, he argued, applied equally to the germs of plants and of animals.

He therefore connected all, both animal and vegetable development, with intelligence, either spent and now unconscious, or still unspent and conscious; and in support of his view as regards vegetable life, he pointed to the way in which all plants have adapted themselves to their habitual environment. Granting that vegetable intelligence at first sight appears to differ materially from animal, yet, he urged, it is like it in the one essential fact that though it has evidently busied itself about matters that are vital to the well-being of the organism that possesses it, it has never shown the slightest tendency to occupy itself with anything else. This, he insisted, is as great a proof of intelligence as any living being can give.

"Plants," said he, "show no sign of interesting themselves in human affairs. We shall never get a rose to understand that five times seven are thirty-five, and there is no use in talking to an oak about fluctuations in the price of stocks. Hence we say that the oak and the rose are unintelligent, and on finding that they do not understand our business conclude that they do not understand their own. But what can a creature who talks in this way know about intelligence? Which shows greater signs of intelligence? He, or the rose and oak?

"And when we call plants stupid for not understanding our business, how capable do we show ourselves of understanding theirs? Can we form even the faintest conception of the way in which a seed from a rose-tree turns earth, air, warmth and water into a rose full-blown? Where does it get its colour from? From the earth, air, &c.? Yes – but how? Those petals of such ineffable texture – that hue that outvies the cheek of a child – that scent again? Look at earth, air, and water – these are all the raw material that the rose has got to work with; does it show any sign of want of intelligence in the alchemy with which it turns mud into rose-leaves? What chemist can do anything comparable? Why does no one try? Simply because every one knows that no human intelligence is equal to the task. We give it up. It is the rose's department; let the rose attend to it – and be dubbed unintelligent because it baffles us by the miracles it works, and the unconcerned businesslike way in which it works them.

"See what pains, again, plants take to protect themselves against their enemies. They scratch, cut, sting, make bad smells, secrete the most dreadful poisons (which Heaven only knows how they contrive to make), cover their precious seeds with spines like those of a hedgehog, frighten insects with delicate nervous systems by assuming portentous shapes, hide themselves, grow in inaccessible places, and tell lies so plausibly as to deceive even their subtlest foes. . . ."

22

It is curious that science seems not once to crop up in any form in the great works of fiction of the Victorian period. A character in Elizabeth Gaskell's Wives and Daughters *and another in George Eliot's* Middlemarch *are said to represent Charles Darwin, but neither utters on the* Origin of Species. *Evidence that science in any sense entered Charles Dickens's consciousness comes only (so far as I am aware) from Sam Weller's declared need for 'a pair of patent double million magnifyin' gas microscopes of hextra power – able to see through a flight of stairs and a deal door.' Across the Channel there was probably a higher level of enlightenment, on the part both of authors and their characters. Here are the eager but dimwitted heroes of Flaubert's last (and unfinished) novel,* Bouvard and Pécuchet, *confronting chemistry.*

To acquire a knowledge of chemistry they procured Regnault's course and first learned that 'simple bodies are perhaps compound'.

Bodies are distinguished as metalloids and metals, a difference which, according to the author, 'is by no means absolute'. Similarly with acids and bases 'since a body may behave like acids or bases according to circumstances'.

They found the notation most fanciful. Multiple proportions worried Pécuchet.

'Since a molecule of A, I suppose, combines with several parts of B, it seems to me that such a molecule ought to divide into as many parts; but if it divides it ceases to be the unit, the original molecule. I don't understand at all!'

'Nor do I!' said Bouvard.

So they resorted to a less difficult work, that of Girardin, where they learned as certain facts that ten litres of air weigh 100 grams, that there is no lead in pencils, that diamonds are nothing but carbon.

What amazed them more than anything was that earth, as an element, does not exist.

They grasped how a blow-lamp works, the facts about gold, silver, washing soda, tinning saucepans; then without the least scruple Bouvard and Pécuchet threw themselves into organic chemistry.

How wonderful to find in living creatures the same substances as those which make up minerals. Nevertheless they felt a sort of humiliation at the idea that their persons contained phosphorus like matches, albumen like white of egg, hydrogen gas like street lamps.

After colours and fats they came on to fermentation.

This led them on to acids, and the law of equivalents upset them once more. They tried to elucidate it with the theory of atoms, and then they were completely lost.

If they were to understand all that, according to Bouvard, they would need instruments.

The expense was considerable, and they had spent too much.

But Dr Vaucorbeil could no doubt enlighten them.

They presented themselves during his consulting hours.

'Yes, gentlemen? What is wrong with you?'

Pécuchet replied that they were not ill, and after explaining the purpose of their visit:

'First we should like to know about superior atomicity.'

The doctor went very red, then criticized them for wanting to learn chemistry.

23

H.G. Wells, who will be encountered again in later chapters, put into Love and Mr Lewisham *what he conceded in his autobiography to have been as precise a description as his memory allowed of his own experiences at the Imperial Institute in South Kensington.*

Lagune's treatment of the exposure was light and vigorous. "The man Chaffery," he said "has made a clean breast of it. His point of view –"

"Facts are facts," said Smithers.

"A fact is a synthesis of impressions," said Lagune; "but that you will learn when you are older. The thing is that we were at cross purposes. I told Chaffery you were beginners. He treated you as beginners – arranged a demonstration."

"It *was* a demonstration," said Smithers.

"Precisely. If it had not been for your interruptions . . ."

"Ah!"

"He forged elementary effects . . ."

"You can't but admit that."

"I don't attempt to deny it. But, as he explained – the thing is necessary – justifiable. Psychic phenomena are subtle, a certain training of the observation is necessary. A medium is a more subtle instrument than a balance or a borax bead, and see how long it is before you can get assured results with a borax bead! In the elementary class, in the introductory phase, conditions are too crude. . . ."

"For honesty."

"Wait a moment. *Is* it dishonest – rigging a demonstration?"

"Of course it is."

"Your professors do it."

"I deny that *in toto*," said Smithers, and repeated with satisfaction, "*in toto*."

"That's all right," said Lagune, "because I have the facts. Your chemical lecturers – you may go downstairs now and ask, if you disbelieve me – always cheat over the indestructibility of matter experiment – always. And then another – a physiography thing. You know the experiment I mean? To demonstrate the existence of the earth's rotation. They use – they use –"

"Foucault's pendulum," said Lewisham. "They use a rubber ball with a pin-hole hidden in the hand, and blow the pendulum round the way it ought to go."

"But that's different," said Smithers.

"Wait a moment," said Lagune, and produced a piece of folded printed paper from his pocket. "Here is a review from *Nature* of the work of no less a person than Professor Greenhill. And see – a convenient pin is introduced in the apparatus for the demonstration of virtual velocities! Read it – if you doubt me. I suppose you doubt me."

Smithers abruptly abandoned his position of denial "*in toto*." "This isn't my point, Mr. Lagune; this isn't my point," he said. "These things that are done in the lecture theatre are not to prove facts, but to give ideas."

"So was my demonstration," said Lagune.

"We didn't understand it in that light."

"Nor does the ordinary person who goes to Science lectures

understand it in that light. He is comforted by the thought that he is seeing things with his own eyes."

"Well, I don't care," said Smithers; "two wrongs don't make a right. To rig demonstrations is wrong."

"There I agree with you. I have spoken plainly with this man Chaffery. He's not a full-blown professor, you know, a highly salaried ornament of the rock of truth like your demonstration-rigging professors here, and so I can speak plainly to him without offence. He takes quite the view they would take. But I am more rigorous. I insist that there shall be no more of this. . . ."

"Next time –" said Smithers, with irony.

"There will be no next time. I have done with elementary exhibitions. You must take the word of the trained observer – just as you do in the matter of chemical analysis."

"Do you mean you are going on with that chap when he's been caught cheating under your very nose?"

"Certainly. Why not?"

Smithers set out to explain why not, and happened on confusion. "I still believe the man has powers," said Lagune.

"Of deception," said Smithers.

"Those I must eliminate," said Lagune. "You might as well refuse to study electricity because it escaped through your body. All new science is elusive. No investigator in his senses would refuse to investigate a compound because it did unexpected things. Either this dissolves in acid or I have nothing more to do with it – eh? That's fine research!"

Then it was the last vestiges of Smithers' manners vanished. "I don't care *what* you say," said Smithers. "It's all rot – it's just rot. Argue if you like – but have you convinced anybody? Put it to the vote?"

"That's democracy with a vengeance," said Lagune. "A general election of the truth half-yearly, eh?"

"That's simply wriggling out of it," said Smithers. "That hasn't anything to do with it at all."

24

We come now to Sinclair Lewis's great novel, Arrowsmith, *published in 1925. The setting (the McGurk Institute of the novel) is the Rockefeller Institute in New York, when it was the pre-eminent centre of biological research in America. Lewis worked closely with his technical advisor, Paul de Kruif, a bacteriologist at the Institute, who eventually forsook science for writing and was the author of a series of popular, highly romantic accounts of science and scientists, now all but unreadable; the best-known was* The Microbe Hunters *(de Kruif always referred to himself as a microbe hunter), widely acclaimed in its day.*

Nearly all the characters in Arrowsmith *appear to have been identifiable denizens of the Rockefeller Institute in the early years of the century. The founder–director, the formidable Simon Flexner, is transmuted in the novel into A. DeWitt Tubbs. According to Sinclair Lewis's biographer, Mark Schorer, the hero, Martin Arrowsmith, is a portrait of R.G. Hussey, later Professor of Pathology at Yale; Professor Gottlieb is a physical picture of Jacques Loeb, although de Kruif in his memoirs states that Gottlieb's opinions are more those of de Kruif's own first patron, the bacteriologist, Frederick G. Novy. Among other portraits in the book is that of Peyton Rous (Rippleton Holabird) a friend of Lewis's – whose name has passed into the language of biology in Rous Sarcoma virus; for this discovery and others Rous was rewarded by the Nobel Prize when he was already in his eighties. And then there is the protein chemist, John Northrop (the main component in Terry Wickett, who also, however, contains elements of another of de Kruif's friends and colleagues, T.J. LeBlanc). Northrop, the last survivor from this splendid* galère, *died in 1987.*

First then, a sample of Paul de Kruif's style, from his memoirs, The Sweeping Wind.

It was hard to say I was leaving him for the Rockefeller because Dr. Novy took a dim view of Rockefellerian bacteriology.

Novy was the Nestor of American microbe hunters. He was the world's authority on the spirochetes of relapsing fever; one of these corkscrew microbes bears his name. He was the first person to cultivate artificially the trypanosome of deadly Nagana. How could I ever have thought of deserting him for the toy science of the Rockefeller? He had got his science working under Robert Koch in Berlin and Emile Roux at the Institut Pasteur in Paris. These shrines of science were far different from the Rockefeller, whose fame was marked by such sad scientific mistakes as that of declaring globoid bodies to be the cause of infantile paralysis.

On hearing my news Dr. Novy was disappointed. For ten years he had tried to beat honesty and accuracy into my head in the tradition of Robert Koch and Emile Roux. Now what would happen to me in the streamlined elegance of the Rockefeller, where sensational scientific events were often portended but seldom came off? What would become of my discipline absorbed from the dedicated, austere Frederick Novy when it was exposed to the atmosphere of a scientific emporium where it didn't very much matter if one made mistakes, what with Rockefeller prestige backed by the big money? What would happen to the scientific conscience of Novy's boy, Paul, when he consorted with the Rockefeller star, Dr. Hideyo Noguchi? Noguchi, a gay little Japanese, had earned fame resting partly on his discovery of a spurious spirochete that turned out, alas, to be definitely not the cause of yellow fever. You forgave Noguchi because he was a nice little guy, but Novy, the master of spirochetes, took Noguchi apart without mercy.

25

And now Sinclair Lewis:

The McGurk Building. A sheer wall, thirty blank stories of glass and limestone, down in the pinched triangle whence New York rules a quarter of the world.

Martin was not overwhelmed by his first hint of New York; after a year in the Chicago Loop, Manhattan seemed leisurely. But when from the elevated railroad he beheld the Woolworth Tower, he was exalted. To him architecture had never existed; buildings were larger or smaller bulks containing more or less interesting objects. His most impassioned architectural comment had been, "There's a cute bungalow; be nice place to live." Now he pondered, "Like to see that tower every day – clouds and storms behind it and everything – so sort of satisfying."

He came along Cedar Street, among thunderous trucks portly with wares from all the world; came to the bronze doors of the McGurk Building and a corridor of intemperately colored terra-cotta, with murals of Andean Indians, pirates booming up the Spanish Main, guarded gold-trains, and the stout walls of Cartagena. At the Cedar Street end of the corridor, a private street, one block long, was the Bank of the Andes and Antilles (Ross McGurk chairman of the board), in whose gold-crusted sanctity red-headed Yankee exporters drew drafts on Quito, and clerks hurled breathless Spanish at bulky women. A sign indicated, at the Liberty Street end, "Passenger Offices, McGurk Line, weekly sailings for the West Indies and South America."

Born to the prairies, never far from the sight of the cornfields, Martin was conveyed to blazing lands and portentous enterprises.

One of the row of bronze-barred elevators was labeled "Express to McGurk Institute." He entered it proudly, feeling himself already a part of the godly association. They rose swiftly, and he had but half-second glimpses of ground glass doors with the signs of mining companies, lumber companies, Central American railroad companies.

The McGurk Institute is probably the only organization for scientific research in the world which is housed in an office building. It has the twenty-ninth and thirtieth stories of the McGurk Building, and the roof is devoted to its animal house and to tiled walks along which (above a world of stenographers and bookkeepers and earnest gentlemen who desire to sell Better-bilt Garments to the golden dons of the Argentine) saunter rapt scientists dreaming of osmosis in Spirogyra.

Later, Martin was to note that the reception-room of the Institute was smaller, yet more forbiddingly polite, in its white paneling and Chippendale chairs, than the lobby of the Rouncefield Clinic, but now he was unconscious of the room, of the staccato girl attendant, of everything except that he was about to see Max Gottlieb, for the first time in five years.

At the door of the laboratory he stared hungrily.

Gottlieb was thin-cheeked and dark as ever, his hawk nose bony, his fierce eyes demanding, but his hair had gone gray, the flesh round his mouth was sunken, and Martin could have wept at the feebleness with which he rose. The old man peered down at him, his hand on Martin's shoulder, but he said only:

"Ah! Dis is good. . . . Your laboratory is three doors down the hall. . . . But I object to one thing in the good paper you send me. You say, 'The regularity of the rate at which the streptolysin disappears suggests that an equation may be found –' "

"But it can, sir!"

"Then why did you not make the equation?"

"Well – I don't know. I wasn't enough of a mathematician."

"Then you should not have published till you knew your math!"

"I – Look, Dr. Gottlieb, do you really think I know enough to work here? I want terribly to succeed."

"Succeed? I have heard that word. It is English? Oh, yes, it is a word that liddle schoolboys use at the University of Winnemac. It means passing examinations. But there are no examinations to pass here. . . . Martin, let us be clear. You know something of laboratory technique; you have heard about dese bacilli; you are not a good chemist, and mathemathics – pfui! – most terrible! But you have curiosity and you are stubborn. You do not accept rules. Therefore I t'ink you will either make a very good scientist or a very bad one, and if you are bad enough, you will be popular with the rich ladies who rule this city, New York, and you can gif lectures for a living or even become, if you get to be plausible enough, a college president. So anyway, it will be interesting."

Half an hour later they were arguing ferociously, Martin asserting that the whole world ought to stop warring and trading and writing and get straightway into laboratories to observe new phenomena; Gottlieb insisting that there were already too many facile scientists, that the one thing necessary was the mathematical analysis (and often the destruction) of phenomena already observed.

It sounded bellicose, and all the while Martin was blissful with the certainty that he had come home.

The laboratory in which they talked (Gottlieb pacing the floor, his long arms fantastically knotted behind his thin back; Martin leaping on and off tall stools) was not in the least remarkable – a sink, a bench with racks of numbered test-tubes, a microscope, a few note-books and hydrogen-ion charts, a grotesque series of bottles connected by glass and rubber tubes on an ordinary kitchen table at the end of the room – yet now and then during his tirades Martin looked about reverently.

Gottlieb interrupted their debate: "What work do you want to do here?"

"Why, sir, I'd like to help you, if I can. I suppose you're cleaning up some things on the synthesis of antibodies."

"Yes, I t'ink I can bring immunity reactions under the mass action law. But you are not to help me. You are to do your own work. What do you want to do? This is not a clinic, wit' patients going through so neat in a row!"

"I want to find a hemolysin for which there's an antibody. There isn't any for streptolysin. I'd like to work with staphylolysin. Would you mind?"

"I do not care what you do – if you just do not steal my staph cultures out of the ice-box, and if you will look mysterious all the time, so Dr. Tubbs, our Director, will t'ink you are up to something big. So! I haf only one suggestion: when you get stuck in a problem, I have a fine collection of detective stories in my office. But no. Should I be serious – this once, when you are just come?

"Perhaps I am a crank, Martin. There are many who hate me. There are plots against me – oh, you t'ink I imagine it, but you shall see! I make many mistakes. But one thing I keep always pure: the religion of a scientist.

"To be a scientist – it is not just a different job, so that a man should choose between being a scientist and being an explorer or a bond-salesman or a physician or a king or a farmer. It is a tangle of ver-y obscure emotions, like mysticism, or wanting to write poetry; it makes its victim all different from the good normal man. The normal man, he does not care much what he does except that he should eat and sleep and make love. But the scientist is intensely religious – he is so religious that he will not accept quarter-truths, because they are an insult to his faith.

"He wants that everything should be subject to inexorable laws. He is equal opposed to the capitalists who t'ink their silly money-grabbing is a system, and to liberals who t'ink man is not a fighting animal; he takes both the American booster and the European aristocrat, and he ignores all their blithering. Ignores it! All of it! He hates the preachers who talk their fables, but he iss not too kindly to the anthropologists and historians who can only make guesses, yet they have the nerf to call themselves scientists! Oh, yes, he is a man that all nice good-natured people should naturally hate!

"He speaks no meaner of the ridiculous faith-healers and chiropractors than he does of the doctors that want to snatch our science before it is tested and rush around hoping they heal people, and spoiling all the clues with their footsteps; and worse than the men like hogs, worse than the imbeciles who have not even heard of science, he hates pseudo-scientists, guess-scientists – like these psycho-analysts; and worse than those comic dream-scientists he hates the men that are allowed in a clean kingdom like biology but know only one text-book and how to lecture to nincompoops all so popular! He is the only real revolutionary, the authentic scientist, because he alone knows how liddle he knows.

"He must be heartless. He lives in a cold, clear light. Yet dis is a funny t'ing: really, in private, he is not cold nor heartless – so much less cold than the Professional Optimists. The world has always been ruled by the Philanthropists: by the doctors that want to use therapeutic methods they do not understand, by the soldiers that want something to defend their country against, by the preachers that yearn to make everybody listen to them, by the kind manufactures that love their workers, by the eloquent statesmen and soft-hearted authors – and see once what a fine mess of hell they haf made of the world! Maybe now it is time for the scientist, who works and searches and never goes around howling how he loves everybody!

"But once again always remember that not all the men who work at science are scientists. So few! The rest – secretaries, press-agents, camp-followers! To be a scientist is like being a Goethe: it is born in you. Sometimes I t'ink you have a liddle of it born in you. If you haf, there is only one t'ing – no, there is two t'ings you must do: work twice as hard as you can, and keep people from using you. I will try to protect you from Success. It is all I can do. So. . . . I should wish, Martin, that you will be very happy here. May Koch bless you!"

26

Again from de Kruif, an impression of Jacques Loeb, the main ingredient in Max Gottlieb, and of some other ornaments of the Rockefeller, including the egregious surgeon, Alexis Carrel, a tireless self-publicist, closely associated with the monkey-gland transplanter, Voronoff, who appears later.

It was thrilling to sit at Jacques Loeb's table listening to that parent of fatherless sea urchins, a veritable eighteenth-century encyclopedist come back to life, making offhand wisecracks about Voltaire and Diderot and pouring sarcasm on contemporary scientists. "As a physicist he is a good husband and father," said Loeb of one eminento, his Mephistophelic eyebrows arched, his black eyes flashing. "Medical science?" said Jacques Loeb, chuckling. "Dat iss a contradiction in terms. Dere iss no such thing." "You should begin with the physical chemistry of proteins, as I do," he admonished his tablemates, who were hard put to it to swap mental punches with the famous founder of the philosophy of a mechanistic conception of life.

What a *Kopf* he was, Jacques Loeb. He gave the Institute a high scientific tone. In addition to his advanced researches, he served as the Institute's front man. I can see him now, piloting Albert Einstein – who looked every inch a concert violinist – about our spotless halls. I adored Jacques Loeb and tried to ape his tart dialectic and I was a thousand miles behind him and would never catch up. In those days I thanked him for encouraging my atheism; he was the peerless leader of the militant godless. But best of all he was the exponent of scientific method as against the prevailing twaddle – *that* was his word – of medical science and I tried secretly to copy him. He was kind to me and backed me up in a semantic hassle over terminology with my chief, Dr. Flexner, reviewing my first manuscript published from the Rockefeller.

Those luncheon sessions were a kindergarten in my stumbling study of character. I never tired of listening to the philosophy of Alexis Carrel, who had won the Nobel Prize in medicine for his ingenious end-to-end anastomosis of arteries, so that he could transplant a kidney, from cat to cat, with its blood supply intact. This worked, too, in men. Carrel, who had been in America a long time, had carefully preserved his French accent, which made him sound to me even more learned than he was. In his lab, in his black surgical gown and cap, Carrel was a magician keeping a bit of chicken heart muscle beating, in a bottle, transplanting it from bottle to bottle ad infinitum over the years, hinting the awesome possibility that physical life is (potentially) immortal. From time to time the birthdays of those bits of chicken in their bottles, tended with devotion by Dr. Ebeling, made newspaper front pages. Dr. Carrel's publicity genius astounded me and I secretly practiced his French accent, hoping it would fortify my own taxicab driver's French if I ever got back to France.

Then at the luncheon table there might be Dr. Peyton Rous, refined, gentle, exquisitely cultured. He had uncovered the virus of a cancer, a sarcoma of chickens; it hinted that all cancers might turn out to be viral, as it now, forty years later, becomes possible that they are. Dr. Rous was so amazed at his own discovery that it was rumoured he couldn't stand the mental strain of going on with it and you couldn't blame him.

27

Miroslav Holub is a Czech microbiologist and poet, whose science nourishes his poetry.

SUFFERING

Ugly creatures, ugly grunting creatures,
Completely concealed under the point of the needle,
  behind the curve of the Research Task Graph,
Disgusting creatures with foam at the mouth,
  with bristles on their bottoms,
One after the other
They close their pink mouths
They open their pink mouths
They grow pale
Flutter their legs
  as if they were running a very
  long distance,

They close ugly blue eyes,
They open ugly blue eyes
  and
  they're
  dead.

But I ask no questions,
no one asks any questions.

And after their death we let the ugly creatures
  run in pieces along the white expanse
  of the paper electrophore
We let them graze in the greenish-blue pool
  of the chromatogram
And in pieces we drive them for a dip
  in alcohol
  and xylol
And the immense eye of the ugly animal god
  watches their every move
  through the tube of the microscope . . .

Einstein once observed, patting the top pocket of his jacket, containing his fountain-pen, that here was his laboratory. Mathematicians and theoretical physicists occupy a different sphere from experimentalists. The great schools of mathematics that thrived in Poland between the wars, appeared to function, in the best Central European tradition, in cafés. Here is the scene as described by Stanislas Ulam in his Adventures of a Mathematician.

The tables had white marble tops on which one could write with a pencil, and, more important, from which notes could be easily erased.

There would be brief spurts of conversation, a few lines would be written on the table, occasional laughter would come from some of the participants, followed by long periods of silence during which we just drank coffee and stared vacantly at each other. The café clients at neighbouring tables must have been puzzled by these strange doings. It is such persistence and habit of concentration which somehow becomes the most important prerequisite for doing genuinely creative mathematical work.

Thinking very hard about the same problem for several hours can produce a severe fatigue, close to a breakdown. I never really experienced a breakdown, but have felt 'strange inside' two or three times during my life. Once I was thinking hard about some mathematical constructions, one after the other, and at the same time trying to keep them all simultaneously in my mind in a very conscious effort. The concentration and mental effort put an added strain on my nerves. Suddenly things started going round and round, and I had to stop.

These long sessions in the cafés with Banach, or more often with Banach and Mazur, were probably unique. Collaboration was on a scale and with an intensity I have never seen surpassed, equaled or approximated anywhere – except perhaps at Los Alamos during the war years.

Banach confided to me once that ever since his youth he had been especially interested in finding proofs – that is, demonstrations of conjectures. He had a subsconscious system for finding hidden paths – the hallmark of his special genius.

After a year or two Banach transferred our daily sessions from the Café Roma to the 'Szkocka' (Scottish Café) just across the street. Stozek was there every day for a couple of hours, playing chess with

Nikliborc and drinking coffee. Other mathematicians surrounded them and kibitzed.

Kuratowski and Steinhaus appeared occasionally. They usually frequented a more genteel teashop that boasted the best pastry in Poland.

It was difficult to outlast or outdrink Banach during these sessions. We discussed problems proposed right there, often with no solution evident even after several hours of thinking. The next day Banach was likely to appear with several small sheets of paper containing outlines of proofs he had completed in the meantime. If they were not polished or even not quite correct, Mazur would frequently put them in a more satisfactory form.

29

E.C. Large was a mycologist at the London Hospital Medical School and a considerable novelist. His best work was Sugar in the Air, *which spins a tale about the pursuit of commercial* in vitro *photosynthesis.*

In something rather less than an hour Pry had the protesting Ackworth making titrations of sugar solutions, and the vacuum oven was going. It irked Pry to have to wait for Ackworth, who was at best excessively deliberate, just because these operations happened to be 'chemical'. He had not Ackworth's chemical qualifications, but he saw no reason why that should prevent him from measuring just how much of a particular solution was required to change another solution from a beautiful ammoniacal blue into a bright red mud. He had done it often enough. He fretted about the laboratory for some time and then said: 'Since you haven't an assistant, don't you think it would be a good idea if you taught me just the elementary parts of some of this, so that I can give you a hand?'

Again Ackworth didn't mind, and the operations on the laboratory bench began to move at a speed determined by Mr. Pry. Ackworth was a sociable person, who began to seek for distractions as soon as he was left to work by himself, and Pry made a mental note of it. In the afternoon, when enough had been done to assure that there would be one or two results which he could work up into a report the next day, Pry, in effect, let Ackworth off, and accompanied him on a tour of the works.

The factory was lofty and spacious, the Cocaine quartz tube installation only occupied one bay, another had been provided for extensions, and yet another for auxiliary machinery, and the great glass storage vats, to contain the sugar that was to have been produced. Near these vats stood hundreds of glass carboys, vessels of ten gallons capacity, packed with straw in steel baskets. Pry called to a rueful looking man with a bulging forehead, who was about the place, and asked him if he could say what they contained.

'That is some of Mr. Cocaine's glucose, sir, wot wasn't sent away.'

'Thanks. Now what is your name, and what do you do here?'

'Name of Plummox, sir, bin here since it was all put in, sir, but I don't know nothink, I'm nobody, you ask that little toad, he treats me like dirt . . .'

Pry looked across at Ackworth. 'Meaning . . .?' he continued.

'That dirty little crawler, Pinks, sir.'

'Ah! I expect we shall come to him presently. Now Plummox, you just get into the laboratory and fetch me a beaker, will you?'

'And pleased to do it, sir, knowing what I do.'

Ackworth stirred up the contents of one of the carboys and drew off a sample into the beaker. 'Put it in two bottles,' said Pry, 'we'll keep one in the safe, and the other we'll analyse . . .'

'Mr. Cocaine, sir, he tests it by holding it up to the light, and when it's like he wants it he says, "Plummox," he says, "if you can make the glucose like that you will know the Cocaine secret", but other times he says, "No! no good yet", and he tells me to pour it down the drain, sir.'

'Ever seen a polarimeter, Plummux?'

'No, sir, not here, sir.'

'Splendid, now tell us some more, what's that?'

'That's the compressor, sir, it hasn't ever bin used, but I heard Mr. Pike, that was the contractor's engineer, sir, telling Mr. MacDuff it was to make liquid air, sir, to get the carbon out.'

'And where do you get your "carbon" from now?'

'Out of them cylinders, sir, it comes out as a sort of gas, like.'

'Carbon dioxide, Plummox, *is* a gas. Except when it's compressed or frozen, when it may be either a liquid or a solid. In those cylinders it's liquid. Ever seen any solid, here; the stuff they use in the icecream carts?'

'No, never, sir.'

'Never mind . . . this the stores?'

'Nobody isn't allowed to go in there, sir, except Mr. Cocaine.'

'Go to Mr. Cocaine, give him my compliments, and tell him I want the key.'

Plummox looked scared, and he hesitated: 'If you say so I suppose it's all right, sir . . .' he hitched his trousers humorously and spat on the floor. 'I'll go!'

'On the contrary,' murmured Pry, as Plummox retreated, 'I think we shall keep *you* on.'

30

The only contemporary literary figures of the first rank in whose work science plays any major part are Primo Levi and John Updike. Here is Updike's description, in his novel Couples, *of a medical school biochemistry laboratory in Boston, and the middle-aged* angst *of a successful biochemist.*

Ken Whitman's field of special competence, after his early interest in echinoid metabolism, was photosynthesis; his doctoral thesis had concerned the 7-carbon sugar sedoheptulose, which occupies a momentary place within the immense chain of reactions whereby the five-sixths of the triosephosphate pool that does not form starch is returned to ribulose-5-phosphate. The process was elegant, and few men under forty were more at home than Ken upon the gigantic ladder, forged by light, that carbon dioxide descends to become carbohydrate. At present he was supervising two graduate students in research concerning the transport of glucose molecules through cell walls. By this point in his career Ken had grown impatient with the molecular politics of sugar and longed to approach the mysterious heart of CO_2 fixation – chlorophyll's transformation of visible light into chemical energy. But here, at this ultimate chamber, the lone reaction that counterbalances the vast expenditures of respiration, that reverses decomposition and death, Ken felt himself barred. Biophysics and electronics were in charge. The grana of stacked quantasomes were structured like the crystal lattices in transistors. Photons excited an electron flow in the cloud of particles present in chlorophyll. Though he had ideas – why chlorophyll? why not any number of equally complex compounds? was the atom of magnesium the clue? – he would have to put himself to school again and, at thirty-two, felt too old. He was wedded to the unglamorous carbon cycle while younger men were achieving fame and opulent grants in such fair fields as neurobiology, virology, and the wonderful new wilderness of nucleic acids. He had a wife, a coming child, a house in need of extensive repair. He had overreached. Life, whose graceful secrets he would have unlocked, pressed upon him clumsily.

As if underwater he moved through the final hour of this heavy gray day. An irreversible, constricted future was brewing in the apparatus of his lab – the fantastic glass alphabet of flasks and retorts, the clamps and slides and tubes, the electromagnetic scales sensitive to the hundredth of a milligram, the dead experiments probably duplicated at Berkeley or across the river. Ken worked on the fourth floor of a monumental neo-Greek benefaction, sooty without and obsolete within, dated 1911. The hall window, whose sill held a dreggish Lily cup, overlooked Boston. Expressways capillariously fed the humped dense center of brick red where the State House dome presided, a gold nucleolus. Dusty excavations ravaged the nearer ground. In the quad directly below, female students in bright spring dresses – dyed trace elements – slid along the paths between polygons of chlorophyll. Ken looked with a weariness unconscious of weariness. There had been rain earlier. The same rain now was falling on Tarbox. The day was so dull the window was partly a mirror in which his handsomeness, that strange outrigger to his career, glanced back at him with a cocked eyebrow, a blurred mouth, and a glint of eye white. Ken shied from this ghost; for most of his life he had consciously avoided narcissism. As a child he had vowed to become a saint of science and his smooth face had developed as his enemy. He turned and walked to the other end of the hall; here, for lack of space, the liquid-scintillation counter, though it had cost the department fifteen thousand, a Packard Tri-Carb, was situated. At the moment it was working, ticking through a chain of isotopically labeled solutions, probably Neusner's minced mice livers. A thick-necked sandy man over forty, Jewish only in the sleepy lids of his eyes. Neusner comported himself with the confidence of the energetically second-rate. His lectures were full of jokes and his papers were full of wishful reasoning. Yet he was liked, and had established forever the spatial configuration of one enzyme. Ken envied him and was not sorry to see, at four-thirty, his lab empty. Neusner was a concertgoer and winetaster and womanizer and mainstay of the faculty supper club; he traveled with the Cambridge political crowd and yesterday had confided to Ken in his hurried emphatic accents the latest Kennedy joke. *One night about three a.m. Jackie hears Jack coming into the White House and she meets him on the stairs. His collar is all rumpled and there's lipstick on his chin and she asks him, Where the hell have you been? and he tells her, I've been having a conference with Madame Nhu, and she says, Oh, and doesn't think any more about it until the next week the same thing happens and this time he says he was sitting up late arguing ideology with Nina Khrushchev . . .* A sallow graduate student was tidying up the deserted labs. A heap of gutted white mice lay like burst grapes on a tray. Pink-eyed cagefuls alertly awaited annihilation. Neusner loved computers and statistical theory and his papers were famous for the sheets of numbers that masked the fantasy of his conclusions. Next door old Prichard, the department's prestigious ornament, was pottering with his newest plaything, the detection and analysis of a memory-substance secreted by the brain. Ken envied the old man his childlike lightness, his freedom to dart through forests of evidence after such a bluebird. Neusner, Prichard – they were both free in a way Ken wasn't. Why? Everyone sensed it, the something wrong with Ken, so intelligent and handsome and careful and secure – the very series expressed it, an unstable compound, unnatural. Prichard, a saint, tried to correct the condition, to give Ken of himself, sawing the air with his papery mottled hands, nodding his unsteady gaunt head, whose flat cheeks seemed rouged, spilling his delicate stammer: *The thing of it, the thing of it is, Wh-Whitman, it's just t-tinkering, you mustn't s-s-suppose life, ah, owes us anything, we just g-get what we can out of the b-bitch, eh? . . .* He wondered why Prichard had never won the Nobel and deduced that his research was like his hobbies, darting this way and that, more enthusiasm than rigor. He thought of photosynthesis and it appeared to him there was a tedious deep flirtatiousness in nature that withheld her secrets while the church burned astronomers and children died of leukemia. That she yielded by whim, wantonly, to those who courted her offhand, with a careless ardor he, Ken, lacked. *The b-b-bitch.*

31

Alex Comfort is a former Director of a Medical Research Council Gerontology Research Unit and a novelist, poet and man of letters. The extract which follows is from Come out to Play *and combines two of his interests – biochemistry and sex (on the pleasures of which he has written abundantly). Reports that human pheromones indeed exist have appeared at intervals, including quite recently.*

"I'm going to put the substance on my own skin. I only want you to record your response. But do exactly what I say: no questions, because I must not inculcate a prejudice before you try."

He put a little plastic wash-hand-basin between us, and filled it from a Winchester. "One per cent citric acid," he explained. "To kill any excess. It acts instantly as a quench bath."

He took off his coat and rolled up his left sleeve. Then he fiddled about with something under the central fume cupboard, and I heard a safe door shut. Marcel came back with a green screw-topped tin. It appeared to be full of sherbet.

"More citric acid. As powder."

He fished out of the sherbet a small phial containing filter-paper strips half the size of a postage stamp, set it down, and took forceps from the drawer. Then he passed me a pencil and a pad of paper.

"This is the procedure," he said. "I will place a strip of paper on my skin here" – he indicated his forearm – "the substance on male skin is slowly hydrolysed. Traces are absorbed of the hydrolysate. No effect is produced on the wearer. What I want you to do is this. Raise your left hand . . ."

"And swear you're telling me the truth, the whole truth and nothing . . ."

"*Soyez sérieux, par exemple.* Raise your left hand and lower it as soon as you notice an effect of any kind upon yourself – mood, thinking, affections, anything."

"Agreed."

"Simultaneously write with your other hand some words describing me. Your reaction towards me, what I am like. Understood?"

"Just reassure me that in France a display of affection between male colleagues isn't a basis for lifelong blackmail," I said.

"There will be nothing of that kind."

He took out a slip of filter-paper. "Now. Left hand up – write." He had his eye fixed on a stopwatch. After thirty seconds I lowered my hand.

"You feel something?"

"I feel a conviction that that was a blank run."

"Correct. On this filter-paper there is nothing at all."

"I've written 'medium height, comes from Midi, voluble, must have a hell of a big research grant'."

"Excellent. Now again." He took another slip. "Go!"

At ten seconds I'd written 'short, dark and handsome' – Marcel's forearm with its little postage stamp was under my nose. I wondered if he'd try two blanks to be sure, and simultaneously began to resent the procedure. In fact I decided, quite suddenly, that I didn't really like this fellow; here was I wasting my time with a bloody little . . . I clapped my left hand down quickly. Marcel, who'd had his eye on my face, put his arm in the bowl. The little stamp drifted off.

"*Hein*?" He looked hard at me.

That stab of furious, contemptuous anger at him had come on slowly enough for me to miss the change in my mood until it had developed, but the return to normal was a definite jolt.

"Yes?"

"I suddenly felt hopping mad with you. I wanted to call you something unforgivable."

Marcel nodded.

"Is that it?" I asked.

"Bitter resentment and jealousy – I am contemptible to you, I have stolen your umbrella: I have stolen your girl! Death to me?"

"More or less."

"What have you written?"

"It's probably something offensive. I don't mean it, you know – it isn't my considered –"

"No, no, of course, but let's see."

He took the paper and read out, "Short, dark and handsome, bloody conceited little – *Qu'est que c'est que ce mot-ci?*"

"It's probably 'frog'," I said, unhappily, "a derogatory name which Englishmen once used of the French."

He was delighted. "Frog, *c'est grenouille? Mais non, c'est encore trop long – regardez . . .*"

The word I had written was 'OTTOPATH'.

"Pathic, psychopath? Or perhaps osteopath – you are a doctor, and this is a deadly insult among doctors."

"Worse," I said. "And I know why I wrote OTTOPATH. There was someone else I didn't like whose name has got in."

"This substance," said Marcel, *con amore*, "is absorbed by the skin. In the female it is unchanged. In the male it is hydrolysed and something is released – of great power. Some women say they can smell it – like cedarwood or cigar boxes. The wearer feels nothing. But *he excites the resentment of all other males*. Is this biologically possible?"

"It's not only possible," I said, slowly, "I've already got some reason to think that a substance of this kind exists in nature. In man. It would explain something – several things. Look, this stuff is dangerous; you may worry about 3-blindmycin, but that could only start an orgy – this stuff is really open to abuse."

He was nodding like a happy Chinaman, delighted to see me looking shaken. "So you see why I am afraid to make your 3-blindmycin."

I told him I saw quite clearly, and I thought his hesitation was wise.

"We will make it, therefore, together. Under your supervision if you wish. After all, it is you who discovered the principle." We shook hands conscientiously.

"I had a very nasty series of experiences with this 1,2-compound," said Marcel. "It took me a long time to realize what was happening. Incidentally, for this one I *have* got a name. It is cocuficin. It makes you feel as if you had been cuckolded. By the wearer. This describes the sensation accurately?"

I agreed that it did, but my mind was really on the word OTTOPATH, and the possibility that cocuficin, or cocuficin-like subliminal stimuli of other kinds, might make the cuckolding an active rather than a passive matter.

32

Vicki Baum was a widely read, now almost forgotten novelist, who occasionally took up scientific themes. The Weeping Wood *is a discursive novel about rubber through the ages. One of its heroes is here seen looking forward to the delights of a scientific conference.*

When Jim received the invitation of the American Chemical Society to the 1939 meeting, which was to take place in Boston "in Commemoration of the Discovery of the Vulcanisation of Rubber by Charles Goodyear in 1839," Janet broke into an Indian war dance.

"Gosh, Jim, that's going to be something! Just what we need to keep us from getting dry rot. It's as if we had been stuck in a swamp all the time, no one to talk our language, not a breath of air – just imagine how you'll wallow in shop talk there. Jimmy, it's too good to be true!" She crept close to him and read the programme with him, greeting every important name, every interesting paper to be read, every exhilarating event, with squeals of excited anticipation. It was as if their membership in the society opened the gates to a chemist's fairyland. Symposiums every day; papers about every conceivable chemical subject; lectures, discussions, gatherings of the lady chemists, amd more symposiums and more papers, and special emphasis on the rubber division, and a reception at Harvard, and a paper on polybutenes and a paper on the catalytic dehydrogenation of mono-olefins to diolefins, which were right up Jim's street; but also dances and banquets and concerts and invitations to the Navy Yard and garden parties and moonlight picnics. "Oh, Jimmy!" Janet sighed, exhausted by contemplating the abundance of pleasures in store for them, "oh, Jim, we've got to be there. I'd die if we missed it. You must make a reservation for us right away. Goodness, I'm so excited I think I'll bust!"

33

Here is what one is exposed to at scientific meetings, from a collection of verse by a Nobel Laureate in Chemistry, Roald Hoffmann.

NEXT SLIDE, PLEASE

there was no question that the reaction worked
but transient colors were seen
in the slurry of sodium methoxide in dichloromethane
and we got a whole lot of products
for which we can't sort out the kinetics
the next slide will show
the most important part
very rapidly
within two minutes
and I forgot to say on further warming
we get in fact the ketone
you can't read it on the slides
but I refer to the structure you saw before
the low temperature infrared spectrum
as I say

gives very direct evidence
so does the NMR
we calculated it
throwing away the geminal coupling
which is of course wrong
there is a difference of 0.9 parts per million
and it is a singlet
and sharp
which means two things
either
you're doing this NMR in excess methoxide
and it's exchanging
or
I would hazard a guess
that certainly in these nucleophilic conditions
there could well be
an alternative path
to the enone you see there
it's difficult to see
you could monitor this quite well in the infrared
I'm sorry in the NMR
my time is up I see
well this is a brief summary of our work
not all of which
I've had time to go into
in as much detail as I wanted
today.

Sydney Smith, divine, Whig, wit and iconoclast (he described the two ancient Universities as 'enormous hulks confined with mooring chains, everything flowing and progressing around them'), was a founder of the Edinburgh Review. *Rebuking his friend, co-founder and editor, Francis Jeffery for being too critical and analytical, he summed up Jeffrey's attitudes thus:*

> Damn the Solar System – Bad Light; planets too distant – pestered with comets – feeble contrivance; could do better myself.

34

V.N. Ipatieff, a famous Russian industrial chemist, reprints in his memoirs, The Life of a Chemist, *this poignant description of the last lecture of a much-admired Professor of Applied Mathematics at the Mikhail Military Academy. It is by one Cherniavsky, and comes from a journal called* Mikhailovtsy, *published in Belgrade in 1937. Professor Budaev had taught at the Academy for thirty-five years, until his seventieth birthday, and throughout this time had never been known to consult a note or write an incorrect line on the blackboard.*

During the regular school days we had two hours of lectures by Professor Budaev. As he entered the class room he carefully wrapped the chalk in a piece of paper and began to write some complicated mathematical equations accompanied by very little explanation. As the lecture drew to a close we noticed that Budaev suddenly stopped and turned to the desk as if he were searching for something. He picked up the eraser and for the first time in his life erased some figures. But the poor old man could not find the mistake that he had made, and he abruptly ended the lecture ten minutes early and went to the professorial room adjacent to the class room. We all remained sitting in silence, realizing that we were witnessing a drama in a human life.

We could see him sitting in the professorial room, smoking and looking at a paper, apparently going through the mathematical proof in which he had made a mistake a short time before. When the second hour began Budaev entered the class room, sat down sidewise in his chair but did not touch the chalk. He remained silent for a few minutes and then said in a very low voice: 'Gentlemen, Budaev made a mistake, so he must be getting old and is no longer any good.' He got up, bowed to the class and left the room. This was his last lecture, and it was also the last time that we ever saw him alive. This man of science and remarkable teacher died shortly afterward, apparently unable to live without teaching and yet no longer physically able to do so. In his memory the students collected a 3,000-ruble fund to bear his name, the income to be given to the student showing the highest proficiency in mathematics in the Mikhail Artillery School.

35

A lecturer of another, probably alas commoner, type was Osborne Reynolds (he of viscous fluid flow and Reynolds's number), of Owen's College in Manchester, as described by J.J. Thomson in Recollections and Reflections*:*

He was one of the most original and independent of men, and never did anything or expressed himself like anybody else. The result was that it was very difficult to take notes at his lectures, so that we had to trust mainly to Rankine's text-books. Occasionally in the higher classes he would forget all about having to lecture and, after waiting for ten minutes or so, we sent the janitor to tell him the class was waiting. He would come rushing into the room pulling on his gown as he came through the door, take a volume of Rankine from the table, open it apparently at random, see some formula or other and say it was wrong. He then went up to the blackboard to prove this. He wrote on the board with his back to us, talking to himself, and every now and then rubbed it all out and said it was wrong. He would then start afresh on a new line, and so on. Generally, towards the end of the lecture, he would finish one which he did not rub out, and say that this proved Rankine was right after all. This, though it did not increase our knowledge of facts, was interesting, for it showed the workings of a very acute mind grappling with a new problem.

36

Arthur Hailey's novel Strong Medicine *is full of remarkably authentic background on research in the pharmaceutical industry and also of the increasingly close relation that has developed in the last decade or so between academic scientists and their industrial brethren. The sub-plot, in which an American pharmaceutical giant sets up a research laboratory in England, because of the lower costs, parallels a number of exactly similar developments in the 1960s and 1970s.*

'And it's likely that big future medical advances will come when we understand the chemistry of DNA better, showing us how genes work and why they sometimes go wrong. That's what I'm researching now, using young and old rats, trying to find differences, varying with age, between the animals' mRNA – messenger ribonucleic acid – which is a template made from their DNA.'

Sam interjected, 'But Alzheimer's disease and the normal ageing process are two separate things, right?'

'It appears so, but there may be overlapping areas.' As Peat-Smith paused, Celia could sense him organizing his thoughts, as a teacher would, into simpler, less scientific words than he was accustomed to using.

'An Alzheimer's victim may have had, at birth, an aberration in his DNA, which contains his coded genetic information. However, someone else, born with more normal DNA, can change that DNA by damaging its environment, the human body. Through smoking, for example, or a harmful diet. For a while, our built-in DNA repair mechanism will take care of that, but as we get older the genetic repair system may slow down or fail entirely. Part of what I'm searching for is a reason for that slowing . . .'

At the end of the explanation, Celia said, 'You're a natural teacher. You enjoy teaching, don't you?'

Peat-Smith appeared surprised. 'Doing some teaching is expected at a university. But, yes, I enjoy it.'

Another facet of this man's interesting personality, Celia thought.

She said, 'I'm beginning to understand the questions. How far are you from answers?'

'Perhaps light-years away. On the other hand we might be close.' Peat-Smith flashed his genuine smile. 'That's a risk that grant givers take.'

The head waiter brought menus and they paused to decide about lunch.

When they had chosen, Peat-Smith said, 'I hope you'll visit my laboratory. I can explain better there what I'm trying to do.'

'We were counting on that,' Sam said. 'Right after lunch.'

While they were eating, Celia asked, 'What is your status at Cambridge, Dr Peat-Smith?'

'I have an appointment as a lecturer; that's more or less equivalent to assistant professor in America. What it means is that I get lab space in the Biochemistry Building, a technician to help me, and freedom to do research of my choice.' He stopped, then added, 'Freedom, that is, if I can get financial backing.'

'About the grant we're speaking of,' Sam said. 'I believe the amount suggested was sixty thousand dollars.'

'Yes. It would be over three years, and is the least I can get by on – to buy equipment and animals, employ three full-time technicians, and conduct experiments. There's nothing in there for me personally.' Peat-Smith grimaced. 'All the same, it's a lot of money, isn't it?'

Sam nodded gravely. 'Yes, it is.'

But it wasn't. As both Sam and Celia knew, sixty thousand dollars was a trifling sum compared with the annual expenditures on research by Felding-Roth Pharmaceuticals or any major drug firm. The question, as always, was: Did Dr Peat-Smith's project have sufficient commercial promise to make an investment worthwhile?

'I get the impression,' Celia told Peat-Smith, 'that you're quite dedicated to the subject of Alzheimer's. Was there some special reason that got you started?'

The young scientist hesitated. Then, meeting Celia's eyes directly, he said, 'My mother is sixty-one, Mrs Jordan. I'm her only child; not surprisingly, we've always been close. She's had Alzheimer's disease for four years and become progressively worse. My father, as best he can, takes care of her, and I go to see her almost every day. Unfortunately, she has no idea who I am.'

Cambridge University Biochemistry Building was a three storeyed red-brick neo-Renaissance structure, plain and unimpressive. It was in Tennis Court Road, a modest lane where no tennis court existed. Martin Peat-Smith, who had come to lunch on a bicycle – a standard form of transportation in Cambridge, it appeared – pedalled energetically ahead while Sam and Celia followed in the Jaguar.

At the building's front door, where they rejoined him, Peat-Smith cautioned, 'I think I should warn you, so you're not surprised, that our facilities here are not the best. We're always crowded, short of space' – again the swift smile – 'and usually short of money. Sometimes it shocks people from outside to see where and how we work.'

Despite the warning, a few minutes later Celia was shocked.

When Peat-Smith left them alone briefly, she whispered to Sam, 'This place is *awful* – like a dungeon! How can anyone do good work here?'

On entering, they had descended a stairway to a basement.

The hallways were gloomy. A series of small rooms leading off them appeared messy, disordered, and cluttered with old equipment. Now they were in a laboratory, not much bigger than the kitchen of a small house, which Peat-Smith had announced was one of two that he worked in, though he shared both with another lecturer who was pursuing a separate project.

While they were talking, the other man and his assistant had come and gone several times, making a private conversation difficult.

The lab was furnished with worn wooden benches, set close together to make the most of available space. Above the benches were old-fashioned gas and electrical outlets, the latter festooned untidily, and probably unsafely, with adapters and many plugs. On the walls were roughly-made shelves, all filled to capacity with books, papers and apparently discarded equipment, amid it, Celia noticed, some outmoded retorts of a type she remembered from her own chemistry work nineteen years earlier. A portion of bench was a makeshift desk. In front was a hard Windsor chair. Several dirty drinking mugs could be seen.

On one bench were several wire cages, inside them, twenty or so rats – two to a cage, and in varying states of activity.

The floor of the laboratory had not been cleaned in some time. Nor had the windows, which were narrow, high up on a wall, and providing a view of the wheels and undersides of cars parked outside. The effect was depressing.

'No matter how it all looks,' Sam told Celia, 'never forget that a lot of scientific history has been made here. Nobel Prize winners have worked in these rooms and walked these halls.'

'That's right,' Martin Peat-Smith said cheerfully; he had returned in time to hear the last remark. 'Fred Sanger was one of them; he discovered the amino acid structure of the insulin molecule in a lab right above us.'

37

Ralph A. Lewin is a biologist with a deft touch as a rhymster.

ELKS, WHELKS, AND THEIR ILK

The monarchs of the Irish bogs
Succumbed to neither men nor dogs

But (most ecologists agree)
To calcium deficiency.
  They scoured the base-deficient peat
For antlers and old shells to eat
Around the Celtic countryside
And, finding all too few, they died.

  Then mourn the passing of the elks
But note the wisdom of the whelks
That roam the shore – their native heath –
With silver-indurated teeth,
  And bore to death their mollusc friends,
Who come to sad, unsuccored ends.
Without the need for extra lime,
The whelks survive to modern time.

  Thus ungulate and gastropod,
And all that live by sea or sod,
Are doomed to be, or not to be,
By biogeochemistry.

38

E.C. Large's second scientific novel, Asleep in the Afternoon, *is not the equal of* Sugar in the Air, *but has pleasant moments. Here he prefigures much later research on anti-noise, in which the sound vibrations are cancelled out by superimposing complementary vibrations on them.*

Then he found a much better affair, the parts of which could be almost completely concealed about his person, so that all that was visible was a thin black cord, attached to a plug which he stuck in his ear. He used that for a long time and was very satisfied with it. He would never have changed it for anything else if it had not been for Professor Hunt-Transom of the Physics Department at Queen Victoria College, London.

That was five years before the events related in Chapter III. Professor Hunt-Transom had been doing some work – for an Asiatic government – on the shriek of underground trains and its effects on Man. The small steel wheels, racing over steel track, through tunnels resonant as organ-pipes, gavc an excessive sensation of speed, and set up very complex and shrill vibrations. These were transmitted through the air, and through the springs and body-work of the coaches, to the nervous systems of the passengers. A project for building underground railways, in the Asiatic country concerned, had been opposed on the ground that 'exposure to shriek', for an hour or more each day, might be bad for Asiatic nervous systems. Professor Hunt-Transom had been commissioned to investigate this absurd contention, and to prepare a report to the effect that it was not so. He did not reach the desired conclusion, but that was beside the point. What did matter was that, as a consequence of Hunt-Transom's report, it was decided that the investigations should continue, to find how the design of the rolling-stock should be modified, so that some of the more excruciating noise and vibration would be 'damped off'. This was a highly scientific job: it meant, amongst other things, opposing vibrations with vibrations so that they nullified each other, and on the mathematical part of this Hugo Boom was called in.

Hunt-Transom's eye was on the plug in Hugo's ear all the time he first talked to him about rolling-stock, and with that boyish curiosity so charming in scientists, he had persuaded Hugo to bring out the concealed parts of the apparatus and let him play with it.

39

The joker in the laboratory is a theme explored only in anthologies. Professor R.V. Jones, who has written in Most Secret War *(see also no. 181) a remarkable autobiographical account of the achievements of scientific intelligence during the Second World War, was a noted* farceur.

Bosch told me that he had worked on an upper floor of a laboratory from which he could see into the windows of a block of flats, and he had found that the occupant of one of them was a newspaper reporter. The telephone in the flat was visible through the window, and Bosch telephoned the reporter pretending to be his own professor. He said that he had just invented a marvellous instrument that could be attached to any ordinary telephone, and which would enable the user to see what was going on at the other end. This was around 1933, when the possibilities of television were just being mooted. The reporter was, of course, incredulous, and the supposed professor offered to give him a demonstration. He told the reporter to point the telephone towards the middle of the room and to stand in front of it and assume any attitude he liked, such as holding one arm up, and when he returned to the telephone he would be told exactly what he had done. Bosch, of course, could see perfectly well what he had done simply by looking through the window. The reporter was appropriately astonished, with the result that the following morning there appeared a most enthusiastic article about Bosch's professor and his marvellous invention, together with a detailed description of the demonstration.

Bosch and I then happily discussed variations on the telephone theme and ultimately I said that it ought to be possible to kid somebody to put a telephone into a bucket of water. I outlined to Bosch the various moves, and we were laughing about the prospect of their success and wondering whom we should select as a victim when one of my colleagues, Gerald Touch, came into the Laboratory and asked why we were so amused. He shared our amusement at the prospect of the bucket of water, and he offered to return to his digs, where several research students resided, and to watch while one or other of them answered the telephone, so as to report whether my plan had been successful.

We therefore waited about twenty minutes and then I telephoned Gerald Touch's digs. Before anyone could answer I rang off again, and repeated this procedure several times, in order to create the impression that someone was trying to ring the number but that something must be wrong. After this spell of induction, I dialled the number again, and heard a voice which I recognized as belonging to a very able research student in chemistry – in fact he had won the Senior Scholarship in Chemistry in the whole University that year. Reverting to the tongue that was my second language, the Cockney that came from my early schooling, I explained that I was the telephone engineer and had just received a complaint from a subscriber who was trying to dial the number and who had failed to get through. From the symptoms that he described I would say that either his dial was running a bit too fast or there was a leak to earth somewhere at the receiving end. I added that we would send a man round in the morning to check the insulation, but it was just possible that the fault could be cleared from the telephone exchange if only we could be quite sure what it was. A few simple tests would check whether this were so, and if the victim would be good enough to help us with these tests, whoever it was who wanted to get through might be able to do so the same evening. Would the victim therefore help with the tests? Immediately, of course, he expressed a readiness to do so, and I explained that I would have to keep him waiting while I got out the appropriate manual so that we could go through the correct test sequence.

I realized that he was so firmly 'hooked' that I could even afford to clown, and I persuaded him to sing loudly into the telephone on the pretext that its carbon granules had seized up. By this time, of course, all the residents of the household had now been alerted, and watched with some amazement the rest of his performance. I told him that his last effort had cleared the microphone and that we were now in a position to trace the leak to earth.

I explained that I would put on a testing signal, and that every time he heard the signal that particular test had proved okay. The appropriate signal was very simply generated by applying my own receiver to its mouthpiece, which resulted in a tremendous squawk. As I had also asked him to listen very carefully for it, he was nearly deafened the first time I did it. I then asked him to place the receiver on the table beside him and touch it. I could, of course, hear the noise of his finger making contact, and immediately I repeated the squawk. When he picked up the receiver I told him that that test had been satisfactory and that we must now try some others, and I led him through a series of antics which involved him holding the receiver by the flex, and as far away from his body as possible, at the same time standing first on one leg and then on the other. When I had given him time to reach each position I duly transmitted the squawk, and thus got him engrossed in listening for it. After this series of tests I told him that we were now getting fairly near the source of the trouble, and that all we now needed was a good 'earth'.

When he asked what that would be I said, 'Well, sir, have you got such a thing as a bucket of water?' He said that he would try to find one, and within a minute or two he came back with the bucket. When he said, 'Well, what do we do now?' I told him to place the bucket on the table beside the telephone and to put his hand into the water to make sure that he was well earthed and then to touch the telephone again. When he did this, he duly heard the appropriate squawk; and when he picked up the receiver again I told him that there was now only one final test and we would have it clinched. When he asked what this was I asked him to pick up the receiver gently by the flex, and hold it over the bucket and then gently lower it into the water. He was quite ready to do so when Gerald Touch, who had been rolling on the floor with agonized laughter, thought the joke had gone far enough, and struggled to his feet. While not wishing to give the game away, he thought that he ought to stop our victim from doing any further damage, and he started to remonstrate, saying that putting the telephone into the water would irretrievably damage it. Our victim then said to me, 'I'm very sorry about this but I'm having difficulty. There is a chap here who is a physicist who says that if I put the telephone into the water it will ruin it!' I could not resist saying, 'Oh, a physicist is he, sir. We know his kind – they think they know everything about electricity. They're always trying to put telephones right by themselves and wrecking them. Don't you worry about him, sir, it's all in my book here.' There was a great guffaw at the other end of the telephone while the victim said to Gerald Touch, 'Ha, ha, you hear that – the engineer said you physicists are always ruining telephones because you think you know all about them.' 'I'm going to do what he tells me.' As he tried to put the telephone into the water Gerald Touch seized his two wrists so as to try to stop him. They stood, swaying in a trial of strength over the bucket and the victim being the stronger man was on the point of succeeding. I heard Touch's voice saying 'It's Jones, you fool!', and our victim, a manifest sportsman, collapsed in laughter.

Bosch and I collaborated on several further occasions. On one we had Leo Szilard go to call on the *Daily Express* in Fleet Street because I had faked a telephone call from the editor asking Szilard to confirm that he had recently invented a radioactive death ray. We were astonished at the strength of Szilard's reaction – it was not until long after World War II that I found that he had just taken out a secret patent on the possibility of a uranium chain reaction and had assigned the patent to the British Admiralty.

40

The following anecdote of the life of geologists in the field (in pursuit of the magnetic South Pole on Shackleton's first journey to the Antarctic), taken from A Geological Miscellany *(editors, G. Y. Craig and E. J. Jones), was recorded by Douglas Mawson, who eventually became a Professor, and indeed a Sir, himself. The hapless Professor is Sir Edgeworth David.*

On 11 December they halted a little earlier than usual to reconnoitre. Mackay started off with the field-glasses for a general look round, Mawson retired into the tent to change photographic plates, and the Professor went out with his sketch-book to get an outline panoramic view of the grand coast ranges then in sight. Afterwards Mawson told the following story:

I was busy changing photographic plates in the only place where it could be done – inside the sleeping bag . . . . Soon after I had done up the bag, having got safely inside, I heard a voice from outside – a gentle voice – calling:

'Mawson, Mawson.'

'Hullo!' said I.

'Oh, you're in the bag changing plates, are you?'

'Yes, Professor.'

There was a silence for some time. Then I heard the Professor calling in a louder tone:

'Mawson!'

I answered again. Well the Professor heard by the sound I was still in the bag, so he said:

'Oh, still changing plates, are you?'

'Yes.'

More silence for some time. After a minute, in a rather loud and anxious tone:

'Mawson!'

I thought there was something up, but could not tell what he was after. I was getting rather tired of it and called out:

'Hullo. What is it? What can I do?'

'Well, Mawson, I am in a rather dangerous position. I am really hanging on by my fingers to the edge of a crevasse, and I don't think I can hold on much longer. I shall have to trouble you to come out and assist me.'

I came out rather quicker than I can say. There was the Professor, just his head showing and hanging on to the edge of a dangerous crevasse.

41

And here is a Professor of Geology, captured in verse by Anon.

EPITAPH ON A GEOLOGIST

Where shall we our great professor inter
That in peace he may rest his bones?

If we hew him a rocky sepulchre,
  He'll rise and break the stones,
And examine each stratum that lies around,
For he's quite in his element under ground.

If with mattock and spade his body we lay
  In the common alluvial soil,
He'll start up and snatch those tools away,
  Of his own geological toil;
In a stratum so young the professor disdains
That embedded should be his organic remains.

Thus expos'd to the drop of some case-hard'ning spring
  His carcase let stalactite cover:
And to Oxford the petrified sage let us bring,
  When he is encrusted all over:
Then 'mid mammoths and crocodiles, high on a shelf,
Let him stand as a monument rais'd to himself.

42

Marianne Moore ruminates on the quartz crystal clocks at the Bell Telephone Laboratories in New Jersey, thermostated in their time vault at 41°C (± 0.01°).

FOUR QUARTZ CRYSTAL CLOCKS

There are four vibrators, the world's exactest clocks;
  and these quartz time-pieces that tell
time intervals to other clocks,
  these worksless clocks work well;
independently the same, kept in
  the 41° Bell
    Laboratory time

vault. Checked by a comparator with Arlington,
  they punctualize the 'radio,
cinema,' and 'presse,' – a group the
  Giraudoux truth-bureau

of hoped-for accuracy has termed
  'instruments of truth'. We know –
    as Jean Giraudoux says

certain Arabs have not heard – that Napoleon
  is dead; that a quartz prism when
the temperature changes, feels
  the change and that the then
electrified alternate edges
  oppositely charged, threaten
    careful timing; so that

this water-clear crystal as the Greeks used to say,
  this 'clear ice' must be kept at the
same coolness. Repetition, with
  the scientist, should be
synonymous with accuracy.
  The lemur-student can see
    that an aye-aye is not

an angwan-tibo, potto, or loris. The seaside burden should not embarrass
the bell-boy with the buoy-ball
  endeavouring to pass
hotel patronesses; nor could a
  practised ear confuse the glass
    eyes for taxidermists

with eye-glasses from the optometrist. And as
  MEridian-7 one-two
one-two gives, each fifteenth second
  in the same voice, the new
data – 'The time will be' so and so –
  you realize that 'when you
    hear the signal', you'll be

hearing Jupiter or jour pater, the day god –
  the salvaged son of Father Time –
telling the cannibal Chronos
  (eater of his proxime
newborn progeny) that punctuality
  is not a crime.

The Illuminati

43

Here I have picked out some pen portraits in biography and fiction of famous scientists of this century and before. I begin with J.B.S. Haldane, who not only towered over British physiology, genetics and biochemistry between the wars and for some time after, but was also unsurpassed as a popular writer about science and was, in the company of J.D. Bernal and the mathematician, Hyman Levy, a political pied-piper, who led many scientists and intellectuals into the Communist Party. Many of his scientific essays appeared in the Daily Worker. *He wrote a fine scientific short story, (which appears below, no. 108). Haldane liked to outrage the stuffier academics. 'I have never gone in really seriously for bestial sodomy' was the sort of remark that he would let slip in his Cambridge College Senior Common Room, before the atmosphere became so hostile that he left for University College, London. As he grew older paranoia often took possession of him. Haldane had assisted his famous physiologist father, J.S., in the laboratory as a boy and had learned early to be his own guinea-pig for dangerous physiological experiments. Haldane admitted to having greatly enjoyed the First World War (his commander called him his bravest and dirtiest officer). In the Second World War his inherited interest in respiratory physiology led him into hazardous under-water experiments with the Admiralty. Haldane ended his life in India. When operated on for the cancer that finally killed him, he wrote a poem, celebrating his experiences. It began: 'I wish I had the voice of Homer./ To sing of rectal carcinoma,' and reached its climax with: 'So now I am like two-faced Janus,/ The only god that sees his anus.' Haldane also wrote a highly successful children's book and was a considerable classical scholar. He was in equal measure loved and feared, and in some quarters detested.*

We begin with Haldane at the Admiralty during the Second World War. This is from Ronald Clark's biography, J.B.S.

Haldane himself was not immune to illusions. He was apt to pursue them vigorously, as shown by the case of Sir Leonard Hill, a distinguished physician and physiologist then in his mid-seventies, who had worked on diving problems for years both with J.B.S.'s father and with Sir Robert Davis, the head of Siebe Gorman. These three older men, all of a generation, were good friends who respected each other's abilities. J.B.S., however, felt antipathy and something like contempt for Hill – for reasons which are unknown but may possibly have had a political basis; being J.B.S., he took little effort to conceal his feelings.

Thus the stage was set for the day when J.B.S., through stupefaction or other causes, gave himself too much oxygen at too high a pressure and succumbed in the chamber to oxygen poisoning. Hill, who was advising Siebe Gorman on purely medical matters and who was watching from outside the chamber, mistakenly attributed Haldane's convulsive movements to asphyxia; he therefore began frantically signalling to Case, who was inside the chamber with Haldane; next he held up written instructions at the window telling Case to increase Haldane's oxygen supply. This Case correctly declined to do – bringing on rage and despair in Hill who thought that Case, too, must be incapacitated.

During decompression Haldane regained consciousness; on complaining of how bad he felt, he was told by Case exactly what had happened. His reaction was fiercely to exclaim: 'Hill was deliberately trying to murder me' – a grotesque claim which he continued to repeat for years.

Oxygen poisoning was, due to one reason or another, not infrequent, and on one occasion Lieutenant – now Professor – Kenneth Donald, one of the young naval officers working with the unit, protested to Haldane that he was suffering from it too often. Surely, it was suggested, this might cause some impairment of intellectual qualities in later life – 'although I admitted he had a great deal to spare I felt that he should reduce these experiments on himself,' says Donald today. 'He was angry in his strange but friendly way and proceeded to shout at me. Inevitably, I shouted back and informed him that he was not only a great scientist but also a 'bloody fakir'. We settled the matter amicably and he did agree to seek my approval of the type and number of experiments he did, if only to ensure that he did not do himself unnecessary damage. A little later Haldane decided that he could get convulsant oxygen poisoning by breathing pure oxygen at atmospheric pressure. He demonstrated this to me up to the stage of twitching but not to the major attack. I informed him with some trepidation that I considered this to be a hysterical elaboration. It was not easy to tell him this and also assure him of my high regard of him as a scientist and a friend. I think, however, the point was made and we heard no more about it.'

Haldane obviously enjoyed the work. He enjoyed it almost blatantly – so much so that he laid himself open to one criticism. 'Some people thought that he was very shrewd and indulging his penchant for melodrama,' says Donald. 'He did take many quite serious risks in my presence on many occasions and this criticism, although it may have had a grain of truth, was entirely unfair. On one occasion, he breathed oxygen at 100 feet (4 atmospheres absolute) in a bath surrounded by blocks of ice. Rather foolishly, he suggested that I breathed oxygen as his attendant as well, to allow immediate decompression if necessary. The result of this was that the wet and frozen professor and the young naval doctor both had oxygen poisoning at the same time; and it is only by good fortune that I did not convulse and Haldane did not drown. Again, Haldane convulsed several times in my arms in the wet pressure pot where he was in a diving suit underwater and I was on a platform above him.'

The work continued through April and May, Haldane taking part as a guinea-pig – quite apart from controlling the tests – in the largest number of experiments. Case was the next most frequently used, followed by Spurway. One result of the whole series was that the routine for submarine-escape was considerably modified. In addition, the foundations were laid for the further work, carried out by Haldane and Spurway a few years later, which helped make possible both the under-water attack on the *Tirpitz* and the clearing by the famous 'P' parties of the underwater mines and time-bombs which blocked the captured Normandy ports.

There were also various *ad hoc* problems which were usually dealt with by Haldane himself and in one of which he suffered an injury to his spinal cord that was to give him intermittent pain for the rest of his life. 'This is due to a bubble of helium formed in this organ while being decompressed while testing in 1940, on behalf of the British Admiralty, a claim by an American firm that a helium-oxygen mixture is safer for divers than air, as being less likely to cause 'bends' and other symptoms during a rapid ascent,' he later wrote. 'I was decompressed according to a time-table on which I had frequently been decompressed without harm after breathing air. I developed fairly intense pain, and have had some discomfort ever since when sitting on a hard surface. I do not complain. I have learned to be sceptical of American salesmanship, even if I learned it the hard way.'

Early in the experiments Haldane discovered that while oxygen is a tasteless gas at atmospheric pressure it begins to acquire, at about five or six atmospheres, the taste of rather stale ginger beer–'a trivial discovery which, for some reason, pleases me greatly', as he described it in his Personal Note for the Royal Society. One result, which he characteristically relished, was that he could now warn young men against believing what they read in the text-books. Oxygen was not, as claimed, quite tasteless; he knew; he had tasted it. Another point, of which he made much, was that this minor discovery illustrated a point in Marxist dialectics; here, after all, the very quantity of a thing altered its quality.

The common coin of experience in all these experiments was loss of consciousness, the onset of fits, or mere nose-bleeding – 'we could usually track down the Professor, paper-chase fashion, by following a trail of small bloody pieces of cotton-wool', say some colleagues. To these were also added the effects of cold. Cold alone, cold and carbon dioxide, and cold and high pressure, had all to be investigated, and Haldane and Case alone took part in the necessary experiments. They were each 'immersed in a bath of melting ice in the pressure chamber, wearing a shirt and trousers. Immersion was not complete but ice was piled so far as practicable, on portions of the body above water level.' Haldane alone subjected himself to the effects of cold, pressure and carbon dioxide, being immersed in a bath of ice for twenty-one minutes, breathing up to 6.5 per cent carbon dioxide and then, when he began to shiver, waiting while the pressure was increased to ten atmospheres. 'He was soon unable to speak coherently (the partial pressure of CO_2 being 6.9 per cent) and lost consciousness after three and a half minutes,' his report commented.

One other unexpected discovery was made after Donald had pointed out that much classical respiratory physiology was founded on physiological idiosyncrasies of J.B.S. and his father.

'I suggested,' says Donald, 'that the "off-effect" of carbon dioxide, which consists of vomiting after the cessation of breathing high carbon dioxide, and which J.B.S. thought to have been important in the *Thetis* disaster, might be a personal idiosyncrasy. As a result of this argument, Professor Haldane, Commander Mould, G.C., G.M., a famous and heroic Australian mine disposal officer, and myself, were sealed in a chamber for nearly eight hours. Haldane analysed, on the Haldane gas apparatus, the fall of oxygen and the rising carbon dioxide, and sprayed mercury and chemicals in various directions as the gaseous environment became more and more desperate. At one stage he called his famous father's famous apparatus a "purple bitch" and threatened to destroy it.

'If I remember rightly, we were breathing something like 11–12 per cent oxygen and about 9 per cent carbon dioxide. Haldane then put on oxygen breathing apparatus and did, in fact, vomit monstrously. I repeated this and it had no effect apart from relieving my somewhat asphyxial sensations. However, Mould also had some nausea. In fact, we were both right; the "off-effect" does occur in some people, but there is extreme variation. After this, we then put on a mask to absorb the carbon dioxide. This was a highly dangerous experiment and both Haldane and I felt we were going to die. Haldane, in fact, said he wanted to die. After these experiments, he dared me to drink three large bottles of beer as he was interested in the effect of alcohol in the hypoxic (low oxygen) state. This I did but felt little difference. I am still wondering why Haldane did not drink the beer.'

44

Here is Haldane, seen by Aldous Huxley; he appears as Professor Shearwater in Antic Hay, *in which Huxley observes the gaudy creatures who made up the social and intellectual whirl of London in the years immediately following the First World War.*

'But why is the physiologue so slow? Up, pachyderm, up! Answer. You hold the key to everything. The key, I tell you, the key. I remember, when I used to hang about the biological laboratories at school, eviscerating frogs – crucified with pins, they were, belly upwards, like little green Christs – I remember once, when I was sitting there, quietly poring over the entrails, in came the laboratory boy and said to the stinks usher: "Please, sir, may I have the key of the Absolute?" And, would you believe it, that usher calmly put his hand in his trouser pocket and fished out a small Yale key and gave it him without a word. What a gesture! The key of the Absolute. But it was only the absolute alcohol the urchin wanted – to pickle some loathsome foetus in, I suppose. God rot his soul in peace! And now, Castor Fiber, out with your key. Tell us about the Archetypal Man, tell us about the primordial Adam. Tell us all about the *boyau rectum.*'

Ponderously, Shearwater moved his clumsy frame; leaning back in his chair he scrutinized Coleman with a large, benevolent curiosity. The eyes under the savage eyebrows were mild and gentle; behind the fearful disguise of the moustache he smiled poutingly, like a baby who sees the approaching bottle. The broad, domed forehead was serene. He ran his hand through his thick brown hair, scratched his head meditatively and then, when he had thoroughly examined, had comprehended and duly classified the strange phenomenon of Coleman, opened his mouth and uttered a little good-natured laugh of amusement.

'Voltaire's question,' he said at last, in his slow, deep voice, 'seemed at the time he asked it an unanswerable piece of irony. It would have seemed almost equally ironic to his contemporaries, if he had asked whether God had a pair of kidneys. We know a little more about the kidneys nowadays. If he had asked me, I should answer: why not? The kidneys are so beautifully organized; they do their work of regulation with such a miraculous – it's hard to find another word – such a positively divine precision, such knowledge and wisdom, that there's no reason why your archetypal man, whoever he is, or any one else for that matter, should be ashamed of owning a pair.'

45

Coleman, who speaks first in that extract, was supposedly recognisable as the composer, Peter Warlock. Two further extracts from Antic Hay *follow. In the first, two conversations twine in counterpoint at a midnight coffee stall by Hyde Park Corner. (Mrs Viveash is Nancy Cunard.)*

'But I know all about love already. I know precious little still about kidneys.'

'But, my good Shearwater, how can you know all about love before you've made it with all women?'

'Off we goes, me and the cop and the 'orse, up in front of the police-court magistrate. . . . '

'Or are you one of those imbeciles,' Mrs Viveash went on, 'who speak of women with a large W and pretend we're all the same? Poor Theodore here might possibly think so in his feebler moments.' Gumbril smiled vaguely from a distance. He was following the man with the teacup into the magistrate's stuffy court. 'And Mercaptan certainly does, because all the women who ever sat on his *dix-huitième* sofa certainly were exactly like one another. And perhaps Casimir does too; all women look like his absurd ideal. But you, Shearwater, you're intelligent. Surely you don't believe anything so stupid?'

Shearwater shook his head.

'The cop, 'e gave evidence against me. "Limping in all four feet," 'e says. "It wasn't," I says, and the police-court vet, 'e bore me out. "The 'orse 'as been very well treated," 'e says. "But 'e's old, 'e's very old." "I know 'e's old," I says. "But where am I goin' to find the price for a young one?" '

'x^2-y^2,' Shearwater was saying, '$=(x+y)(x-y)$. And the equation holds good whatever the values of x and y. . . . It's the same with your love business, Mrs Viveash. The relation is still fundamentally the same, whatever the value of the unknown personal quantities concerned. Little individual tics and peculiarities – after all, what do they matter?'

'What indeed!' said Coleman. 'Tics, mere tics. Sheep ticks, horse ticks, bed bugs, tape worms, taint worms, guinea worms, liver flukes. . . .'

' "The 'orse must be destroyed," says the beak. "'E's too old for work." "But I'm not," I says. "I can't get a old age pension at thirty-two, can I? 'Ow am I to earn my living if you take away what I earns my living by?" '

Mrs Viveash smiled agonizingly. 'Here's a man who thinks personal peculiarities are trivial and unimportant,' she said. 'You're not even interested in people, then?'

' "I don't know what you can do," 'e says. "I'm only 'ere to administer the law." "Seems a queer sort of law," I says. "What law is it?" '

Shearwater scratched his head. Under his formidable black moustache he smiled at last his ingenuous, childish smile. 'No,' he said. 'No, I suppose I'm not. It hadn't occurred to me, until you said it. But I suppose I'm not. No.' He laughed, quite delighted, it seemed, by this discovery about himself.

' "What law is it?" ' 'e says. "The Croolty to Animals law. That's what it is," 'e says.'

The smile of mockery and suffering appeared and faded. 'One of these days,' said Mrs Viveash, 'you may find them more absorbing than you do now.'

'Meanwhile,' said Shearwater . . .

'I couldn't find a job 'ere, and 'aving been workin' on my own, my own master like, couldn't get unemployment pay. So when we 'eard of jobs at Portsmouth, we thought we'd try to get one, even if it did mean walkin' there.'

'Meanwhile, I have my kidneys.'

He was half-way down the last flight, when with a rattle and a squeak of hinges the door of the house, which was only separated by a short lobby from the foot of the stairs, opened, revealing, on the doorstep, Shearwater and a friend, eagerly talking.

'. . . I take my rabbit,' the friend was saying – he was a young man with dark, protruding eyes, and staring, doggy nostrils; very eager, lively and loud. 'I take my rabbit and I inject into it the solution of eyes, pulped eyes of another dead rabbit. You see?'

Gumbril's first instinct was to rush up the stairs and hide in the first likely-looking corner. But he pulled himself together at once. He was a Complete Man, and Complete Men do not hide; moreover, he was sufficiently disguised to be quite unrecognizable. He stood where he was, and listened to the conversation.

'The rabbit,' continued the young man, and with his bright eyes and staring, sniffing nose, he looked like a poacher's terrier ready to go barking after the first white tail that passed his way; 'the rabbit naturally develops the appropriate resistance, develops a specific anti-eye to protect itself. I then take some of its anti-eye serum and inject it into my female rabbit; I then immediately breed from her.' He paused.

'Well?' asked Shearwater, in his slow, ponderous way. He lifted his great round head inquiringly and looked at the doggy young man from under his bushy eyebrows.

The doggy young man smiled triumphantly. 'The young ones,' he said, emphasizing his words by striking his right fist against the extended palm of his left hand, 'the young ones are born with defective sight.'

Thoughtfully Shearwater pulled at his formidable moustache. 'H'm,' he said slowly. 'Very remarkable.'

'You realize the full significance of it?' asked the young man. 'We seem to be affecting the germ-plasm directly. We have found a way of making acquired characteristics . . .'

'Pardon me,' said Gumbril. He had decided that it was time to be gone. He ran down the stairs and across the tiled hall, he pushed his way firmly but politely between the talkers.

'. . . heritable,' continued the young man, imperturbably eager, speaking through and over and round the obstacle.

'Damn!' said Shearwater. The Complete Man had trodden on his toe. 'Sorry,' he added, absent-mindedly apologizing for the injury he had received.

Gumbril hurried off along the street. 'If we really have found out a technique for influencing the germ-plasm directly . . .' he heard the doggy young man saying; but he was already too far away to catch the rest of the sentence. There are many ways, he reflected, of spending an afternoon.

The doggy young man refused to come in, he had to get in his game of tennis before dinner. Shearwater climbed the stairs alone. He was taking off his hat in the little hall of his own apartment, when Rosie came out of the sitting-room with a trayful of tea-things.

'Well?' he asked, kissing her affectionately on the forehead. 'Well? People to tea?'

'Only one,' Rosie replied. 'I'll go and make you a fresh cup.'

She glided off, rustling in her pink kimono towards the kitchen.

Shearwater sat down in the sitting-room. He had brought home with him from the library the fifteenth volume of the *Journal of Biochemistry*. There was something in it he wanted to look up. He turned over the pages. Ah, here it was. He began reading. Rosie came back again.

'Here's your tea,' she said.

He thanked her without looking up. The tea grew cold on the little table at his side.

Lying on the sofa, Rosie pondered and remembered. Had the events of the afternoon, she asked herself, really happened? They seemed very improbable and remote, now, in this studious silence. She couldn't help feeling a little disappointed. Was it only this? So simple and obvious? She tried to work herself up into a more exalted mood. She even tried to feel guilty; but there she failed completely. She tried to feel rapturous; but without much more success. Still, he certainly had been a most extraordinary man. Such impudence, and at the same time such delicacy and tact.

It was a pity she couldn't afford to change the furniture. She saw now that it wouldn't do at all. She would go and tell Aunt Aggie about the dreadful middle-classness of her Art and Craftiness.

She ought to have an Empire *chaise longue*. Like Madame Récamier. She could see herself lying there, dispensing tea. 'Like a delicious pink snake.' He had called her that.

Well, really, now she came to think of it all again, it had been too queer, too queer.

'What's a hedonist?' she suddenly asked.

Shearwater looked up from the *Journal of Biochemistry*. 'What?' he said.

'A hedonist.'

'A man who holds that the end of life is pleasure.'

A 'conscientious hedonist' – ah, that was good.

'This tea is cold,' Shearwater remarked.

'You should have drunk it before,' she said. The silence renewed and prolonged itself.

47

P. B. Medawar in Memoirs of a Thinking Radish *has left this vignette of his trying colleague at University College, London.*

My Zoology Department was a rambling gormenghastly structure with many rooms in unexpected places and I was happy to be able to put my predecessor, Professor D. M. S. Watson, into a suite of rooms where he could continue working on his fossils with a secretary and assistant.

The teaching staff I inherited at University College were a quite exceptionally intelligent and likeable lot, enriched by a number of visiting workers with unusual capability: from the United Kingdom, Alex Comfort, the biochemist, poet, and novelist; Anne McLaren, who became one of the few women Fellows of the Royal Society; Donald Michie, now England's foremost authority on 'machine intelligence'; and a number of American guests – Paul Terasaki, Paul Russell, Jerry Lawrence, and Bill Hildeman. In addition the Department of Biometry, of which J. B. S. Haldane was the head, was physically housed in my Department, though not administratively a part of it, thank God.

The American guests were friendly and well liked and wonderfully patient with the teasing they had to put up with from the department, the worst offender being Haldane whose favourite provocation was to argue that civil liberties were every bit as much endangered in the USA as in Russia and East Europe. One of my American guests one December came to complain to me formally that there was no daytime in England: night fell with commendable regularity but he got up in the dark and night fell, it seemed to him, almost immediately after. I couldn't help sympathizing with him because University College was surrounded by the principal main-line railway stations at a time when steam trains were still darkening the sky with soot and grime.

This company, at least two of whom were notable conversationalists, made our tea-table the liveliest and the wittiest I have ever attended, but we trod very warily when we came to politics: Haldane was a card-carrying Party member who was a frequent contributor to the *Daily Worker*, on which he had served as chairman of the editorial board. He was a complete innocent politically and believed everything he was told. I can remember asking him if he did not think it very strange that the lately disgraced Beria, in spite of the exalted position he held in the state, had all the while been in the pay of the Americans. Haldane was not in the least surprised. 'People in high-up positions sometimes get careless, you know.' Haldane also heartily approved of the murder in Czechoslovakia of the politicians Clementis and Slansky.

Haldane worked in a room which was never tidied or cleaned. The fossils upon which he was proposing one day to undertake some biometric studies were clearly undergoing a second interment, the sediments consisting this time of atmospheric dust and soot, together with manuscripts, committee papers, and official communications of the College. When anything got lost, Haldane always claimed it was something of *their* doing. His liking and admiration for the working classes was purely notional and his colleagues soon formed the opinion that he couldn't bear the sight of them; it was the laugh of the Department when electricians whose duty it was to rewire his office demanded danger money for working in it.

48

Aldous Huxley has also left a memorable fictional portrait – Lord Edward in Point Counter Point *– of J.B.S. Haldane's father, J.S., Professor of Physiology of Oxford. J.S. was apparently considerably annoyed.*

That evening he told his father that he was not going to stand for Parliament. Still agitated by the morning's revelations of Parnellism, the old gentleman was furious. Lord Edward was entirely unmoved; his mind was made up. The next day he advertised for a tutor. In the spring of the following year he was in Berlin working under Du Bois Reymond.

Forty years had passed since then. The studies of osmosis, which had indirectly given him a wife, had also given him a reputation. His work on assimilation and growth was celebrated. But what he regarded as the real task of his life – the great theoretical treatise on physical biology – was still unfinished. 'The life of the animal is only a fragment of the total life of the universe.' Claude Bernard's words had been his life-long theme as well as his original inspiration. The book on which he had been working all these years was but an elaboration, a quantitative and mathematical illustration of them.

Upstairs in the laboratory the day's work had just begun. Lord Edward preferred to work at night. He found the daylight hours disagreeably noisy. Breakfasting at half-past one, he would walk for an hour or two in the afternoon and return to read or write till lunch-time at eight. At nine or half-past he would do some practical work with his assistant, and when that was over they would sit down to work on the great book or to discussion of its problems. At one, Lord Edward had his supper, and at about four or five he would go to bed.

Diminished and in fragments the B minor Suite came floating up from the great hall to the ears of the two men in the laboratory. They were too busy to realize that they were hearing it.

'Forceps,' said Lord Edward to his assistant. He had a very deep voice, indistinct and without, so to speak, a clearly defined contour. 'A furry voice,' his daughter Lucy had called it, when she was a child.

Illidge handed him the fine bright instrument. Lord Edward made a deep noise that signified thanks and turned back with the forceps to the anaesthetized newt that lay stretched out on the diminutive operating table. Illidge watched him critically, and approved. The Old Man was doing the job extraordinarily well. Illidge was always astonished by Lord Edward's skill. You would never have expected a huge, lumbering creature like the Old Man to be so exquisitely neat. His big hands could do the finest work; it was a pleasure to watch them.

'There!' said Lord Edward at last and straightened himself up as far as his rheumatically bent back would allow him. 'I think that's all right, don't you?'

Illidge nodded. 'Perfectly all right,' he said in an accent that had certainly not been formed in any of the ancient and expensive seats of learning. It hinted of Lancashire origins. He was a small man, with a boyish-looking freckled face and red hair.

The newt began to wake up. Illidge put it away in a place of safety. The animal had no tail; it has lost that eight days ago, and tonight the little bud of regenerated tissue which would normally have grown into a new tail had been removed and grafted on to the stump of its amputated right foreleg. Transplanted to its new position, would the bud turn into a foreleg, or continue incongruously to grow as a tail? Their first experiment had been with a tail-bud only just formed; it had duly turned into a leg. In the next, they had given the bud time to grow to a considerable size before they transplanted it; it had proved too far committed to tailhood to be able to adapt itself to the new conditions; they had manufactured a monster with a tail where an arm should have been. Tonight they were experimenting on a bud of intermediate age.

Lord Edward took a pipe out of his pocket and began to fill it, looking meditatively meanwhile at the newt. 'Interesting to see what happens this time,' he said in his profound indistinct voice. 'I should think we must be just about on the border line between . . .' He left the sentence unfinished: it was always difficult for him to find the words to express his meaning. 'The bud will have a difficult choice.'

'To be or not to be,' said Illidge facetiously, and started to laugh; but seeing that Lord Edward showed no signs of having been amused, he checked himself. Almost put his foot in it again. He felt annoyed with himself and also, unreasonably, with the Old Man.

Lord Edward filled his pipe. 'Tail becomes leg,' he said meditatively. 'What's the mechanism? Chemical peculiarities in the neighbouring . . . ? It can't obviously be the blood. Or do you suppose it has something to do with the electric tension? It does vary, of course, in different parts of the body. Though why we don't all just vaguely proliferate like cancers . . . Growing in a definite shape is very unlikely, when you come to think of it. Very mysterious and . . .' His voice trailed off into a deep and husky murmur.

Illidge listened disapprovingly. When the Old Man started off like this about the major and fundamental problems of biology, you never knew where he'd be getting to. Why, as likely as not he'd begin talking about God. It really made one blush. He was determined to prevent anything so discreditable happening this time. 'The next step with these newts,' he said in his most briskly practical tone, 'is to tinker with the nervous system and see whether that has any influence on the grafts. Suppose, for example, we excised a piece of the spine . . .'

But Lord Edward was not listening to his assistant. He had taken his pipe out of his mouth, he had lifted his head and at the same time slightly cocked it on one side. He was frowning, as though making an effort to seize and remember something. He raised his hand in a gesture that commanded silence; Illidge interrupted himself in the middle of his sentence and also listened. A pattern of melody faintly traced itself upon the silence.

'Bach?' said Lord Edward in a whisper.

Pongileoni's blowing and the scraping of the anonymous fiddlers had shaken the air in the great hall, had set the glass of the windows looking on to it vibrating; and this in turn had shaken the air in Lord Edward's apartment on the further side. The shaking air rattled Lord Edward's *membrana tympani*; the interlocked *malleus, incus* and stirrup bones were set in motion so as to agitate the membrane of the oval window and raise an infinitesimal storm in the fluid of the labyrinth. The hairy endings of the auditory nerve shuddered like weeds in a rough sea; a vast number of obscure miracles were performed in the brain, and Lord Edward ecstatically whispered 'Bach!' He smiled with pleasure, his eyes lit up. The young girl was singing to herself in solitude under the floating clouds. And then the cloud-solitary philosopher began poetically to meditate. 'We must really go downstairs and listen,' said Lord Edward. He got up. 'Come,' he said. 'Work can wait. One doesn't hear this sort of thing every night.'

'But what about clothes,' said Illidge doubtfully. 'I can't come down like this.' He looked down at himself. It had been a cheap suit at the best of times. Age had not improved it.

'Oh, that doesn't matter.' A dog with the smell of rabbits in his nostrils could hardly have shown a more indecent eagerness than Lord Edward at the sound of Pongileoni's flute. He took his assistant's arm and hurried him out of the door, and along the corridor towards the stairs. 'It's just a little party,' he went on. 'I seem to remember my wife having said . . . Quite informal. And besides,' he added, inventing new excuses to justify the violence of his musical appetite, 'we can just slip in without . . . Nobody will notice.'

Illidge had his doubts. 'I'm afraid it's not a very small party,' he began; he had seen the motors arriving.

'Never mind, never mind,' interrupted Lord Edward, lusting irrepressibly for Bach.

Illidge abandoned himself. He would look like a horrible fool, he reflected, in his shiny blue serge suit. But perhaps, on second thoughts, it was better to appear in shiny blue – straight from the laboratory, after all, and under the protection of the master of the house (himself in a tweed jacket), than in that old and, as he had perceived during previous excursions into Lady Edward's luscious world, deplorably shoddy and ill-made evening suit of his. It was better to be totally different from the rich and smart – a visitor from another intellectual planet – than a fourth-rate and snobbish imitator. Dressed in blue, one might be stared at as an oddity; in badly cut black (like a waiter) one was contemptuously ignored, one was despised for trying without success to be what one obviously wasn't.

Illidge braced himself to play the part of the Martian visitor with firmness, even assertively.

Their entrance was even more embarrassingly conspicuous than Illidge had anticipated. The great staircase at Tantamount House comes down from the first floor in two branches which join, like a pair of equal rivers, to precipitate themselves in a single architectural cataract of Verona marble into the hall. It debouches under the arcades, in the centre of one of the sides of the covered quadrangle, opposite the vestibule and the front door. Coming in from the street, one looks across the hall and sees through the central arch of the opposite arcade the wide stairs and shining balustrades climbing up to a landing on which a Venus by Canova, the pride of the third marquess's collection, stands pedestalled in an alcove, screening with a modest but coquettish gesture of her two hands, or rather failing to screen, her marble charms. It was at the foot of this triumphal slope of marble that Lady Edward had posted the orchestra; her guests were seated in serried rows confronting it. When Illidge and Lord Edward turned the corner in front of Canova's Venus, tiptoeing, as they approached the music and the listening crowd, with steps ever more laboriously conspiratorial, they found themselves suddenly at the focus of a hundred pairs of eyes. A gust of curiosity stirred the assembled guests. The apparition from a world so different from theirs of this huge bent old man, pipe-smoking and tweed-jacketed, seemed strangely portentous. He had a certain air of the skeleton in the cupboard – broken loose; or of one of those monsters which haunt the palaces of only the best and most aristocratic families. The Beastie of Glamis, the Minotaur itself could hardly have aroused more interest than did Lord Edward. Lorgnons were raised, there was a general craning to left and right, as people tried to look round the well-fed obstacles in front of them. Becoming suddenly aware of so many inquisitive glances, Lord Edward took fright. A consciousness of social sin possessed him; he took his pipe out of his mouth and put it away, still smoking, into the pocket of his jacket. He halted irresolutely. Flight or advance? He turned this way and that, pivoting his whole bent body from the hips with a curious swinging motion, like the slow ponderous balancing of a camel's neck. For a moment he wanted to retreat. But love of Bach was stronger than his terrors.

49

The following passage from Nicholas Wade's The Nobel Prize Duel *will give an inkling of how the pursuit of science has changed in seventy years. A. Schally and R. Guillemin shared the Nobel Prize for the discovery of the pituitary releasing factors.*

Matsuo's other surprise was Schally himself. The bluntness which Schally had developed almost to an art form in his adopted country was the polar opposite of the delicacy and indirection with which personal relations are conducted in Japanese society. Schally straightaway handed Matsuo a tiny test tube containing a smidgeon of dried powder and started talking. "This is LH-RH," said Schally, using his term for LRF. "This is 800 micrograms which I collected from 160,000 pig brains. This is all there is. I have spent ten years and I have found it." Matsuo marveled at the monotony of anyone spending so long extracting the material he had just been given. "Be careful, please don't drop it," Schally kept saying.

Schally lost no time in describing the situation created by his rivalry with Guillemin. "I hate the group of Guillemin," he explained to Matsuo: "They have purified more than several milligrams of LH-RH from sheep brains. I cannot lose to Guillemin. I would like you to determine the structure of LH-RH as soon as possible – I want to beat Guillemin." Matsuo was astonished to be told of Schally's private feelings at their first meeting. Little did he understand the purpose of this strange Westerner who was baring his breast to him. Matsuo was simply receiving the routine indoctrination for each new recruit to Schally's forces, the pep talk given to the soldier going into battle. Hatred of the enemy was the central theme in Schally's motivational course.

50

Nikolai Timofeyev-Ressovsky was a noted Russian geneticist, who went to Berlin to do postdoctoral work in the 1920s, wisely decided to shun Russia during the Terror and survived the war, carrying on his work on Drosophila genetics more or less undisturbed. Following the capture of Berlin by the Russian Army, the Gulag inevitably became his lot. Released, he later fell foul of Lysenko (see below, no. 201). Daniil Granin has written a biographical novel, The Geneticist, *subtitled* The Life of Nikolai Timofeyev-Ressovsky, known as Ur, *which caused a considerable stir in scientific circles when it was published in* Novy Mir *in 1987.*

Seeing Ur with Lyapunov and others left an unexpected impression. Academician Lev Alexandrovich Senkevich was older than Ur and had known him as a student. 'Both were mighty Colossi; they walked together in near-silence, because they needed no words to understand each other,' S. Schnol recalls. 'They seemed to me monsters from the primeval swamp. To splash around in the water with them was frightening. They were like the philosophers in the Nekrasov painting; their thoughts revolved around the universe, belief, consciousness; they strode along in silence, immersed in their deliberations.'

Ur and Senkevich discussed loudly at a banquet why strokes were now so common. They arrived at the conclusion that in earlier days, when they were children, people in inns and guest-houses were bitten by bed-bugs, which injected anticoagulants, that is to say clot-inhibitors, into the blood, so strokes were prevented. With Ur's jokes one never quite knew what to think – it sounded like nonsense, but there was something in it all the same.

At Ur's jubilee celebration, there was Boris Stepanovich Matveyev, one of his teachers. The young people were amazed of course to set eyes on a teacher of their teacher. Boris Stepanovich had been Demonstrator in Kolzov's course on the vertebrates. Suddenly he asked Ur, so that all could hear: 'Kolyusha, did we ever teach you anything?' 'That you did, Boris Stepanovich.' 'Then tell me, please, Kolyusha, what are the rudimentary veins in mammals, that remain from the reptiles?'

All sat, captivated. It was Ur's seventieth birthday party. Boris Stepanovich was over eighty, but for the young people both were antideluvian relics. Ur started to snuffle, furrowed his brow and roared: 'Vena azygos and vena hemiazygos.' It was too much for Boris Stepanovich; his tears started and Ur too was moved.

Sukachev, Pryanishnikov, Asturos, Vavilov, Kolzov, Senkevich – names like these form a mountain range. They are the towering heights, the standards against which integrity was measured. They were feared – what would they say? There was no true, fixed collective opinion – what one understands by collective opinion did not exist: there was no academic milieu to uphold ethical standards, to condemn anyone for plagiarism, exploitation of students or dishonourable conduct, or to bestow praise for civic courage and decency. Collective opinion was replaced by individual scientists, in whom moral and professional authority were happily united. But their day was done, the giants had, as the saying goes, departed into darkness, and no one had replaced them. At least that is how it seemed to us.

Those whose word was feared grew even fewer. There was no one left before whom to feel shame. Some were dead, others were banned, silenced or had despaired. Their code of honour had become too demanding, so they were called old-fashioned. They themselves – the prophets, the knights of truth, guardians of honour – became legends.

51

Elie Metchnikoff, a Russian eccentric and one of the great originals of nineteenth-century biology, a founder of the science of immunology, was the subject of a biography by his wife, Olga Metchnikoff. Here is a description of Metchnikoff's final decline.

M. Roux then proposed that we should be transferred to Pasteur's old flat; the rooms were spacious and much cooler. This idea rejoiced and touched Elie very much. As he thanked M. Roux, he said to him: "See how my life is bound with the Pasteur Institute. I have worked here for years; I am nursed here during my illness; in order to complete the connection I ought to be incinerated in the great oven where our dead animals are burnt, and my ashes could be kept in an urn in one of the cupboards in the library." "What a gruesome joke!" answered M. Roux, really taking those words for a joke. But directly after he was gone Elie turned to me with an anxious look and said, "Well, what do you think of my idea?" I saw by his earnest expression that he meant what he said, and I answered that I thought it a very good idea. The Pasteur Institute had become his refuge, the centre of all his scientific interests; he loved it; he had spent his best years there. Let his ashes be laid there some day; it would be in perfect harmony with his past.

52

The great Belgian biologist, Jules Bordet, one of the many European luminaries who joined the Rockefeller Institute in its early years, was a pupil of Metchnikoff; his anecdote is recorded in Paul de Kruif's memoirs, The Sweeping Wind.

The frail, famous small man with his drooping, pathetic mustache sat chain-smoking cigarettes that he rolled in brown paper. He answered my eager questions as to how he had developed his complement-fixation reaction that had resulted in the Wassermann test. He was a progenitor of the science of serology.

He recounted how he had isolated the bacillus of *coqueluche* – the whooping cough – from his own little son's dirty handkerchief that had lain neglected on a lab bench for days. He told how he had been the first to spy the spirochete of syphilis, before Schaudinn, how he didn't dare publish such a tremendous discovery without confirmation by his master, Metchnikoff, at the Pasteur Institute. "I sent my stained preparations – Giemsa method – to Metchnikoff," Bordet explained. "But Metchnikoff was getting old," he said smiling, "and couldn't see too well through the microscope. He wrote me he wasn't sure about my new spirochetes. He warned me against publishing. So I lost that one," said Bordet with a shrug and a sad smile.

53

Some of the great Victorian worthies of science appear in a Novel of Ideas of the period, W.H. Mallock's The New Republic. *Stockton is the physicist, Tyndall, and Mr Storks is T.H. Huxley – Darwin's Bulldog.*

'Well', said Lawrence, 'that man by himself, turning over the books on the table – the man with the black whiskers, spectacles and bushy eyebrows – is Mr Storks of the Royal Society, who is great on the physical basis of life and the imaginative basis of God. The man with long locks in the window, explaining a microscope in so eager a way to that dark-haired girl, is Professor Stockton – of the Royal Society also, and member and president of many societies more.'

54

And here is a sample of their conversation:

During this speech Mr Storks had remained with his face buried in his hands, every now and then drawing his breath through his teeth, as if he were in pain. When it was over he looked up with a scared expression, as if he hardly knew where he was, and seemed quite unable to utter a syllable.

'Of course,' said Mr Stockton, 'mere science, as science, does not deal with moral rights and wrongs.'

'No,' said Mr Saunders, 'for it has been shown that right and wrong are terms of a bygone age, connoting altogether false ideas. Mere automata, as science shows we are – clockwork machines, wound up by meat and drink –'

'As for that,' broke in Mr Storks, who had by this time recovered himself – and his weighty voice at once silenced Mr Saunders, 'I would advise our young friend not to be too confident. We may be automata, or we may not. Science has not yet decided. And upon my word,' he said, striking the table, 'I don't myself care which we are. Supposing the Deity – if there be one – should offer to make me a machine, if I am not one, on condition that I should always go right, I, for one, would gladly close with the proposal.'

'But you forget,' said Allen, 'that in the moral sense there would be no going right at all, if there were not also the possibility of going wrong. If your watch keeps good time you don't call it virtuous, not if it keeps bad time do you call it sinful.'

'Sin! Lord Allen,' said Mr Storks, 'is a word that has helped to retard moral and social progress more than anything. Nothing is good or bad, but thinking makes it so; and the superstitious and morbid way in which a number of entirely innocent things have been banned as sin has caused more than half the tragedies of the world. Science will establish an entirely new basis of morality, and the sunlight of national approbation will shine on many a thing, hitherto overshadowed by the cause of a hypothetical God.'

55

Charles Darwin too could sometimes show his claws. Here, in The Voyage of the Beagle, *he is being catty about Lamarck. The tucutuco is a mole-like creature, with an onomatopoetic name which describes the sound that it emits under ground.*

The man who caught them asserted that very many are invariably found blind. A specimen which I preserved in spirits was in this state; Mr. Reid considers it to be the effect of inflammation in the nictitating membrane. When the animal was alive I placed my finger within half an inch of its head, and not the slightest notice was taken: it made its way, however, about the room nearly as well as the others. Considering the strictly subterranean habits of the tucutuco, the blindness, though so common, cannot be a very serious evil; yet it appears strange that any animal should possess an organ frequently subject to be injured. Lamarck would have been delighted with this fact, had he known it, when speculating (probably with more truth than usual with him) on the gradually *acquired* blindness of the Aspalax, a Gnawer living under ground, and of the Proteus, a reptile living in dark caverns filled with water; in both of which animals the eye is in an almost rudimentary state, and is covered by a tendinous membrane and skin. In the common mole the eye is extraordinarily small but perfect, though many anatomists doubt whether it is connected with the true optic nerve; its vision must certainly be imperfect, though probably useful to the animal when it leaves its burrow. In the tucutuco, which I believe never comes to the surface of the ground, the eye is rather larger, but often rendered blind and useless, though without apparently causing any inconvenience to the animal; no doubt Lamarck would have said that the tucutuco is now passing into the state of the Aspalax and Proteus.

56

The heroine of H.G. Wells's feminist novel, Ann Veronica, *is an aspiring zoologist. The atmosphere of the Imperial Institute is as he knew it; Russell is T.H. Huxley; Capes may be his demonstrator, Howes.*

The biological laboratory had an atmosphere that was all its own. It was at the top of the building, and looked clear over a clustering mass of inferior buildings towards Regent's Park. It was long and narrow, a well-lit, well-ventilated, quiet gallery of small tables and sinks, pervaded by a thin smell of methylated spirit and of a mitigated and sterilized organic decay. Along the inner side was a wonderfully arranged series of displayed specimens that Russell himself had prepared. The supreme effect for Ann Veronica was its surpassing relevance; it made every other atmosphere she knew seem discursive and confused. The whole place and everything in it aimed at one thing – to illustrate, to elaborate, to criticize and illuminate, and make ever plainer and plainer the significance of animal and vegetable structure. It dealt from floor to ceiling and end to end with the theory of the forms of life; the very duster by the blackboard was there to do its share in that work, the very washers in the taps; the room was more simply concentrated in aim even than a church. To that, perhaps, a large part of its satisfyingness was due. Contrasted with the confused movements and presences of a Fabian meeting, or the inexplicable enthusiasm behind the suffrage demand, with the speeches that were partly egotistical displays, partly artful manoeuvres, and partly incoherent cries for unsoundly formulated ends, compared with the comings and goings of audiences and supporters that were like the eddy-driven drift of paper in the street, this long, quiet, methodical chamber shone like a star seen through clouds.

Day after day for a measured hour in the lecture theatre, with elaborate power and patience, Russell pieced together difficulty and suggestion, instance and counter-instance, in the elaborate construction of the family tree of life. And then the students went into the long laboratory and followed out these facts in almost living tissue with microscope and scalpel, probe and microtome, and the utmost of their skill and care, making now and then a raid into the compact museum of illustration next door, in which specimens and models and directions stood in disciplined ranks, under the direction of the demonstrator Capes. There was a couple of blackboards at each end of the aisle of tables, and at these Capes, with quick and nervous speech that contrasted vividly with Russell's slow, definitive articulation, directed the dissection and made illuminating comments on the structures under examination. Then he would come along the laboratory, sitting down by each student in turn, checking the work and discussing its difficulties, and answering questions arising out of Russell's lecture.

Ann Veronica had come to the Imperial College obsessed by the great figure of Russell, by the part he had played in the Darwinian controversies, and by the resolute effect of the grim-lipped, yellow, leonine face beneath the mane of silvery hair. Capes was rather a discovery. Capes was something superadded. Russell burnt like a beacon, but Capes illuminated by darting flashes and threw light, even if it was but momentary light, into a hundred corners that Russell left steadfastly in the shade.

Capes was an exceptionally fair man of two or three and thirty, so ruddily blond that it was a mercy he had escaped light eyelashes, and with a minor but by no means contemptible reputation of his own. He talked at the blackboard in a pleasant, very slightly lisping voice with a curious spontaneity, and was sometimes very clumsy in his exposition, and sometimes very vivid. He dissected rather awkwardly and hurriedly, but, on the whole, effectively, and drew with an impatient directness that made up in significance what it lacked in precision. Across the blackboard the coloured chalks flew like flights of variously tinted rockets as diagram after diagram flickered into being.

57

Huxley's own memoir of his early life and the uncomfortable beginnings of his career as the greatest lecturer of his time, are succinct, even telegraphic.

During the four years of our absence, I sent home communication after communication to the 'Linnean Society,' with the same result as that obtained by Noah when he sent the raven out of his ark. Tired at last of hearing nothing about them, I determined to do or die, and in 1849 I drew up a more elaborate paper and forwarded it to the Royal Society. This was my dove, if I had only known it. But owing to the movements of the ship, I heard nothing of that either until my return to England in the latter end of the year 1850, when I found that it was printed and published, and that a huge packet of separate copies awaited me. When I hear some of my young friends complain of want of sympathy and encouragement, I am inclined to think that my naval life was not the least valuable part of my education.

Three years after my return were occupied by a battle between my scientific friends on the one hand and the Admiralty on the other, as to whether the latter ought, or ought not, to act up to the spirit of a pledge they had given to encourage officers who had done scientific work by contributing to the expense of publishing mine. At last the Admiralty, getting tired, I suppose, cut short the discussion by ordering me to join a ship, which thing I declined to do, and as Rastignac, in the *Père Goriot*, says to Paris, I said to London, '*à nous deux*.' I desired to obtain a Professorship of either Physiology or Comparative Anatomy, and as vacancies occurred I applied, but in vain. My friend, Professor Tyndall, and I were candidates at the same time, he for the Chair of Physics and I for that of Natural History in the University of Toronto, which, fortunately, as it turned out, would not look at either of us. I say fortunately, not from any lack of respect for Toronto, but because I soon made up my mind that London was the place for me, and hence I have steadily declined the inducements to leave it, which have at various times been offered. At last, in 1854, on the translation of my warm friend Edward Forbes to Edinburgh, Sir Henry De la Beche, the Director-General of the Geological Survey, offered me the post Forbes vacated of Paleontologist and Lecturer on Natural History. I refused the former point blank, and accepted the latter only provisionally, telling Sir Henry that I did not care for fossils, and that I should give up Natural History as soon as I could get a physiological post. But I held the office for thirty-one years, and a large part of my work has been paleontological.

At that time I disliked public speaking, and had a firm conviction that I should break down every time I opened my mouth. I believe

I had every fault a speaker could have (except talking at random or indulging in rhetoric), when I spoke to the first important audience I ever addressed, on a Friday evening at the Royal Institution, in 1852. Yet, I must confess to having been guilty, *malgré moi*, of as much public speaking as most of my contemporaries, and for the last ten years it ceased to be so much of a bugbear to me. I used to pity myself for having to go through this training, but I am now more disposed to compassionate the unfortunate audiences, especially my ever-friendly hearers at the Royal Institution, who were the subjects of my oratorical experiments.

58

Here is another, factual description of T.H. Huxley, from Wells's Experiment in Autobiography.

As I knew Huxley he was a yellow-faced, square-faced old man, with bright little brown eyes, lurking as it were in caves under his heavy grey eyebrows, and a mane of grey hair brushed back from his wall of forehead. He lectured in a clear firm voice without hurry and without delay, turning to the blackboard behind him to sketch some diagram, and always dusting the chalk from his fingers rather fastidiously before he resumed. He fell ill presently, and after some delay Howes, uneasy, irritable, brilliant, took his place, lecturing and drawing breathlessly and leaving the blackboard a smother of graceful coloured lines. At the back of the auditorium were curtains, giving upon a museum devoted to the invertebrata. I was told that while Huxley lectured Charles Darwin had been wont at times to come through those very curtains from the gallery behind and sit and listen until his friend and ally had done. In my time Darwin had been dead for only a year or so (he died in 1882).

These two were very great men. They thought boldly, carefully and simply, they spoke and wrote fearlessly and plainly, they lived modestly and decently; they were mighty intellectual liberators. It is a pity that so many of the younger scientific workers of today, ignorant of the conditions of mental life in the early nineteenth century and standing for the most part on the ground won, cleared and prepared for them by these giants, find a perverse pleasure in belittling them.

59

Samuel Butler's mock-heroic poem, Hudibras, *is a seventeenth-century satire on Presbyterianism and humbug.*

SIR HUDIBRAS, HIS PASSING WORTH

He was in *Logick* a great Critick,
Profoundly skill'd in Analytick.
He could distinguish, and divide
A Hair 'twixt *South* and *South-West* side:
On either which he would dispute,
Confute, change hands, and still confute.
He'd undertake to prove by force
Of Argument, a Man's no Horse.
He'd prove a Buzard is no Fowl,
And that a *Lord* may be an Owl;
A Calf an *Alderman*, a Goose a *Justice*,
And Rooks *Committee-men* and *Trustees*.
He'd run in Debt by Disputation,
And pay with Ratiocination,
All this by Syllogism, true
In Mood and Figure, he would do.

In *Mathematicks* he was greater
Than *Tycho Brahe*, or *Erra Pater:*
For he by *Geometrick* scale
Could take the size of *Pots of Ale;*
Resolve by Signes and Tangents straight,
If *Bread* or *Butter* wanted weight;
And wisely tell what hour o'th day
The Clock does strike, by *Algebra*.

60

Conan Doyle's Professor Challenger – this extract is from one of the stories, The Poison Belt *– is generally held to have been based on the zoologist (Professor at University College, London and at Oxford, and Director of the Natural History Museum) and populariser, Sir Ray*

Lankester. An admixture of the Scottish physiologist, William Rutherford, has also been detected.

It is imperative that now at once, while these stupendous events are still clear in my mind, I should set them down with that exactness of detail which time may blur. But even as I do so, I am overwhelmed by the wonder of the fact that it should be our little group of the 'Lost World'–Professor Challenger, Professor Summerlee, Lord John Roxton, and myself–who have passed through this amazing experience.

When, some years ago, I chronicled in the *Daily Gazette* our epoch-making journey in South America, I little thought that it should ever fall to my lot to tell an even stranger personal experience, one which is unique in all human annals, and must stand out in the records of history as a great peak among the humble foothills which surround it. The event itself will always be marvellous, but the circumstances that we four were together at the time of this extraordinary episode came about in a most natural and, indeed, inevitable fashion. I will explain the events which led up to it as shortly and as clearly as I can, though I am well aware that the fuller the detail upon such a subject the more welcome it will be to the reader, for the public curiosity has been and still is insatiable.

It was upon Friday, the twenty-seventh of August – a date for ever memorable in the history of the world – that I went down to the office of my paper and asked for three days' leave of absence from Mr. McArdle, who still presided over our news department. The good old Scotchman shook his head, scratched his dwindling fringe of ruddy fluff, and finally put his reluctance into words.

'I was thinking, Mr. Malone, that we could employ you to advantage these days. I was thinking there was a story that you are the only man that could handle as it should be handled.'

'I am sorry for that,' said I, trying to hide my disappointment. 'Of course if I am needed, there is an end of the matter. But the engagement was important and intimate. If I could be spared –'

'Well, I don't see that you can.'

It was bitter, but I had to put the best face I could upon it. After all, it was my own fault, for I should have known by this time that a journalist has no right to make plans of his own.

'Then I'll think no more of it,' said I, with as much cheerfulness as I could assume at so short a notice. 'What was it that you wanted me to do?'

'Well, it was just to interview that deevil of a man down at Rotherfield.'

'You don't mean Professor Challenger?' I cried.

'Aye, it's just him that I do mean. He ran young Alec Simpson, of the *Courier*, a mile down the high road last week by the collar of his coat and the slack of his breeches. You'll have read of it, likely, in the police report. Our boys would as soon interview a loose alligator in the Zoo. But you could do it, I'm thinking – an old friend like you.'

'Why,' said I, greatly relieved, 'this makes it all easy. It so happens that it was to visit Professor Challenger at Rotherfield that I was asking for leave of absence. The fact is, that it is the anniversary of our main adventure on the plateau three years ago, and he has asked our whole party down to his house to see him and celebrate the occasion.'

'Capital!' cried McArdle, rubbing his hands and beaming through his glasses. 'Then you will be able to get his opeenions out of him. In any other man I would say it was all moonshine, but the fellow has made good once, and who knows but he may again!'

'Get what out of him?' I asked. 'What has he been doing?'

'Haven't you seen his letter on "Scientific Possibeelities" in today's *Times*?'

'No.'

McArdle dived down and picked a copy from the floor. . . .

This was the letter which I read to the news editor of the *Gazette*:

'SCIENTIFIC POSSIBILITIES'

'Sir,–I have read with amusement, not wholly unmixed with some less complimentary emotion, the complacent and wholly fatuous letter of James Wilson MacPhail, which has lately appeared in your columns upon the subject of the blurring of Frauenhofer's lines in the spectra both of the planets and of the fixed stars. He dismisses the matter as of no significance. To a wider intelligence it may well seem of very great possible importance – so great as to involve the ultimate welfare of every man, woman, and child upon this planet. I can hardly hope, by the use of scientific language, to convey any sense of my meaning to those ineffectual people who gather their ideas from the columns of a daily newspaper. I will endeavour, therefore, to condescend to their limitation, and to indicate the situation by the use of a homely analogy which will be within the limits of the intelligence of your readers.'

'Man, he's a wonder – a living wonder!' said McArdle, shaking his head reflectively. 'He'd put up the feathers of a sucking-dove and set up a riot in a Quaker's meeting. No wonder he has made London too hot for him. It's a peety, Mr. Malone, for it's a grand brain! Well, let's have the analogy.'

'We will suppose,' I read, 'that a small bundle of connected corks was launched in a sluggish current upon a voyage across the Atlantic. The corks drift slowly on from day to day with the same conditions all round them. If the corks were sentient we could imagine that they would consider these conditions to be permanent and assured. But we, with our superior knowledge, know that many things might happen to surprise the corks. They might possibly float up against a ship, or a sleeping whale, or become entangled in seaweed. In any case, their voyage would probably end by their being thrown up on the rocky coast of Labrador. But what could they know of all this while they drifted so gently day by day in what they thought was a limitless and homogeneous ocean?

'Your readers will possibly comprehend that the Atlantic, in this parable, stands for the mighty ocean of ether through which we drift, and that the bunch of corks represents the little and obscure planetary system to which we belong. A third-rate sun, with its rag-tag and bobtail of insignificant satellites, we float under the same daily conditions towards some unknown end, some squalid catastrophe which will overwhelm us at the ultimate confines of space, where we are swept over an etheric Niagara, or dashed upon some unthinkable Labrador. I see no room here for the shallow and ignorant optimism of your correspondent, Mr. James Wilson MacPhail, but many reasons why we should watch with a very close and interested attention every indication of change in those cosmic surroundings upon which our own ultimate fate may depend.'

'Man, he'd have made a grand meenister,' said McArdle. 'It just booms like an organ. Let's get doun to what it is that's troubling him.'

'The general blurring and shifting of Frauenhofer's lines of the spectrum point, in my opinion, to a widespread cosmic change of a subtle and singular character. Light from a planet is the reflected light of the sun. Light from a star is a self-produced light. But the spectra both from planets and stars have, in this instance, all undergone the same change. Is it, then, a change in those planets and stars? To me such an idea is inconceivable. What common change could simultaneously come upon them all? Is it a change in our own atmosphere? It is possible, but in the highest degree improbable, since we see no signs of it around us, and chemical analysis has failed to reveal it. What, then, is the third possibility? That it may be a change in the conducting medium, in that infinitely fine ether which extends from star to star and pervades the whole universe. Deep in that ocean we are floating upon a slow current. Might that current not drift us into belts of ether which are novel and have properties of which we have never conceived? There is a change somewhere. This cosmic disturbance of the spectrum proves it. It may be a good change. It may be an evil one. It may be a neutral one. We do not know. Shallow observers may treat the matter as one which can be disregarded, but one who like myself is possessed of the deeper intelligence of the true philosopher will understand that the possibilities of the universe are incalculable and that the wisest man is he who holds himself ready for the unexpected. To take an obvious example, who would undertake to say that the mysterious and universal outbreak of illness, recorded in your columns this very morning as having broken out among the indigenous races of Sumatra, has no connection with some cosmic change to which they may respond more quickly than the more complex peoples of Europe? I throw out the idea for what it is worth. To assert it is, in the present stage, as unprofitable as to deny it, but it is an unimaginative numskull who is too dense to perceive that it is well within the bounds of scientific possibility.

Yours faithfully

GEORGE EDWARD CHALLENGER.

THE BRIARS, ROTHERFIELD.'

'It's a fine, steemulating letter,' said McArdle, thoughtfully, fitting a cigarette into the long glass tube which he used as a holder. 'What's your opeenion of it, Mr. Malone?'

I had to confess my total and humiliating ignorance of the subject at issue. What, for example, were Frauenhofer's lines? McArdle had just been studying the matter with the aid of our tame scientist at the office, and he picked from his desk two of those many-coloured spectral bands which bear a general resemblance to the hat-ribbons of some young and ambitious cricket club. He pointed out to me that there were certain black lines which formed cross-bars upon the series of brilliant colours extending from the red at one end, through gradations of orange, yellow, green, blue, and indigo, to the violet at the other.

'Those dark bands are Frauenhofer's lines,' said he. 'The colours are just light itself. Every light, if you can split it up with a prism, gives the same colours. They tell us nothing. It is the lines that count, because they vary according to what it may be that produces the light. It is these lines that have been blurred instead of clear this last week, and all the astronomers have been quarrelling over the reason. Here's a photograph of the blurred lines for our issue tomorrow. The public have taken no interest in the matter up to now, but this letter of Challenger's in *The Times* will make them wake up, I'm thinking.'

'And this about Sumatra?'

'Well, it's a long cry from a blurred line in a spectrum to a sick nigger in Sumatra. And yet the chiel has shown us once before that he knows what he's talking about. There is some queer illness down yonder, that's beyond all doubt, and today there's a cable just come in from Singapore that the lighthouses are out of action in the Straits of Sudan, and two ships on the beach in consequence. Anyhow, it's good enough for you to interview Challenger upon. If you get anything definite, let us have a column by Monday.'

I was coming out from the news editor's room, turning over my new mission in my mind, when I heard my name called from the waiting-room below. It was a telegraph-boy with a wire which had been forwarded from my lodgings at Streatham. The message was from the very man we had been discussing, and ran thus:

'Malone, 17, Hill Street, Streatham. – Bring oxygen. – *CHALLENGER*.'

'Bring oxygen!' The Professor, as I remembered him, had an elephantine sense of humour capable of the most clumsy and unwieldy gambollings. Was this one of those jokes which used to reduce him to uproarious laughter, when his eyes would disappear, and he was all gaping mouth and wagging beard, supremely indifferent to the gravity of all around him? I turned the words over, but could make nothing even remotely jocose out of them. Then surely it was a concise order – though a very strange one. He was the last man in the world whose deliberate command I should care to disobey. Possibly some chemical experiment was afoot; possibly – Well, it was no business of mine to speculate upon why he wanted it. I must get it. There was nearly an hour before I should catch the train at Victoria. I took a taxi, and having ascertained the address from the telephone book, I made for the Oxygen Tube Supply Company in Oxford Street.

61

*The titan of twentieth-century biochemistry, Warburg, is the subject of a biography (*Otto Warburg*) by one of several of his students to win a Nobel Prize of his own, H.A. Krebs. The flavour of Warburg's forbidding personality comes through in an anecdote, stemming from his first visit to the USA after the Second World War (in which he had remained in his laboratory in Berlin, with the status – for he was half Jewish – of Honorary Aryan). How, a journalist asked him, did he react to the allegation that he was a great scientist but a rotten human being? 'I am glad,' Warburg replied in his most chilling manner, 'they do not say the other way round.' Here, quoted in Krebs's book, is a description of life in Warburg's lab by one of his collaborators, Karlfried Gawehn.*

The official working hours in the 1950s were from 9 a.m. to 6 p.m. Monday to Friday, with a mid-day break from 1 to 2 p.m.; on Saturdays we worked from 9 a.m. to 1 p.m. Warburg usually arrived shortly before 9 a.m. and it was a matter of course that everyone else was there by 8.45 a.m. at the latest. We usually extended our luncheon break a little because Warburg was normally away for 2 ½ hours. He left with Heiss at about 1 p.m. for his home (1 minute away) and after lunch would take a brief nap and not come back before 3.30 p.m. Nevertheless, our working day was at least 8 hours because Warburg did not leave the Institute before 6.30 p.m. Here again it was accepted that we did not leave before him, if only because he often required some of our latest results for his evening work at home. His working week was regularly about 60 hours.

Warburg failed to appreciate that people might occasionally need an hour or two's leave of absence to deal with personal matters – for example to visit a doctor or bank or child's school. When asked for permission he would say, 'You have enough time in the evening to attend to such things and besides, you have at your disposal 7 to 8 weeks' leave each year.' In the early 1960s the Directors of the Max Planck Institutes were discussing a revision of working hours. Warburg's comments were characteristically humorous-cum-aggressive: 'Can you imagine,' he said, 'those people' (meaning his fellow-directors) 'no longer wish to work on Saturdays? When I asked them why they considered it no longer necessary, they said that employees needed time for shopping. So I told them that perhaps they should also give their employees Friday off so that they could do still more shopping.' To me this answer sums up Warburg's attitude very well: for him there were no reasonable grounds, apart from death, for not working. Illnesses such as influenza were no excuse for absence.

Warburg himself adhered rigidly to this spartan style of working. When he was awarded a decoration by the Federal Government and was invited to the ceremonial investiture at the Town Hall, he telephoned the official: 'Send me my decoration by mail. My experiments allow me no time to leave my laboratory.'

At the end of the 1960s however, Warburg had to give up the fight and allow a five-day working week to be introduced.

The summer vacation was usually from the beginning of August until mid-September although Warburg himself preferred not to return before the middle of October in order to avoid celebrations and congratulations from colleagues on his birthday, 8 October. But there was no let-up in the laboratory between mid-September and mid-October. Work proceeded at full speed and results had to be reported to Warburg by telephone at least twice a day, in the morning at 9 a.m. and in the evening at 5 p.m.

When experiments were successful and the principles of his theories were not contradicted by experimental results, Warburg would be relaxed and would often reminisce about interesting happenings during the Hitler regime, during the War, and during his encounters with Russians and Americans after the War. He also talked about his private life. Mostly his talk was spiced with witty comments on politicians, officers, or his scientific opponents.

62

A figure of rather similar character and impact on German science was the great physicist, Walther Nernst. An overbearing and autocratic personality, he was by no means universally popular. A story was told about Nernst after his death, which occurred in 1945 on his estate in East Prussia. Nernst was buried there and two eminent physicists, Bonhoeffer and Bodenstein, stood as pall-bearers. When the Russians annexed East Prussia, Nernst's widow decided that he should be reinterred in Berlin and the services of Bonhoeffer and Bodenstein were called on again. Some years later, another decision was taken that Nernst should lie in that Valhalla of great physicists, the municipal cemetery in Göttingen. Bonhoeffer and Bodenstein at the graveside looked at each other morosely. 'This,' said Bodenstein to Bonhoeffer 'is getting to be tiresome.' Bonhoeffer wagged his finger. 'You can't,' he said, 'bury Nernst too often.'

Nernst and his times are the subject of an absorbing study by Kurt Mendelssohn, The World of Walther Nernst, *subtitled* The Rise and Fall of German Science. *Here is just one anecdote from it.*

A probably apocryphal story records Nernst's surprise on a cold winter morning to find the cowshed pleasantly warm and wondering whether it was heated. On being told that this was due entirely to the metabolic heat of the cows, he sold the cows and invested the proceeds in carp since he felt that he did not intend to spend his money on meat production attended with so much waste heat when the same thing could be done isothermally. Whatever the truth of this anecdote, Nernst forever praised the thermodynamic efficiency of fish who were able to put on weight at ambient temperature.

63

Chaim Weizmann – the chemist, who came from Russia to Manchester, and whose manufacturing process for acetone made a critical contribution to the course of the First World War, became the first President of Israel and brought science to the desert in the shape of the Weizmann (at that time the Sieff) Institute – is the posthumous hero of a scientific thriller, The Sun Chemist, *by Lionel Davidson.*

Chaim Weizmann was born in 1874 in Motol, a small village in Byelorussia, and as a boy moved with his family to Pinsk, a few miles away, which was bigger and even nastier. His father was a timber merchant, not prosperous, but he managed to put all of his large family through university. There was a rather sound family way of doing this. As each child completed university, he got a job and began contributing to the tuition fees of the next in line. (Years later, while Chaimchik was pawning his compasses, or scraping a living with Verochka in Manchester, he still managed to send a pound or two a month to keep two sisters going in Switzerland.)

He soon got away from Pinsk and went to Germany to study chemistry, ultimately to the Technische Hochschule at Berlin Charlottenburg where he worked under the immediate direction of a Dr Bistrzycki. When Bistrzycki was called to a professorship at Fribourg in Switzerland in 1896, Weizmann followed him. He picked up his D.Sc. there in 1899, and went to the University of Geneva as a junior lecturer.

Dyestuff chemistry was much the thing at the time, and this is what he had been doing with Bistrzycki. He immediately began researching and publishing at a great rate. In a single year he produced three very extensive papers and took out four well-documented patents. But he was busy in a bewildering number of directions.

Switzerland was at the time a hotbed of political activity. There were numerous groups of impassioned émigrés, mainly Russian, covering a wide spectrum of contrary opinion. There were simple Socialists, not so simple Socialists, Communists (including incipient Bolsheviks and Mensheviks), Anarchists, Bundists, Zionists. The wild object of many of them was to create a revolution in unchanging Russia, and of the last group to alter an equally unchanging situation, the dispersion of the Jews.

Zionism as a political movement was of later vintage than the others. Its basis was that the millions of Jews scattered about the world were not simply religious minorities in their different countries, as might be Protestants, Catholics, or Muslims, but a single people exiled from a particular land. The proposition was to repurchase the land, and the movement's organizer, a Viennese journalist called Theodor Herzl (whose dignified portrait today appears on Israeli hundred-pound notes), in fact tried to do this by offering the Sultan of Turkey several million pounds for a 'charter' to it. The deal fell through, to the Sultan's regret, but there were very many alternative proposals, hotly contested by the impecunious polemicists and students who made up the active body.

Weizmann had been a Zionist for years, and in Switzerland found fertile ground and much unattached or even downright errant Jewish youth. He decided to collect what he could of it for Zionism, and with half a dozen friends arranged a meeting in the Russian library. This was a rash thing to do without securing the prior approval of G. V. Plekhanov, doyen of the squabbling émigré society and founder of Marxism in Russia. (In later life, Verochka recalled often seeing his two juniors Lenin and Trotsky meeting in a flat across the street.) Plekhanov's disapproval could virtually be guaranteed for interlopers to his scene, so that when the founding seven arrived at their venue they found, with no surprise, that all the furniture had been removed. They held the meeting, nonetheless, standing up, voted on a Hebrew name for themselves, *Ha-Shachar*, The Dawn, and then voted to call a mammoth conference to recruit membership.

This tremendous affair, addressed by representatives of all factions, lasted for three and a half days and ended at four in the morning with a great personal triumph for Weizmann – 118 new members – and a trembling confrontation with Plekhanov.

'What do you mean by bringing dissension into our ranks?' the offended Marxist demanded.

'Monsieur Plekhanov,' Weizmann grandly informed him, 'you are not the Czar!'

Apart from these heated public affairs, the young experimenter was having a couple of private ones. He was living with a young lady, Sophia Getzova, to whom he was engaged, and carrying on with another, Vera Khatzmann, a medical student from Rostov-on-Don. By about 1904, things began to get on top of him, and he decided to narrow his activities to the scientific and try his luck elsewhere.

His fancy fell on Manchester, center of the British textile industry, which had an excellent university department of organic chemistry presided over by the distinguished Professor Perkin. Perkin's father, many years before, had made his name in the field of dyestuff chemistry by synthesizing aniline blue (thus heralding in the 'mauve decade') – a fact that Weizmann thought might dispose him in favor of another dyestuff chemist. But Perkin greatly took to the animated young Russian anyway. In an affable conversation in German (Weizmann as yet had no English), he pointed out that while no staff jobs were open at the moment, he could offer him the use of a little basement laboratory at a nominal sum of six pounds, with the services of a lab boy thrown in. Weizmann accepted, and while Perkin went off on vacation – it was the summer of 1904 – installed himself in the empty university.

By the time Perkin returned, he had quite a lot of English. He had learned it from the chemistry department stores book, the Bible, and the purchased works of Macaulay and Gladstone; and also – as he wrote to his sweet darling, the joy from Rostov-on-Don

– from conversations with a young demonstrator of Perkin's who had arrived back from vacation. With this young man he instituted a series of experiments, so that when Perkin took up his chair again, much refreshed by his holiday, Weizmann was able to show him quite a lot. By the winter he was on the staff, with students of his own.

He couldn't, however, stay away from Zionism. Pogroms in Russia brought a mass protest rally in Manchester, which the young Russian was asked to address. At the rally was the prospective Liberal candidate for the constituency, a Mr W. S. Churchill, keeping his eye on the electorate. He couldn't understand a word of Weizmann's fiery oration in Yiddish, but was much impressed by the effect on the audience, and made haste to wring the orator's hand and to hint that he could be of great service in swaying the Jewish vote. Weizmann declined: he said he was interested only in Zionist issues. Anyway, the January, 1906, elections were at hand, with other politicians astir. Contesting them was the Prime Minister himself, Mr Balfour, also with a Manchester constituency. Came January and Mr Balfour, and Mr Balfour's agent, who thought he ought to have fifteen minutes with the intriguing young man who knew so much about Russia and the state of the Jews there. The fifteen minutes stretched to an hour and a quarter, and ended with both men knowing rather more about the state of everything.

He was determined to stick to science, however, and he did. He brought over Miss Vera Khatzmann, his Verochka, married her, and slogged on with his chemistry. An interesting new problem had appeared. The world's supply of rubber was unequal to the demand: a task for the synthetic chemist. Perkin interested himself, and put teams on it, including Weizmann's.

Chaimchik's approach was novel. He had become interested in fermentation. Verochka's sister had married a scientist who lived in Paris on the Left Bank. The Weizmanns visited from time to time, and Chaimchik picked up a free-lance assignment that involved work at the Pasteur Institute, kingdom of the great fermenter himself.

The basis of fermentation was that micro-organisms, bacteria, could by creating a ferment in one substance change it into another. He looked up the literature and found that the essential substance of rubber was the five-carbon compound isoprene. Further study showed that a Russian called Winogradsky had recently observed a five-carbon compound in nature. It could be isolated by fermenting sugar with certain bacteria to produce a volatile compound exhibiting the odor of fusel oil.

Weizmann repeated the experiment in Manchester, and found that Winogradsky had got it wrong. The substance produced, though smelling of fusel oil, was something else. It was not a five-carbon molecule, either. It was a four-carbon one, and it was butyl alcohol. He was a very dogged experimenter, and he tried it many times with a variety of bacteria. It always turned out the same, but in the end he got more butyl alcohol.

Professor Perkin, to whom he showed the results, permitted himself one of his rare puns. He said, 'Your butyl alcohol is a very futile alcohol,' and advised him to pour it down the sink. Chaimchik didn't do this. He kept on refining the produce with a variety of treatments. He got a very large yield of butyl alcohol and smaller quantities of other substances, including methyl alcohol and acetone, the latter about 30 percent of the total.

He kept on doing this, and the First World War broke out, and a new problem asserted itself. It impinged on Manchester in the form of a Dr Rintoul, from the Scottish branch of the Nobel explosives firm, whose problem was most acute. His firm was supplying the British fleet with cordite for its large naval guns. Strategic considerations made it imperative that the location of these guns should be concealed from the enemy. The solution was to propel the shells by smokeless gunpowder, made possible by the chemical solvent acetone, previously, but no longer, obtainable in generous supply from the forests of Europe as a by-product of charcoal. Not all the forests of Britain could supply the present need for acetone. Was there some other method of producing it?

'Walk this way, Dr Rintoul,' said Dr Weizmann, and showed him a method.

Dr Rintoul made haste to the telephone, and the night train from Scotland brought the managing director of Nobel's, together with several of the senior scientific staff. They went carefully through Chaimchik's lab books and repeated his experiments: acetone in abundance. Terms were stated for this valuable patent, to which Chaimchik and the university readily agreed. And then occurred a strange accident. The Nobel works blew up. They were unable to take up the patent. The problem was no less urgent for the accident, and Chaimchik, placed in charge of it, was sent by express train to London and ushered into the office of the First Lord of the Admiralty. He found the First Lord of the Admiralty was Mr

W.S. Churchill, last encountered wringing his hand while trying to get him to nobble the Jewish vote in Manchester.

The two men got on famously, and Churchill asked him what he required. Weizmann replied that existing fermentation plants were largely operated by distillers of whisky and gin. Churchill told him to take his pick, and he picked the Nicolson gin distillery in Bow, which was immediately sequestered for his use.

For the next two years he took on a most daunting, almost mind-boggling task, the one-man creation of a completely new industry: industrial fermentation. The government built him a factory and took over the largest distilleries in the country. He himself took over the laboratory of the Lister Institute in Chelsea, to train teams of chemists to go out and operate the plants.

The process depended on a bacterium that he had isolated and in countless experiments improved, *Clostridium acetobutylicum weizmann*. It worked on starch-containing products, and the method he had perfected demanded large quantities of maize. When U-boat warfare interrupted overseas supplies, he switched to horse chestnuts in Britain, and the process crossed the Atlantic to be employed on Canadian maize. A plant was taken over in Toronto, and soon turning out acetone; and when the United States entered the war, the process was also adopted there, at two big distilleries in Terre Haute, Indiana. His operations had spread to Asia before the war ended; but by that time he had transferred his energies.

While he had been keeping the Navy's guns firing, the Army's were blowing the Turks out of their old Ottoman Empire, which included Palestine. The wartime coalition government was headed by Lloyd George, very keen on his Old Testament, and the Foreign Secretary was Mr A. J. Balfour, whose mind had been so enlarged on the Jewish question in 1906. Weizmann became more tremendously busy. He had never ceased to be active in Zionism, had attended all the big prewar European conferences. But now the war had cut off the European societies, and from being a well-fancied middleweight in the movement he had become its senior statesman. Much negotiation brought about the Balfour Declaration, which declared: 'His Majesty's Government view with favour the establishment of a Jewish National Home in Palestine'; and while his ancient sparring partners Lenin and Trotsky were raising hell making their incredible dream come true in Russia, Chaimchik sped off to Palestine to attend to his.

He found fighting still going on, but lost no time in getting down to his first scheme, a project planned more than half his life, the laying of a cornerstone for a Hebrew University in Jerusalem. He invited the victorious General Allenby to the ceremony, and, recalling the prophetic utterance that the Word should go forth from Jerusalem, Allenby was both happy and moved. Gunfire could still be heard rolling in the Jerusalem hills while the future center of learning was founded, and everybody was very moved.

But things were moving everywhere, and in a variety of directions. On the scientific front, acetone was phasing out, and was anyway being produced as a by-product of the rising petroleum industry. Keeping pace with the petroleum industry was the rising automobile industry. The automobiles needed painting, and mass-production methods required fast-drying varnishes. The best solvent was found to come from Chaimchik's butyl alcohol, which Professor Perkin had advised him to pour down the sink – still obtainable and in large quantities, by his patent method.

He let his patent agents attend to that one, together with Commercial Solvents, and threw himself into politics.

He was the leader of world Zionism, the builder-up of the national home, the settler of the people in it, the raiser of the money to do the job. The university was his pet project and he raised that. For the next thirteen years, almost every minute was accounted for. He traveled, exhorted, pleaded, presided; and in 1931, thoroughly exhausted, found himself kicked out of the job owing to factional differences.

I was sitting in his chair and staring out at his grave as I pondered this, his presence strong in the room, so that when the hand fell on my shoulder I silently rose and almost went through the ceiling.

'Igor, we are going down to lunch now,' Connie said. 'And Meyer wants you to call him afterwards. He didn't want you disturbed while you were reading.'

64

Here now is Weizmann himself, from his autobiography, Trial and Error, *during the First World War.*

Mr. Churchill was brisk, fascinating, charming and energetic. Almost his first words were: 'Well, Dr. Weizmann, we need thirty thousand tons of acetone. Can you make it?' I was so terrified by this lordly request that I almost turned tail. I answered: 'So far I have succeeded in making a few hundred cubic centimetres of acetone at a time by the fermentation process. I do my work in a laboratory. I am not a technician, I am only a research chemist. But, if I were somehow able to produce a ton of acetone, I would be able to multiply that by any factor you chose. Once the bacteriology of the process is established, it is only a question of brewing. I must get hold of a brewing engineer from one of the big distilleries, and we will set about the preliminary task. I shall naturally need the support of the Government to obtain the people, the equipment, the emplacements and the rest of it. I myself can't even determine what will be required.'

I was given a *carte blanche* by Mr. Churchill and the department, and I took upon myself a task which was to tax all my energies for the next two years, and which was to have consequences which I did not then foresee.

65

We now go back three centuries to Isaac Newton, a rather unlikeable man ('I dreamed,' wrote the Astronomer Royal, John Flamsteed and an admirer of Newton, 'that Newton was dead'), who appears often in poetry, most familiarly in Wordsworth's lines, inspired by contemplation of Newton's statue in St. John's College, Cambridge. The famous passage concludes:

And from my pillow, looking forth by light
Of moon or favouring stars, I could behold
The antechapel where the statue stood
Of Newton with his prism and silent face,
The marble index of a mind for ever
Voyaging through strange seas of Thought, alone.

66

And now Alexander Pope, with a modern riposte by Sir John Squire:

Nature, and Nature's laws, lay hid in night,
God said, *Let Newton be!* and all was light.

It did not last: the Devil howling *Ho,*
Let Einstein be, restored the status quo.

67

Here is how George Bernard Shaw imagines him in his play, In Good King Charles's Golden Days. *('Good King Charles' – Charles II – was something of a patron of science and founded the Royal Society, but its meetings at Gresham College in the City of London, according to Pepys, 'he mightily laughed at for spending time only in the weighing of ayre'.)*

Newton, aged 38, comes in from the garden, hatless, deep in calculation, his fists clenched, tapping his knuckles together to tick off the stages of the equation. He stumbles over the mat.

MRS BASHAM. Oh, do look where youre going, Mr Newton. Someday youll walk into the river and drown yourself. I thought you were out at the university.

NEWTON. Now dont scold, Mrs Basham, dont scold. I forgot to go out. I thought of a way of making a calculation that has been puzzling me.

MRS BASHAM. And you have been sitting out there forgetting everything else since breakfast. However, since you have one of your calculating fits on I wonder would you mind doing a little sum for me to check the washing bill. How much is three times seven?

NEWTON. Three times seven? Oh, that is quite easy.

MRS BASHAM. I suppose it is to you, sir; but it beats me. At school I got as far as addition and subtraction; but I never could do multiplication or division.

NEWTON. Why, neither could I: I was too lazy. But they are quite unnecessary: addition and subtraction are quite sufficient. You add the logarithms of the numbers; and the antilogarithm of the sum of the two is the answer. Let me see: three times seven? The logarithm of three must be decimal four seven seven or thereabouts. The logarithm of seven is, say, decimal eight four five. That makes one decimal three two two, doesnt it? What's the antilogarithm of one decimal three two two? Well, it must be less than twentytwo and more than twenty. You will be safe if you put it down as–

Sally returns.

SALLY. Please, maam, Jack says it's twentyone.

NEWTON. Extraordinary! Here was I blundering over this simple problem for a whole minute; and this uneducated fish hawker solves it in a flash! He is a better mathematician than I.

68

Newton's absent-mindedness is well attested. A handed-down story, recorded by Thomas Moore in his Journal, *goes like this:*

Inviting a friend to dinner and forgetting it; the friend arriving and finding the philosopher in a fit of absorption. Dinner brought up for one: the friend (without disturbing Newton) sitting down and despatching it, and Newton, after recovering from his reverie, looking at the empty dishes and saying, "Well really, if it weren't for the proof before my eyes, I could have sworn that I had not yet dined."

And again: At some seldom times when he designed to dine in hall, would turn to the left hand and go out into the street, when making a stop he found his mistake, would hastily turn back, and then sometimes instead of going into hall, return to his chamber again.

It was also related that at home in Woolsthorpe he insisted that his breakfast egg must cook exactly five minutes; on one occasion the maid entered the kitchen to find Newton before the stove, thoughtfully looking at the egg, which rested in his hand, while his watch lay in the saucepan of boiling water.

Shaw's picture of Newton, confronted with the accounts, also has a historical base. It was indeed written (by Alexander Pope) that Sir Isaac Newton, though so deep in Algebra and Fluxions, could not readily make up a common account; and when he was Master of the Mint, used to get somebody to make up his accounts for him.

Something of the quality of Newton's genius is caught up by the Cambridge scholar, William Whewell, here quoted by E.N. da C. Andrade, writing in a Royal Society symposium on Newton and the Science of his Age (1943):

As we read the *Principia* we feel as when we are in an ancient armoury where the weapons are of gigantic size; and as we look at them we marvel what manner of man he was who could use them as a weapon what we can scarcely lift as a burden.

We leave Newton with Byron (from Don Juan, *Canto X):*

When Newton saw an apple fall, he found
  In that slight startle from his contemplation –
'Tis *said* (for I'll not answer above ground
  For any sage's creed or calculation) –
A mode of proving that the earth was round
  In a most natural whirl, called 'gravitation';
And this is the soul mortal who could grapple,
  Since Adam, with a fall, or with an apple.

69

The quality of James Clerk Maxwell is caught in this accomplished expression of his craft in verse. The title makes allusion to Maxwell's contemporary, John Tyndall (see no. 53), and to the Assyrian harp or nabla, which is in the form of an inverted delta, ∇*, del, the Hamiltonian operator* $\left(i\frac{d}{dx} + j\frac{d}{dy} + k\frac{d}{dz}\right)$*. 'The Chief Musician upon Nabla' is the mathematician, Tait.*

TO THE CHIEF MUSICIAN UPON NABLA
A TYNDALLIC ODE

I come from fields of fractured ice,
  Whose wounds are cured by squeezing,
Melting they cool, but in a trice,
  Get warm again by freezing.
Here, in the frosty air, the sprays
  With fern-like hoar-frost bristle,
There, liquid stars their watery rays
  Shoot through the solid crystal.

I come from empyrean fires –
  From microscopic spaces,
Where molecules with fierce desires,
  Shiver in hot embraces.

The atoms clash, the spectra flash,
  Projected on the screen,
The double D, magnesian *b*,
  And Thallium's living green.

We place our eye where these dark rays
  Unite in this dark focus,
Right on the source of power we gaze,
  Without a screen to cloak us.
Then where the eye was placed at first,
  We place a disc of platinum,
It glows, it puckers! will it burst?
  How ever shall we flatten him!

This crystal tube the electric ray
  Shows optically clean,
No dust or haze within, but stay!
  All has not yet been seen.
What gleams are these of heavenly blue?
  What air-drawn form appearing,
What mystic fish, that, ghostlike, through
  The empty space is steering?

I light this sympathetic flame,
  My faintest wish that answers,
I sing, it sweetly sings the same,
  It dances with the dancers.
I shout, I whistle, clap my hands,
  And stamp upon the platform,
The flame responds to my commands,
  In this form and in that form.

What means that thrilling, drilling scream,
  Protect me! 'tis the siren:
Her heart is fire, her breath is steam,
  Her larynx is of iron.
Sun! dart thy beams! in tepid streams,
  Rise, viewless exhalations!
And lap me round, that no rude sound
  May mar my meditations.

Here let me pause. – These transient facts,
These fugitive impressions,
Must be transformed by mental acts,
To permanent possessions.
Then summon up your grasp of mind,
Your fancy scientific,
Till sights and sounds with thought combined,
Become of truth prolific.

Go to! prepare your mental bricks,
Fetch them from every quarter.
Firm on the sand your basement fix
With best sensation mortar.
The top shall rise to heaven on high –
Or such an elevation,
That the swift whirl with which we fly
Shall conquer gravitation.

70

Ernest (Lord) Rutherford was probably the greatest experimental physicist of this century. 'Always,' one of his contemporaries told him, 'on the crest of a wave'. 'Well, I made the wave, didn't I?' Rutherford replied. Then he added modestly, 'or at least some of it.' A former student, M.L. Oliphant, describes Rutherford's exuberant character (from Notes and Records of the Royal Society, *Vol. 27 No. 1).*

Rutherford's participation in the experiments was limited to discussion of what to do next, and deep interest in the results. He gave us a completely free hand in the design of experiments and running of the equipment, but kept us on our toes all the time. Usually he came to see us twice each day, late in the morning and shortly before six o'clock in the evening. Sometimes, he would turn up at other times, uncannily aware that something interesting was happening. This was when things were almost bound to go wrong. With Rutherford looking over the operator's shoulder, impatiently awaiting the outcome of an observation, silly mistakes were apt to be made. On two occasions, Rutherford pushed something through the thin mica window through which the products of transformations emerged, creating panic as air flowed into the vacuum and we rushed to shut off and cool down the oil diffusion pumps. He apologized profusely, but disappeared for days while we cleaned up and got going again.

If Rutherford appeared just at the end of a run, he insisted that the record be developed as rapidly as possible, barely allowed it to be dipped in the fixing bath, and sat at the table in the next room, dripping fixing solution upon our papers and his own clothes, as he examined the tracing. His pipe dribbled ash all over the wet and sticky photographic paper. He damaged it irreparably with the stump of a pencil from his pocket, with which he attempted to mark the soft, messy material. Searching impatiently for the interesting parts of the long record, he pulled it from the coil in Crowe's hands to fall to the dirty stone floor, trampling on it as he got up in the end. We had then to do our best to finish fixing, washing and drying the paper strips, often damaged beyond repair. When it was possible, we concealed records from him till they had been properly processed and measured up by us, but this was impossible when he was present while the record was being taken. Once, at the end of a particularly heavy day, when the experiments had gone well, we decided to postpone development till next morning when we were fresh and we could handle the long strip in new developer and fixer without damage. Just as we were leaving Rutherford came in. He became extremely angry when he heard what we had decided, and insisted that we develop the film at once. 'I can't understand it,' he thundered. 'Here you have exciting results and you are too damned lazy to look at them tonight.'

We did our best, but the developer was almost exhausted, and the fixing bath yellowed with use. The result was a messy record which even Rutherford could not interpret. In the end, he went off, muttering to himself that he did not know why he was blessed with such a group of incompetent colleagues. After dinner that night, he telephoned me at home: 'Er! Er! Is that you Oliphant? I'm er, er, sorry to have been so bad tempered tonight. Would you call in to see me at Newnham Cottage as you go to the Laboratory in the morning?'

Next day he was even more contrite: 'Mary says I've ruined my suit. Did you manage to salvage the record?'

He drove us mercilessly, but we loved him for it.

In 1933, G. N. Lewis, from Berkeley, visited the Cavendish and presented Rutherford with about half a gram of almost pure heavy water which he had concentrated electrolytically. We made arrangements to recover a small volume of deuterium gas produced from this, after it had passed through the apparatus, and to purify it for re-use. A small portion of the heavy water was converted, by ion exchange, into a solid salt which was spread on our target. Immediately, we obtained exciting results when a beam of deuterium ions fell on this deuterium containing target, even at very low energies. There was a copious emission of two groups of particles which we identified as long range protons and short range tritium ions which originated in the same process and therefore escaped in pairs opposite one another. It was impressive to experience Rutherford's enthusiasm and the approximate arithmetic by which he calculated the range-energy relationship of tritium nuclei from the known range-energy curves for alpha-particles and protons. We were able to show, by passing the particles through a magnetic and an electric field at right angles, that both particles had the same momentum, and we obtained a value for the mass of tritium very close to that now accepted. We showed also that there was a copious emission of neutrons, using a helium ionization counting chamber at high pressure. I managed, with Crowe's help, to split a piece of mica of uniform thickness, and stopping power equivalent to only 1.5 millimetres of air, to cover the window on our equipment. We at once observed an emission of particles which clearly carried a double charge and appeared to be alpha-particles, in numbers equal to that of the protons or tritons. The equality of fluxes suggested strongly that all three particles originated in the same nuclear process. Rutherford produced hypothesis after hypothesis, covering reams of paper with abortive arithmetic, and finally we went home to think about it.

I went all over the afternoon's work again, telephoned Cockcroft who had no new ideas to offer, and went to bed tired out. At 3 a.m., the telephone rang. Fearing bad news, for a call at that time is always ominous, my wife, who wakens instantly, answered it and came back to tell me that 'the Professor' wanted to speak to me. Still drugged with sleep, I heard an apologetic voice express sorrow for wakening me, then excitedly say: 'I've got it. Those short-range particles are helium of mass three.'

Shocked into attention, I asked on what possible grounds could he conclude that this was so, as no combination of twice two could give two particles of mass three and one of mass unity. Rutherford roared: 'Reasons! Reasons! I feel it in my water!'

He then told me that he believed the helium particle of mass 3 to be the companion of a neutron, produced in an alternative reaction which just happened to occur with the same probability as the reaction producing protons and tritons.

I went back to bed, but not to sleep. I called in to see Rutherford at Newnham Cottage after breakfast, and went through his approximate calculations with him. We agreed that the way to clinch the conclusion was to measure, as accurately as we could, the range of the doubly charged group of particles, and the energy of the neutrons. I went through our records from the helium pressure chamber, measuring the amplitudes of the most energetic of the helium recoils, and obtaining a maximum neutron energy of about 2 million electron-volts, while my colleagues estimated more accurately the range of the short group. Of course, Rutherford was right. By the end of the morning we had satisfied ourselves that an alternative reaction of two deuterons, produced a neutron and a helium particle of mass 3, the energy released being close to that in the other reaction. The mass of helium three worked out to be a little less than that of tritium.

We all shared Rutherford's excitement. We had found two new isotopes and measured their masses, and we understood the remarkable deuterium reactions. That evening, I wrote a note describing our work. This was pencilled all over by Rutherford in the morning, retyped and sent off to *Nature*. Only in the war was I to experience such a hectic few days of work, but at no other time have I felt the same sense of accomplishment or such comradeship as Rutherford radiated that day.

71

Sir Edward Bullard has the following recollection; 'An early breakthrough in spending money was the purchase of a doughnut magnet designed by Cockroft, which was used to focus α particles. Rutherford, when showing it to a visitor said "That cost as much as a research student for a year – but it does twice as much work." '

Here is another picture of Rutherford, a universally esteemed figure, taken from C. P. Snow's collection of character sketches, Variety of Men.

The difficulty is to separate the inner man from the Rutherfordiana, much of which is quite genuine. From behind a screen in a Cambridge tailor's, a friend and I heard a reverberating voice: 'That shirt's too tight round the neck. Every day I grow in girth. *And* in mentality.' Yet his physical make-up was more nervous than it seemed. In the same way, his temperament, which seemed exuberantly powerful, massively simple, rejoicing with childish satisfaction in creation and fame, was not quite so simple as all that. His was a personality of Johnsonian scale. As with Johnson, the façade was overbearing and unbroken. But there were fissures within.

No one could have enjoyed himself more, either in creative work or the honours it brought him. He worked hard, but with immense gusto; he got pleasure not only from the high moments, but also from the hours of what to others would be drudgery, sitting in the dark counting the alpha particle scintillations on the screen. His insight was direct, his intuition, with one curious exception, infallible. No scientist has made fewer mistakes. In the corpus of his published work, one of the largest in scientific history, there was nothing he had to correct afterwards. By thirty he had already set going the science of nuclear physics – single handed, as a professor on £500 a year, in the isolation of late-Victorian Montreal. By forty, now in Manchester, he had found the structure of the atom, on which all modern nuclear physics depends.

It was an astonishing career, creatively active until the month he died. He was born very poor, as I have said. New Zealand was, in the 1880s, the most remote of provinces, but he managed to get a good education; enough of the old Scottish tradition had percolated there, and he won all the prizes. He was as original as Einstein, but unlike Einstein he did not revolt against formal instruction; he was top in classics as well as in everything else. He started research – on the subject of wireless waves – with equipment such as one might rustle up today in an African laboratory. That did not deter him: 'I could do research at the North Pole,' he once proclaimed, and it was true. Then he was awarded one of the 1851 overseas scholarships (which later brought to England Florey, Oliphant, Philip Bowden, a whole series of gifted antipodeans). In fact, he got the scholarship only because another man, placed above him, chose to get married: with the curious humility that was interwoven with his boastfulness, he was grateful all his life. There was a proposal, when he was Lord Rutherford, President of the Royal Society, the greatest of living experimental scientists, to cut down these scholarships. Rutherford was on the committee. He was too upset to speak: at last he blurted out:

'If it had not been for them, I shouldn't have been.'

That was nonsense. Nothing could have stopped him. He brought his wireless work to Cambridge, anticipated Marconi, and then dropped it because he saw a field – radio-activity – more scientifically interesting.

If he had pushed on with wireless, incidentally, he couldn't have avoided becoming rich. But for that he never had time to spare. He provided for his wife and daughter, they lived in comfortable middle-class houses, and that was all. His work led directly to the atomic energy industry, spending, within ten years of his death, thousands of millions of pounds. He himself never earned, or wanted to earn, more than a professor's salary – about £1,600 a year at the Cavendish in the thirties. In his will he left precisely the value of his Nobel prize, then worth £7,000. Of the people I am writing about, he died much the poorest: even G. H. Hardy, who by Rutherford's side looked so ascetic and unworldly, happened not to be above taking an interest in his investments.

As soon as Rutherford got on to radio-activity, he was set on his life's work. His ideas were simple, rugged, material: he kept them so. He thought of atoms as though they were tennis balls. He discovered particles smaller than atoms, and discovered how they moved or bounced. Sometimes the particles bounced the wrong way. Then he inspected the facts and made a new but always simple picture. In that way he moved, as certainly as a sleepwalker, from unstable radio-active atoms to the discovery of the nucleus and the structure of the atom.

In 1919 he made one of the significant discoveries of all time: he broke up a nucleus of nitrogen by a direct hit from an alpha particle. That is, man could get inside the atomic nucleus and play with it if he could find the right projectiles. These projectiles could either be provided by radio-active atoms or by ordinary atoms speeded up by electrical machines.

The rest of that story leads to the technical and military history of our time. Rutherford himself never built the great machines which have dominated modern particle physics, though some of his pupils, notably Cockcroft, started them. Rutherford worked with bizarrely simple apparatus: but in fact he carried the use of such apparatus as far as it would go. His researches remain the last supreme single-handed achievement in fundamental physics. No one else can ever work there again – in the old Cavendish phrase – with sealing wax and string.

It was not done without noise: it was done with anger and

storms, but also with an overflow of creative energy, with abundance and generosity, as though research were the easiest and most natural avocation in the world. He had deep sympathy with the creative arts, particularly literature; he read more novels than most literary people manage to do. He had no use for critics of any kind. He felt both suspicion and dislike of the people who invested scientific research or any other brach of creation with an aura of difficulty, who used long, methodological words to explain things which he did perfectly by instinct. 'Those fellows,' he used to call them. 'Those fellows' were the logicians, the critics, the metaphysicians. They were clever; the were usually more lucid than he was; in argument against them he often felt at a disadvantage. Yet somehow they never produced a serious piece of work, whereas he was the greatest experimental scientist of the age.

I have heard larger claims made for him. I remember one discussion in particular, a year or two after his death, by half a dozen men, all of whom had international reputations in science. Darwin was there; G. I. Taylor; Fowler and some others. Was Rutherford the greatest experimental scientist since Michael Faraday? Without any doubt. Greater than Faraday? Possibly so. And then – it is interesting, as it shows the anonymous Tolstoyan nature of organized science – how many years' difference would it have made if he had never lived? How much longer before the nucleus would have been understood as we now understand it? Perhaps ten years. More likely only five.

Rutherford's intellect was so strong that he would, in the long run, have accepted that judgement. But he would not have liked it. His estimate of his own powers was realistic, but if it erred at all, it did not err on the modest side. 'There is no room for this particle in the atom as designed by *me*,' I once heard him assure a large audience. It was part of his nature that, stupendous as his work was, he should consider it ten per cent more so. It was also part of his nature that, quite without acting, he should behave constantly as though he were ten per cent larger than life. Worldly success? He loved every minute of it: flattery, titles, the company of the high official world. He said in a speech: 'As I was standing in the drawing-room at Trinity, a *clergyman* came in. And I said to him: "I'm Lord Rutherford." And he said to me: "I'm the Archbishop of York." And I don't suppose either of us believed the other.'

He was a great man, a very great man, by any standards which we can apply. He was not subtle: but he was clever as well as creatively gifted, magnanimous (within the human limits) as well as hearty. He was also superbly and magnificently vain as well as wise – the combination is commoner than we think when we are young. He enjoyed a life of miraculous success. On the whole he enjoyed his own personality. But I am sure that, even quite late in his life, he felt stabs of a sickening insecurity.

Somewhere at the roots of that abundant and creative nature there was a painful, shrinking nerve. One has only to read his letters as a young man to discern it. There are passages of self-doubt which are not to be explained completely by a humble colonial childhood and youth. He was uncertain in secret, abnormally so for a young man of his gifts. He kept the secret as his personality flowered and hid it. But there was a mysterious diffidence behind it all. He hated the faintest suspicion of being patronized, even when he was a world figure. Archbishop Lang was once tactless enough to suggest that he supposed a famous scientist had no time for reading. Rutherford immediately felt that he was being regarded as an ignorant roughneck. He produced a formidable list of his last month's reading. Then, half innocently, half malevolently: 'And what do you manage to read, your Grice?' 'I am afraid,' said the Archbishop, somewhat out of his depth, 'that a man in my position really doesn't have the leisure . . .' 'Ah, yes, your Grice,' said Rutherford in triumph, 'it must be a dog's life! It must be a dog's life!'

Once I had an opportunity of seeing that diffidence face to face. In the autumn of 1934 I published my first novel, which was called *The Search* [see no. 213] and the background of which was the scientific world. Not long after it came out, Rutherford met me in King's Parade. 'What have you been doing to us, young man?' he asked vociferously. I began to describe the novel, but it was not necessary; he announced that he had read it with care. He went on to invite, or rather command, me to take a stroll with him round the Backs. Like most of my scientific friends, he was good-natured about the book, which has some descriptions of the scientific experience which are probably somewhere near the truth. He praised it. I was gratified. It was a sunny October afternoon. Suddenly he said: 'I didn't like the erotic bits. I suppose it's because we belong to different generations.'

The book, I thought, was reticient enough. I did not know how to reply.

In complete seriousness and simplicity, he made another suggestion. He hoped that I was not going to write all my novels about scientists. I assured him that I was not – certainly not another for a long time.

He nodded. He was looking gentler than usual, and thoughtful. 'It's a small world, you know,' he said. He meant the world of science. 'Keep off us as much as you can. People are bound to think that you are getting at some of us. And I suppose we've all got things that we don't want anyone to see.'

72

While the Cavendish Laboratory in Cambridge was creating twentieth-century physics under Rutherford, F. Lindemann, Lord Cherwell, was doing considerably less for physics in Oxford. Lindemann's rise during the Second World War had several ill-consequences. C. P. Snow in Science and Government *has described vividly (if, as often appears, inaccurately) his rise to power and his famous feud with his one-time friend, Sir Henry Tizard. Snow saw in Lindemann "a character who made a novelist's fingers itch."*

A great deal else of Lindemann's personality struck them also as uncomfortable and strained. About him there hung an air of indefinable malaise – so that, if one was drawn to him at all, one wanted to alleviate it. He was formidable, he was savage, he had a suspicious malevolent sadistic turn of what he would have called humour, though it was not really that. But he did not seem, when it came to the most fundamental things, to understand his own life, and despite his intelligence and will, he did not seem good at grappling with it. He enjoyed none of the sensual pleasures. He never drank. He was an extreme and cranky vegetarian, who lived largely on the whites of eggs, Port Salut cheese, and olive oil. So far as is known, he had no sexual relations. And yet he was a man of intense emotions.

Tizard, whose emotions were also deep and difficult to control, had an outgoing nature, which, luckily for him, found him wife and family and friends. Lindemann's passions were repressed and turned in upon himself. You could hear the difference in their kind of joke. Tizard, as I mentioned, had a tongue which was harsh, which could be rough with pretentious persons, but which was in the long run good-natured. Lindemann's had the bitter edge of repression.

I remember being in Oxford one morning when the Honours List had been published. I think this must have been during the war. I was talking to Lindemann. I happened to remark that the English honours system must cause far more pain than pleasure: that every January and June the pleasure to those who got awards was nothing like so great as the pain of those who did not. Miraculously Lindemann's sombre, heavy face lit up. His brown eyes were usually sad, but now they were glowing. With a gleeful sneer he said: 'Of course it is. It wouldn't be any use getting an award if one didn't think of all the people who were miserable because they hadn't managed it.'

73

Lindemann apparently got on well with Einstein, whose political naïveté he however deplored. Ronald Clark, in his biography of Einstein, suggests that they were united in the view that human beings counted for little, compared to the truths of physics. 'Lindemann', he continues, 'according to a colleague, "had time for a few dukes, but regarded the rest of mankind as furry little animals." Substituting "pacifists" for "dukes", much the same was true of Einstein.'

Lindemann was a surprise appointment to the Chair in Oxford in 1924. His introduction to the University was disastrous and is divertingly described by the economist and friend of Lindemann, Roy Harrod, in his biography, The Prof. *The occasion was a 'debate' organised by undergraduates, and Lindemann was opposed by Harrod's tutor, the redoubtable philosopher, H.W.B. Joseph.*

To the dismay of the audience he proceeded to extract from the satchel a thick wad of manuscript. Balancing a pair of *pince-nez* on the lower end of his nose, he made ready to read from his papers. We steeled ourselves to a long session.

In one respect Joseph was at one with J. A. Smith; it appeared from an early stage that he was setting out to prove that the Theory of Relativity was 'wrong'. But he did not, like Smith, proceed at a technical level; what he had to offer was, in its own way, genuine philosophy. Joseph had a most remarkable style, which will always be remembered by, and indeed sometimes had an important influence on, those who heard him. His philosophy was influenced by that of the already deceased J. Cook Wilson; it was a philosophy in which a rather crude realism had been grafted on the doctrines of the ancient Greek philosophers, especially Plato. What was peculiar was, not the corpus of the doctrine, but the manner of its elucidation. It was a tradition of this school, a very sound and excellent tradition, that one should eschew polysyllables and all technical philosophical jargon, and confine oneself to short English, preferably Anglo-Saxon, words, the meaning of which was plain from their habitual use in ordinary discourse. These philosophers were careful also to keep their sentences, in so far as the subject allowed, crisp and clear. Thus, if one listened to one sentence, one had the idea that this was a most beautifully lucid exposition and expressed a thought that commended itself to plain common sense. But Joseph had a certain cleverness, which prevented his remaining quite in the austere Cook Wilson tradition. He was very fluent; and he had learnt the trick, which was a truly remarkable one, of running his sentences together into a discourse of interminable length. Thus, although each sentence seemed all right by itself, it was exceedingly difficult to grasp the meaning of the totality. He achieved what was almost an acrobatic trick. He rushed on and on; one could not complain that any single word was of obscure meaning, or even that any single clause was. None the less the whole was difficult to follow, and sometimes one could not help having the feeling, as sentence succeeded sentence, that he had cheated somehow in the process of leading us up to his clear and well-defined conclusion.

Professor Joad, in his wireless talks in the Brains Trust, has habituated large British audiences to the idea that a philosopher is apt to say, 'It all depends on what you mean by that'. Joad's technique was a feeble caricature of Joseph's. The latter chose his ground carefully, and when he turned to you and gave you a piercing look and said, 'What do you mean by that?', your heart sank; you thought hastily how to explain your meaning; you realized that you could not explain it; you realized finally that you had meant just nothing at all.

Joseph had the idea that certain words that are commonly used express some genuine apprehension by the mind. On this particular evening he was much concerned with such words as 'greater than', 'less than', 'before', 'after', 'simultaneously', 'moving relatively to'. What was meant by these words was based on a definite intellectual apprehension. One must not import different meanings, which violated these original apprehensions, into them. There they were, tokens of the mind's power of grasping certain things. And so he proceeded forward, in a lengthy and elaborate demonstration, to show that among the mind's original powers to grasp certain things, powers indicated by the use of words, powers which you could only challenge by using words in senses that were manifestly improper, was the mind's knowledge that space was Euclidean in character. Therefore the Theory of Relativity must be wrong.

Undergraduates, hand-picked although they are for the great seats of learning, do not always get things quite right. There was current in New College a limerick about Joseph:

> There was an old person called Joseph
> Whom nobody knows if he knows if
> He knows what he knows, which accounts
> I suppose
> For the mental condition of Joseph.

They were right in thinking that the question of what he knew or did not know was quite essential to the inner personality of Joseph. But the whole point about him was quite the opposite to that suggested by the limerick. What was peculiar was his very great degree of assurance that he *did* know certain things. For instance, he knew, absolutely and unshakably, that the space in which we live is in fact Euclidean.

At long last his discourse ended. One felt, as he proceeded, that the smooth polished surface of his phrases ought really to be interminable; and yet they did in fact terminate, no one quite knew why. All eyes were turned upon Professor Lindemann. What on earth was he to say? He had had lengthy discussions with Einstein, Max Planck, Broglie and other great men of thought. But I do not suppose that he had ever heard anything like this before. It was a hot-house Oxford product. He had been fairly and squarely challenged. He had been told that the Theory of Relativity was quite wrong, and this great chain of argument had been furnished.

There is a further point that must be made about Joseph. His paper against Relativity must have caused him much arduous work and taken much time to compose. But he did not show the slightest sign of his ever having seriously tried to understand either what the theoretical considerations were, or what the experimental results had been, that had led these distinguished physicists to feel the need to expound these tiresome theories of the relativity of space and time. It was evident that, in the ordinary sense of language, he knew nothing whatever about the Theory of Relativity. Since it was so evident to him that the conclusion reached was untrue, on quite different and sufficient philosophical grounds, there was no need for him to bother with the reasons why certain persons had been induced to frame such a theory. Indeed, I would go further. I have doubts myself as to whether Joseph, who had very limited intellectual capacity, despite his quite extraordinary linguistic acrobatics, would ever have been capable of understanding the Theory of Relativity.

So what was Professor Lindemann to do about it? He resumed, in his previous style of brief staccato sentences. He reiterated certain points. He gave some further illustrations. Then, with the corners of his lips turned down and an ironic expression on his face, he said, 'Well, if you really suppose that you have private inspiration enabling you to know that. . . .' But that was precisely what Joseph claimed. When there was a pause in Professor Lindemann's utterances, Joseph began unwinding again. And so each time. The Prof. really never got to grips with his argument; he did not know how to do so; none of the real points of interest in relation to relativity had been touched on; the whole game must have seemed to him to be perfectly futile. None the less there was this distinguished audience listening, and he was by no means winning in debate.

I suppose that some others among the learned persons present must have contributed something to the discussion. If so, that is entirely effaced from my memory. The spotlight was upon the Lindemann-Joseph interaction.

I mingled among the audience as they finally left the room. The Wykehamist Greats men were jubilant; a scientific professor had been torn to pieces; the Theory of Relativity had been shown to be untrue. But I was reluctant to join in their jubilation. I had already had unhappy experience with Joseph. Unlike those Wykehamists, I had read a great deal of philosophy at my school (Westminster), and had come up to Oxford full of theories and of earnestness to learn more. My arguments with Joseph had led to nothing but frustration. He had successfully shown that I was unable to express my thoughts in clear English and that sometimes what I had written for him meant nothing at all. But he seemed to be totally indifferent to what I had *tried* to mean, of the thought behind my words, just as he had been totally indifferent to the question of what were the theoretical considerations and the empirical facts that had led to the Theory of Relativity. Thus I had a certain fellow feeling with the unfortunate Professor Lindemann. I remember that I turned to an old friend, N. A. Beechman, a Balliol Greats man, afterwards

President of the Union, and still later Minister of the Crown. He had a certain worldly shrewdness such as is necessary to those who interest themselves in politics. I asked him, 'Which was right?' He answered at once, 'Of course Professor Lindemann was right'. That little bit of worldly wisdom saved him from the illusions of the Wykehamists. In retrospect it is easy enough to perceive that Professor Lindemann was indeed right. But many of these young men were genuinely thinking otherwise; they believed then, in 1920, or whenever precisely it was, that relativity had been disproved by these verbal tricks. I had given up classics for history in my last two years at school and won a history scholarship for New College; but New College insisted on my doing Greats, the crown of all humane studies, before reverting to history. But what was this Greats? What was the significance, in relation to the pressing problems thrown up by science on the one hand and by the evolution of social affairs on the other, of this linguistic dialectic purveyed by Greats? Professor Lindemann remained in my mind after that evening as a sort of symbol of the free advance of the human spirit. I did not suppose that I should ever see him again.

74

In Einstein, *Clark recounts the following anecdote of Abraham Pais's about an encounter between those two giants, Einstein and Niels Bohr. The year was 1949 and Bohr was busy with an article about their epistemological arguments, which he was completing for a* Festschrift *in celebration of Einstein's seventieth birthday. Bohr called Pais into his office:*

We went there [says Pais] . . . After we had entered, Bohr asked me to sit down ('I always need an origin for the co-ordinate system') and soon started to pace furiously around the oblong table in the centre of the room. He then asked me if I could put down a few sentences as they would emerge during his pacing. It should be explained that, at such sessions, Bohr never had a full sentence ready. He would often dwell on one word, coax it, implore it, to find the continuation. This could go on for many minutes. At that moment the word was 'Einstein'. There Bohr was, almost running around the table and repeating: 'Einstein . . . Einstein . . .' It would have been a curious sight for someone not familiar with Bohr. After a little while he walked to the window, gazed out, repeating every now and then: 'Einstein . . . Einstein . . .'

At that moment the door opened very softly and Einstein tiptoed in.

He beckoned to me with a finger on his lips to be very quiet, his urchin smile on his face. He was to explain a few minutes later the reason for his behaviour. Einstein was not allowed by his doctor to buy any tobacco. However, the doctor had not forbidden him to steal tobacco, and this was precisely what he set out to do now. Always on tiptoe he made a bee-line for Bohr's tobacco pot which stood on the table at which I was sitting. Meanwhile Bohr, unaware, was standing at the window, muttering 'Einstein . . . Einstein . . .' I was at a loss what to do, especially because I had at that moment not the faintest idea what Einstein was up to.

Then Bohr, with a firm 'Einstein', turned around. There they were face to face, as if Bohr had summoned him forth. It is an understatement to say that for a moment Bohr was speechless. I myself, who had seen it coming, had distinctly felt uncanny for a moment, so I could well understand Bohr's own reaction. A moment later the spell was broken when Einstein explained his mission and soon we were all bursting with laughter.

75

And here is John Updike on the neutrino, the particle that emerged as one of the great triumphs of twentieth-century theoretical and experimental physics. Updike was plainly impressed by a sentence that he read in the American Scientist, *and which he made the preamble to the poem: 'Every second, hundreds of billions of these neutrinos pass through each square inch of our bodies, coming from above during the day and from below at night, when the sun is shining on the other side of the earth!'*

COSMIC GALL

Neutrinos, they are very small.
  They have no charge and have no mass
And do not interact at all.
The earth is just a silly ball
  To them, through which they simply pass,
Like dustmaids down a drafty hall
  Or photons through a sheet of glass.

They snub the most exquisite gas,
Ignore the most substantial wall,
Cold-shoulder steel and sounding brass,
Insult the stallion in his stall,
And, scorning barriers of class,
Infiltrate you and me! Like tall
And painless guillotines, they fall
Down through our heads into the grass.
At night, they enter at Nepal
And pierce the lover and his lass
From underneath the bed – you call
It wonderful; I call it crass.

76

The most entertaining chronicler of post-war physics is the theoretical physicist and writer, Jeremy Bernstein. Here, from his collection of essays, Experiencing Science, *is the story of how the peerless Julian Schwinger came to be discovered by I.I. Rabi (see above, no. 14), then already the Czar of Physics at Columbia University.*

'It was sort of romantic, in a way,' Rabi recalls. 'You can fix it to a year – 1935. Einstein, Boris Podolsky, and Nathan Rosen had just published their famous paper' – on an apparent paradox in the quantum theory of measurement. 'I was reading the paper, and my way of reading a paper was to bring in a student and explain it to him. In this case, the student was Lloyd Motz, who's now a professor of astronomy at Columbia. We were arguing about something, and after a while Motz said there was someone waiting outside the office, and asked if he could bring him in. He brought in this kid.' Schwinger was then sixteen. 'So I told him to sit down someplace, and he sat down. Motz and I were arguing, and this kid pipes up and settles the argument by the use of the completeness theorem' – an important mathematical theorem used frequently in the quantum theory. 'And I said, "Who the hell is this?" Well, it turned out that he was a sophomore at City College, and he was doing very badly – flunking his courses, not in physics, but doing very badly. I talked to him for a while and was deeply impressed. He had already written a paper on quantum electrodynamics. So I asked him if he wanted to transfer, and he said yes. He gave me a transcript, and I looked at it. He was failing – English, and just about everything else. He spoke well. I said, "What's the matter with you? You're flunking English. You speak well, and you sound like an educated person." He said, "I have no time to do the themes." I tried to get him admitted to Columbia on a scholarship. I saw the director of admissions. He looked at the transcript and said, "A scholarship?" He wouldn't even admit him. Well, Hans Bethe was passing through, and I asked him if he would read Schwinger's paper. He read it and thought well of it. So I asked Bethe to write me a letter. He did. And, armed with this letter, I got Schwinger admitted. He entered Columbia as a junior and actually made Phi Beta Kappa. He turned over a new leaf.' (George Uhlenbeck also remembers being asked by Rabi to plead Schwinger's case.)

Rabi went on to tell me about Schwinger's graduation. 'One Sunday morning, the dean calls me up and says, "Schwinger has all his course requirements to graduate, but he doesn't have enough maturity credits." ' At that time, Columbia gave grades for performance in the course work and also something called a maturity credit, which was a measure of the degree of difficulty of the course. It was an attempt to stop students who had taken nothing but snap courses from getting a degree. 'I was sore that the dean had called me on this thing. So I said, "I suppose if Schwinger doesn't have enough maturity credits, he can't graduate." Silence. Then the dean said, "I'll be damned if I am going to stop Schwinger from graduating just because he doesn't have enough maturity credits." That's what I wanted to hear. I didn't want to beg the dean to graduate Schwinger, but if he hadn't he would have been a damn fool.' At the time of his graduation, Schwinger had just finished completing the material for his Ph.D. thesis.

77

And here from Bernstein's recollections of his career in physics, The Life it Brings, *is a description of the mature Schwinger in action.*

When the war broke out Rabi and Schwinger both went to the Radiation Laboratory in Cambridge, where radar was being developed. According to Rabi, Schwinger worked all night at Radiation Laboratory and slept all day. 'At five o'clock when everybody was leaving, you'd see Schwinger coming in.' People who had left mathematical problems on their desks or on blackboards would often find them solved anonymously the next morning. Rabi recalled, 'The problems he solved were just fantastic. He lectured twice a week on his current work.' Following the lectures, the more experimentally inclined physicists in the audience would use Schwinger's ideas to invent new radars. In 1946 Schwinger was appointed an associate professor in physics at Harvard and a year later, at twenty-nine, he became one of the youngest full professors in the history of the university.

Schwinger gave courses in all branches of theoretical physics. Like Mackey's, these courses were completely original. It was a standing joke in the profession at the time that almost any unsolved problem in theoretical physics could be found solved in Schwinger's unpublished classroom notes. As it happened – and this was a piece of luck – when I entered graduate school, in the fall of 1951, Schwinger was starting a cycle of quantum mechanics lectures that, in the end, would last for a year and a half. These lectures were a kind of theater. Like any great performer, Schwinger would always be slightly late. Since he still worked all night, the classes were scheduled for the late morning. He drove a small blue Cadillac – the make of car also set him apart from the herd – and a scout would watch out for its arrival, at which point we would all rush into the classroom to find seats. In those days Schwinger was averse to most forms of physical exercise. When I got to know him better, he would occasionally kid me about my tennis playing and bicycle riding. Later, after he left Harvard and went to the University of California at Los Angeles, Schwinger became, I am told, an avid tennis player. But at the time I speak of he had a slightly oblate look, accentuated by the stark white pallor of his skin, the result of his nocturnal hours. There was a remarkable Indian graduate student in the class who said to me one day, apropos of nothing evident, '*Our* Schwinger is very fat.' By the standards of rural India, he certainly was.

To add to the general sense of theater, Schwinger never used, at least in those days, any lecture notes. As one who has spent much of the last thirty years giving lectures in theoretical physics, I find this nearly incomprehensible. He did incredibly complicated calculations on the blackboard without the shred of a note. This goes beyond memorization, although that may have played a part. The individual steps of these calculations were, one gathers, so trivial to Schwinger that he could carry them out more or less spontaneously. However, he did not like to be interrupted, and questions in class were not encouraged. You had the feeling that you were assisting in a remarkable performance, and just as you would not stand up in a theater and ask a question of an actor while he was working, so you did not ask questions of Schwinger. Schwinger also had some speech mannerisms that many of his students began unconsciously to imitate. By 1975, when Schwinger left Harvard, he had produced sixty-eight Ph.D.s, an incredibly high number for a theorist when you consider that each Ph.D. represents at least one publishable research idea. I don't know how many of these acolytes began to say 'nucular' and 'we can effectively regard,' two of the Schwinger locutions. One of my contemporaries constructed what I thought was the perfect model of a Schwinger sentence. It began 'Although "one" is not perfectly "zero," we can effectively regard . . .' In short, Julian was another legend.

78

From the same source, a remarkable examination at close quarters of how another great theoretical physicist, Murray Gell-Mann, worked.

I had come to Paris with a second physics problem that had been nagging at me ever since I left Princeton. It was a sort of pedagogical problem, too technical to describe here, that was connected to the theory of radioactive decays that Feynman and Gell-Mann had proposed. I explained my question to Michel, who said it sounded good enough to work on. It was by now early November, about the time that Gell-Mann had said he would be arriving in Paris. In fact, no sooner had Michel and I begun to formulate what it was that we wanted to study, and how we might go about it, than Gell-Mann himself walked into our basement. He asked us what we were doing and I explained. He said something about the problem's sounding interesting and then disappeared. The next day, at about the same hour, he reappeared and said in his perfect French, *'Messieurs, le problème est rèsolu.'* He must have spent all night on it. What he had done was produce a general framework that could accommodate our problem along with one that he had been working on. Michel and I were stupefied. I decided that this would be a real opportunity to see how creative physics of the highest order is done – a little like playing Van Voorhies in tennis.

Unlike several of the first-rank physicists, Gell-Mann genuinely likes to collaborate. He seems to need someone to bounce ideas off. My colleague, the late John Sakurai, once asked Dirac, over dinner at the Institute, whether he had ever had a priority fight over any of his ideas. Dirac answered, succinctly, 'The really good ideas are had by only one person.' He almost never collaborated. Many of Gell-Mann's most outstanding papers were done by himself, and there is the one great paper with Feynman. But when it comes to working out the ramifications of ideas, Gell-Mann, more often than not, works with other people. In Paris there were basically four of us, Michel and I and Gell-Mann and Maurice Lévy. We saw each other almost daily and exchanged all kinds of ideas, and I had the chance to see whether I could figure out what made Gell-Mann tick. What was the secret of that kind of scientific creativity? I finally decided, after almost a year, that I hadn't learned the secret and that, in some sense, it could not be understood.

In a way I pity the historians of science and others who are attempting to put together a reliable history of the postwar development of elementary particle physics. At first sight the problem seems almost trivial; all you have to do, you may think, is ask the participants for their accounts of what happened. But such stories can be woefully misleading. I've been astonished to read descriptions of events I had witnessed whose meaning at the time seemed completely different from these later versions. This disparity may reflect my faulty memory – or it may reflect a not unnatural desire by the participants to make things appear more coherent than they actually were. Having issued this caveat, I will try to describe the work done by Gell-Mann in Paris that led to his creation, the following fall, of what he called the Eightfold Way and then the quark.

Gell-Mann had a particular method of working. Periodically, he summarized what he had been doing. The summaries were, in effect, manuscripts of papers, though he often did not publish them. Writing the papers was his way of collecting his thoughts. He let me read some of these handwritten manuscripts. Their theme and the theme of the work that Gell-Mann told me about when we talked together several times a week was the same – it was an attempt to find a unifying symmetry principle that would give meaning to the newly discovered strange particles. The problem was that the experiments did not suggest such a symmetry. Indeed, the events that occurred in the discovery of this principle were, as far as I am concerned, a perfect counter to Le Prince Ringuet's Baconian notion of going into the laboratory with an open mind and no theoretical guidance. [See no. 13 above.] If his method had been followed, no one would have gotten anywhere in the physics of elementary particles.

Since experiment was of little or no help, what took place, as far as I could make out, was a kind of theoretical 'playing around.' Gell-Mann had various mathematical objects and was playing with them like toys. He would say to me such things as 'The currents are commuting like angular momenta.' Given the currents in question, this seemed a true statement, but one whose significance was beyond me. I did not see why it was of any interest. Occasionally he would come up with a new form of current that had even 'cuter' algebraic properties, which he would be very pleased with and which meant even less to me. I am afraid I tried his patience with my obtuseness. It got to the point where I would hide out for a few days just to try to make sense of what was going on. When I called Gell-Mann, he would say, 'Where have you been. I've discovered millions of things.' As far as I could see, there was not the remotest hint that any of this was leading anywhere. It seemed to me like open-ended speculation, but in this I was entirely wrong. When Gell-Mann returned to Cal Tech in the fall of 1960, and began talking to a mathematician named R. Block, he came to realize that there was a name for what he was doing – generating Lie groups – and that it had been, since its invention by the Norwegian mathematician Sophus Lie at the end of the nineteenth century, a well-formulated branch of pure mathematics. Viewed this way, and in retrospect, a program of how to proceed is quite clear. You would make a systematic study of the simple Lie groups, looking for one with properties that resemble those of the elementary particles. Indeed this is how Gell-Mann did it after Block pointed out to him that these groups had been classified by the French mathematician E. Cartan. As it happened one of the most elementary Lie groups worked, although it wasn't until 1964 that a group of thirty-three physicists at the Brookhaven Laboratory found the key particle whose existence Gell-Mann had predicted, along with its essential properties. It is impossible to believe that this very odd particle – known as the omega-minus – would have been found if no one had been looking for it; and if it had been found without the theory, its existence would have seemed incomprehensible. In 1969 Gell-Mann was awarded the Nobel Prize in Physics for this work. I doubt whether there have been many more deserving recipients.

79

Richard Feynman was one of the transcendent originals of our time. His opinions and experiences were dictated and transcribed by a friend under the title, Surely You're Joking, Mr Feynman! *Here is the flavour of the man.*

I don't know why, but I'm always very careless, when I go on a trip, about the address or telephone number or anything of the people who invited me. I figure I'll be met, or somebody else will know where we're going; it'll get straightened out somehow.

One time, in 1957, I went to a gravity conference at the University of North Carolina. I was supposed to be an expert in a different field who looks at gravity.

I landed at the airport a day late for the conference (I couldn't make it the first day), and I went out to where the taxis were. I said to the dispatcher, 'I'd like to go to the University of North Carolina.'

'Which do you mean,' he said, 'the State University of North Carolina at Raleigh, or the University of North Carolina at Chapel Hill?'

Needless to say, I hadn't the slightest idea. 'Where are they?' I asked, figuring that one must be near the other.

'One's north of here, and the other is south of here, about the same distance.'

I had nothing with me that showed which one it was, and there was nobody else going to the conference a day late like I was.

That gave me an idea. 'Listen,' I said to the dispatcher. 'The main meeting began yesterday, so there were a whole lot of guys going to the meeting who must have come through here yesterday. Let me describe them to you: They would have their heads kind of in the air, and they would be talking to each other, not paying attention to where they were going, saying things to each other, like "G-mu-nu. G-mu-nu." '

His face lit up. 'Ah, yes,' he said. 'You mean Chapel Hill!' He called the next taxi waiting in line. 'Take this man to the university at Chapel Hill.'

'Thank you,' I said, and I went to the conference.

80

Robert Oppenheimer was a tormented spirit, about whom much has been written. The following two sharp sketches are by Jeremy Bernstein, from The Life it Brings.

It was more or less expected that everyone would give a colloquium, or part of a colloquium, during his stay. The experience could be traumatic because of Oppenheimer's often maddening behavior. He smoked incessantly – the habit finally killed him – and emitted, from time to time, a racking cough. He often interrupted the speaker with a Delphic remark, after which he would look around, I guess to see whether the others had gotten the point. He could also be brutal. Someone told me that at a seminar at Berkeley, when a pneumatic drill was being used outside the window, he interrupted the speaker to say 'How can I follow this with all that noise out there, especially when that noise is making more sense that you are?' The only time he ever went after me, I thoroughly deserved it. I had asked an idiotic question about mu mesons in astrophysics, and Oppenheimer said, in a scornful tone, 'Mu mesons are found at the University at Chicago and not in the stars.' At this point Cohen, who was sitting next to me, whispered, 'Bernstein, this is a good day to keep down in the trenches with the rest of the troops.' I kept my mouth shut for the rest of the seminar.

81

Sanders Theatre, where the lectures were given, was filled to capacity. It must be difficult for physicists of the present generation to imagine what a presence Oppenheimer was in our field, and what an electrifying public figure. His use of language was somewhat opaque, often poetic, and he had an odd, clipped diction that commanded attention. Yet I do not have the foggiest recollection of what he said except that in the last lecture he dealt with the revolution wrought in our thinking by quantum mechanics. I was seated directly behind two of those classic blue-haired Boston dowagers who look fragile but, I am sure, will outlive us all. A gentle aroma of violets floated around them. Oppenheimer had a small blackboard on the stage. At one point he said that to make things precise he would write down the fact that in quantum mechanics the position and momentum don't 'commute.' It is not important to know what that means; very few people in his audience knew. He took a chalk and carefully wrote qp – pq = ih The two ladies clutched each other for reassurance. Perhaps they thought that the formula was going to explode. I was tempted to lean over and tell them that it was all right – that Oppenheimer was just showing off a bit – but I didn't, because it would have destroyed the beauty of the moment.

82

The reaction of the Boston matrons echoes the insect-fanciers listening to Professor Poulton. The extract is from The Journal of a Disappointed Man *by W.N.P. Barbellion (real name Bruce Frederick Cummings), a frustrated marine biologist and a fine writer, whose short and bitter life ended in 1919.*

March 4.

The Entomological Society

There were a great many Scarabees present who exhibited to one another poor little pinned insects in collecting boxes. . . . It was really a one-man show, Prof. Poulton, a man of very considerable scientific attainments, being present, and shouting with a raucous voice in a way that must have scared some of the timid, unassuming collectors of our country's butterflies and moths. Like a great powerful sheep-dog, he got up and barked, 'Mendelian characters', or 'Germ plasm', what time the obedient flock ran together and bleated a pitiful applause. I suppose, having frequently heard these and similar phrases fall from the lips of the great man at these reunions, they have come to regard them as symbols of a ritual which they think it pious to accept without any question. So every time the Professor says, 'Allelomorph', or some such phrase, they cross themselves and never venture to ask him what the hell it is all about.

Next, another vignette by Jeremy Bernstein, this one of the great Swiss theoretical physicist, Wolfgang Pauli, a man noted for brilliance, intolerance, overweening arrogance and an acerbic style of utterance. None were spared. Even as a young man he was supposed to have engaged in the following exchange with the eminent theoretician, Paul Ehrenfest. 'I like your papers better than you', said Ehrenfest at their first encounter. 'That's strange,' replied Pauli, 'because I like you better than your papers.' Of a graduate student's seminar, Pauli was supposed to have remarked 'So young and already he has achieved so little.' To his great contemporary Lev Davidovich Landau, when the latter took umbrage at Pauli's rudeness and asked 'You think perhaps that all I have said is nonsense?', Pauli replied 'Not at all, not at all; what you have been saying is so confused that I cannot tell whether it is nonsense or not.' Pauli died at the age of fifty-seven. The physicist Charles Enz has told of the final days of Pauli's illness, which were spent in room 137 of a Zurich hospital. Now 137 is a 'magic number', delivered by Dirac's wave equation, and relating to the fine structure of the hydrogen spectrum. The experimental constant, introduced by Sommerfeld, became a fetish for A.S. Eddington, who would always seek to hang his hat on a peg of that number in the cloakroom. The conjunction troubled Pauli's last days.

Pauli had been engaged in a peculiar enterprise with his former collaborator Werner Heisenberg, another of the great architects of the quantum theory. For a while they claimed to have solved all the unsolved problems in elementary particle theory; they reduced everything to a single equation. When calmer spirits examined the matter, they concluded that the whole thing was a chimera. The dénouement, for Pauli, came at a lecture he delivered at Columbia University in the large lecture hall in Pupin Laboratory. Even though there had been an attempt to keep the talk secret, the room was filled to capacity. The audience was studded with past, present, and future Nobel Prize winners, including Niels Bohr. After Pauli delivered his lecture, Bohr was asked to comment. There then occurred one of the most unusual, and in its unearthly way most moving, demonstrations I have ever witnessed. Bohr's basic point was that as a fundamental theory it was crazy, but not crazy enough. This was a very important observation. The great advances, like relativity and the quantum theory, do seem – at first sight, and especially if one has been brought up in the physics that preceded them – to be crazy, to violate common sense in a fundamental way. On the other hand, Pauli's theory was just bizarre, a strange-looking equation that stared at you like a hieroglyph. Pauli objected to Bohr's assessment; he said the theory *was* crazy enough.

At this point these two monumental figures in modern physics began moving in a conjoined circular orbit around the long lecture table. Whenever Bohr faced the audience from the front of the table, he repeated that the theory was not crazy enough, and whenever Pauli faced the group, he would say it was. I recall wondering what anyone from the other world – the nonphysicists' world – would make of this. Dyson was asked to comment and refused. Afterward he remarked to me that it was like watching the 'death of a noble animal.' He was prescient. Pauli died not many months later, in 1958, at the age of fifty-eight, of a previously undetected cancer. Before that, he had renounced 'Heisenberg's theory,' as he now referred to it, in the most acidulous manner. One could only wonder whether Pauli's brief love affair with it was a sign that he was already ill.

84

Another celebrated attribute of Pauli's – the so-called Second Pauli Principle (the first being the Exclusion Principle of atomic structure) – was that his proximity was death to all mechanical devices, especially laboratory apparatus. It was said that when an explosion devastated the Physics Department at the University of Berne, it was discovered that Pauli had at that instant been in a train, halted at Berne station en route *to Zürich. H.G.B. Casimir, the Dutch physicist and a vastly successful Director of Research at Philips, has left the following story in his recollections,* Haphazard Reality. *L. Rosenfeld was another distinguished theoretician.*

Much of Casimir's early career in theoretical physics revolved around Niels Bohr's Institute in Copenhagen; Bohr was a fervent rationalist; when taxed with the unreason of displaying a horseshoe on the door of his country cottage he is supposed to have said: 'Of course it is nonsense, but they say it works even if you don't believe in it.' The passage from Casimir's book continues with an example of Bohr's sly humour.

The [second] passage relates to an almost unbelievable demonstration of the Pauli effect. Rosenfeld and I compared our recollections and they agreed exactly, so this is definitely an authentic and well-established case. Let me first explain about the Pauli effect. It was alleged that Pauli was so utterly remote from any kind of experiment that his very presence in the vicinity of a research laboratory was sufficient to cause apparatus to break down in the most unaccountable way. Pauli's friends assiduously collected – and possibly invented – examples, and he himself thoroughly enjoyed such stories. He may even have been inclined to believe the effect really existed, but I shall come back to that later on. Of course, because of the legend, experimental physicists tended to get nervous whenever Pauli entered their laboratory, so they may have made errors. But now for Rosenfeld's story.

Pauli, so far as I remember, was rather subdued, except on one spectacular occasion. Heitler, by lecturing on the theory of the homopolar bond, unexpectedly excited his wrath: for, as it turned out, he had a strong dislike to this theory. Hardly had Heitler finished, than Pauli moved to the blackboard in a state of great agitation, pacing to and fro he angrily started to voice his grievance, while Heitler sat down on a chair at the edge of the podium. 'At long distance,' Pauli explained, 'the theory is certainly wrong, since we have there van der Waals attraction; at short distance, obviously, it is also entirely wrong.' At his point he had reached the end of the podium opposite so that where Heitler was sitting. He turned round and was now walking towards him, threateningly pointing in his direction the piece of chalk he was holding in his hand: 'Und nun,' he exclaimed, 'gibt es eine an den guten Glauben der Physiker appellierende Aussage, die behauptet, das diese Näherung, die falsch ist in grossen Abständen und falsch in kleinen Abständen, trotzdem in einem Zwischengebiet qualitativ richtig sein soll!' (And now there is a statement, invoking the credulity of the physicists, that claims that this approximation which is wrong at large distance and is wrong at short distance yet is qualitatively true in an intermediate region.) He was now quite near to Heitler. The latter leaned back suddenly, the back of the chair gave way with a great crash, and poor Heitler tumbled backwards (luckily without hurting himself too much).

Casimir who also remembers the incident notes that Gamow was the first to shout: 'Pauli effect.' And as an afterthought he adds: 'Sometimes I wonder whether Gamow had not done something to the chair beforehand.'

I have nothing to add to this story, apart from the remark that unlike Pauli himself, Heitler was definitely a lightweight, which made the demonstration all the more impressive.

For me, the most important consequence of this conference was that Bohr invited me to stay on, and for the next two years I spent more than half of my time in Copenhagen. My father, not too well acquainted with the world of physics, may have had some doubt whether the man with whom I was working was really as famous as I said he was. So he addressed one of the first letters he wrote me: Casimir c/o Niels Bohr, Denmark. Of course, the letter arrived without delay. The Danish Post Office had not even troubled to add an address; they had only scrawled an ø on the envelope. After that I think my parents felt more convinced that I was in good hands; they were even more convinced when they met Mrs. Bohr. I found it so evident that the letter arrived that I did not keep the envelope. A pity; by now it would be a nice collector's item.

During my stay in Denmark I often acted as a kind of private secretary to Bohr, who liked to have someone to talk to about his ideas. Moreover, writing a paper, or rather trying to write a paper, was for him a way of thinking. For that kind of assistance shorthand was not required. Sentences came haltingly, hesitatingly, were often broken off and restarted. But one had to become accustomed to Bohr's voice, which was soft and rather indistinct, whether he spoke German, English, or Danish; and his habit of walking around the room all the time did not help. I put quite a bit of energy into learning Danish (I am still reasonably fluent) and I believe I got fairly good at understanding Bohr.

Working with Bohr in that way was a unique experience. Ehrenfest taught me the importance of clear, crisp formulations and was a master at finding simple examples illustrating the gist of a physical theory. Later Pauli forced me not to shun elaborate mathematical analysis. Bohr was both more profound and closer to reality. As a young man he had done beautiful experiments on surface tension and had built most of the apparatus with his own hands, and his grasp of orders of magnitude went all the way from the atomic nucleus to engineering problems of daily life.

The following anecdote illustrates that point. Close to Bohr's Institute there is a body of water – I hesitate whether to call it a lake or a pond – about three kilometers long and between 150 and 200 meters wide, the Sortedamsø. It is crossed by several bridges. One day Bohr took me on a stroll along that lake and across one of the bridges. 'Look,' he said, 'I'll show you a curious resonance phenomenon.' The parapet of that bridge was built in the following way. Stone pillars, about four feet high and ten feet apart, were linked near their tops by stout iron bars (or rather more likely, tubes) let into the stone. Halfway between each two pillars an iron ring was anchored in the stonework of the bridge, and two heavy chains, one on each side, were suspended between shackles welded to the top bar close to the stone pillars and that ring. Bohr grasped one chain near the top bar and set it swinging, and to my surprise the chain at the other end of the top bar began to swing too. 'A remarkable example of resonance,' Bohr said. I was much impressed, but suddenly Bohr began to laugh. Of course, resonance was quite out of the question; the coupling forces were extremely small and the oscillations were strongly damped. What happened was that Bohr, when moving the chain, was rotating the top bar, which was let into, but not fastened to, the stone pillars, and in that way he had moved the two chains simultaneously. I was crestfallen that I had shown so little practical sense, but Bohr consoled me, saying that Heisenberg had also been taken in; he had even given a whole lecture on resonance!

That bridge [was] known in Bohr's Institute as the Resonance Bridge.

85

The age of Big Science – of radiotelescopes and of particle accelerators that cost as much as the Gross National Product of an average country in the Southern Hemisphere – brought with it a new breed of scientific entrepreneurs, of whom Carlo Rubbia, the first Nobel Laureate of the European Nuclear Research Centre, CERN, is an archetype. Gary Taubes is a science journalist with a Ph.D. in physics, and in his book Nobel Dreams *has penetrated to the heart of an undertaking that is not without its seamy side. The theoretician (theoreticians, it could be argued, have less difficulty in remaining morally pure), Marty Perl, has offered the following judgement on his experimental brethren: 'This generation of high-energy physicists could also have done very well in the retail garment trade.' Here is an illustration of how the members of this fraternity run their laboratories.*

A woman physicist had been waiting for a couple of weeks to steal just a few minutes of Rubbia's time and discuss what she considered a crucial and highly important piece of physics. Rubbia also thought it was important, but he had been flying around the world, coming and going, and the physicist had just about given up hope.

Finally, one morning she gets a call from Rubbia. She picks up the phone and Rubbia says, 'Okay, I have exactly twenty minutes to talk to you about your work.' This is great, she thinks. She slams down the phone, runs full speed to Rubbia's office, making it there in about ten seconds, only to find that his door is locked. She turns to Rubbia's secretary and says, 'Carlo's door is locked?'

'Yes,' the secretary replies. 'Carlo was calling from the airport in Zurich.'

Meanwhile Rubbia has called back and is saying to his secretary, 'What the hell is the matter with that woman? I tell her I can talk to her about her work, and she hangs up on me.'

86

And this is how their subjects talk about them:

After Glashow's talk, the physicists broke for coffee. Out in the lobby, DiLella had a copy of the morning edition of *Corriere della Sera*, the Milan paper. He was translating an article to an audience of physicists, who stood around laughing and cheering him on. Hans Hoffmann, of UA1, Gordon Kane, and Steve Ellis were with him, and four or five others.

'Physicist Rubbia Now Wants a Second Nobel Prize,' DiLella said, translating the headline. 'The Italian scientist is trailing a strange phenomenon of missing energy. . . .

'Carlo Rubbia was asked by this guy here,' DiLella explained, ' "What would you like to do next?" And he said, "Get a second Nobel." That's what he said. Huh? Quote, unquote.'

DiLella stopped for a moment to let the laughter die down.

'Then it says that the Nobel Effect, which is generally quoted by scientists as being a fearful form of self-satisfaction, taking away any driving force or will, in Carlo Rubbia is working to the contrary and is trying to push him toward higher and higher enterprises.

'Then it says that Carlo Rubbia has spoken to the four hundred physicists who were listening to him like they were entranced.

'Two hundred twenty,' DiLella said, correcting the article.

'But it always doubles,' Hoffmann said.

'Like the number of Nobel prizes,' added a third.

87

A generation earlier, similar characters were operating, if on a comparatively minute scale, in the higher power strata of physics. R.A. Millikan, President of the California Institute of Technology, is here portrayed by Nuel Pharr Davis in his dual biography, Lawrence and Oppenheimer.

Birge once called at his office for a conference. It was disturbed by a secretary.

'I'm sorry, Dr Millikan,' she said, 'but there are two reporters outside who insist on seeing you.'

'Oh, those damned reporters,' said Millikan. 'How they pester me!'

Birge later learned Millikan had made appointments for them to come. Millikan's anonymous subordinates saw him in a light different from the public one. 'I couldn't stand him,' said Thomas O'Donnell, his laboratory mechanic. Always precise, Birge earned his dislike by exposing inaccuracies in his work. A puckish student of Birge's named Edward Condon once wrote a paper for the purpose. 'How dare you let such a thing come out of your laboratories? I'll kill Condon if he publishes that,' Millikan said to another of the Berkeley faculty.

Birge's attitude toward Oppenheimer furnished the key to Millikan's. A merely bright person hobbling after a genius: that was the way Birge thought of himself when he and Oppenheimer talked physics. Inhumanly predisposed though he was to activity in research, Birge does not pretend to have relished the involuntary self-comparison. Was there not something alien, repulsive, perhaps even peculiarly Jewish in the way Oppenheimer could comprehend and define so much faster? As for Millikan, on this point he had no doubts.

'Millikan loathed Oppenheimer, wouldn't match the promotions we gave him here, and harassed him maliciously,' said Birge. Oppenheimer did not complain and may not have known enough about colleges to understand what he had to complain of. Birge remembers a complaint from another source, Niels Bohr, the greatest physicist in Europe: 'Visiting from Denmark, Bohr sized things up at Pasadena and talked with me and with the Berkeley administration. Oppenheimer began to stay longer into the spring here. Later he taught the full session at Berkeley and cut his stay at Pasadena to four summer weeks. Millikan just left his name in the faculty register and made him miserable when the chance came.'

According to a story of Walter Elsasser's, a fence around a half-built laboratory on the Caltech campus was adorned with a sign proclaiming

JESUS SAVES. Below it a student graffito artist had added: BUT MILLIKAN GETS THE CREDIT.

88

Back then to the theoreticians: Sir Richard Blackmore (d. 1729) seems to have known his cosmology. The famous Dane was of course Tycho Brahe (see no. 17).

Copernicus, who rightly did condemn
This eldest system, form'd a wiser scheme;
In which he leaves the sun at rest, and rolls
The orb terrestial on its proper poles;
Which makes the night and day by this career,
And by its flow and crooked course the year.
The famous Dane, who oft the modern guides,
To earth and sun their provinces divides:
The earth's rotation makes the night and day;
The sun revolving through th' ecliptic way
Effects the various seasons of the year,
Which in their turn for happy ends appear.
This scheme or that, which pleases best, embrace,
Still we the Fountain of their motion trace.
  Kepler asserts these wonders may be done
By the magnetic virtue of the sun,
Which he, to gain his end, thinks fit to place
Full in the centre of that mighty space,
Which does the spheres, where planets roll, include,
And leaves him with attractive force endued.
The sun, thus seated, by mechanic law,
The earth and every distant planet draws;
By which attraction all the planets, found
Within his reach, are turn'd in aether round.

89

C.P. Snow, during his Cambridge days, was a friend of the Professor of Mathematics, G.H. Hardy. Hardy was modest about his attainments; his pride, he said, was that he had been able to talk on equal terms with

Littlewood and with Ramanujan. He rated himself as 40%, relative to Littlewood's 60%, with Ramanujan as 100%. The latter is now recognised as one of the mathematical geniuses of all time, and the magnitude of some of his discoveries is still dawning on the academic world. Here Snow tells how Hardy brought Ramanujan out of India and recounts a wonderful anecdote.

In the late afternoon, a stroll back to his rooms. That particular day, though, while the timetable wasn't altered, internally things were not going according to plan. At the back of his mind, getting in the way of his complete pleasure in his [real tennis] game, the Indian manuscript nagged away. Wild theorems. Theorems such as he had never seen before, nor imagined. A fraud of genius? A question was forming itself in his mind. As it was Hardy's mind, the question was forming itself with epigrammatic clarity: is a fraud of genius more probable than an unknown mathematician of genius? Clearly the answer was no. Back in his rooms in Trinity, he had another look at the script. He sent word to Littlewood (probably by messenger, certainly not by telephone, for which, like all mechanical contrivances including fountain pens, he had a deep distrust) that they must have a discussion after hall.

When the meal was over, there may have been a slight delay. Hardy liked a glass of wine, but, despite the glorious vistas of 'Alan St Aubyn' which had fired his youthful imagination, he found he did not really enjoy lingering in the combination-room over port and walnuts. Littlewood, a good deal more *homme moyen sensuel*, did. So there may have been a delay. Anyway, by nine o'clock or so they were in one of Hardy's rooms, with the manuscript stretched out in front of them.

That is an occasion at which one would have liked to be present. Hardy with his combination of remorseless clarity and intellectual panache (he was very English, but in argument he showed the characteristics that Latin minds have often assumed to be their own): Littlewood, imaginative, powerful, humorous. Apparently it did not take them long. Before midnight they knew, and knew for certain. The writer of these manuscripts was a man of genius. That was as much as they could judge, that night. It was only later that Hardy decided that Ramanujan was, in terms of *natural* mathematical genius, in the class of Gauss and Euler: but that he could not expect, because of the defects of his education, and because he had come on the scene too late in the line of mathematical history, to make a contribution on the same scale.

It all sounds easy, the kind of judgement great mathematicians should have been able to make. But I mentioned that there were two persons who do not come out of the story with credit. Out of chivalry Hardy concealed this in all that he said or wrote about Ramanujan. The two people concerned have now been dead, however, for many years, and it is time to tell the truth. It is simple. Hardy was not the first eminent mathematician to be sent the Ramanujan manuscripts. There had been two before him, both English, both of the highest professional standing. They had each returned the manuscripts without comment. I don't think history relates what they said, if anything, when Ramanujan became famous. Anyone who has been sent unsolicited material will have a sneaking sympathy with them.

Anyway, the following day Hardy went into action. Ramanujan must be brought to England, Hardy decided. Money was not a major problem. Trinity has usually been good at supporting unorthodox talent (the college did the same for Kapitsa a few years later). Once Hardy was determined, no human agency could have stopped Ramanujan, but they needed a certain amount of help from a superhuman one.

Ramanujan turned out to be a poor clerk in Madras, living with his wife on twenty pounds a year. But he was also a Brahmin, unusually strict about his religious observances, with a mother who was even stricter. It seemed impossible that he could break the proscriptions and cross the water. Fortunately his mother had the highest respect for the goddess of Namakkal. One morning Ramanujan's mother made a startling announcement. She had had a dream on the previous night in which she saw her son seated in a big hall among a group of Europeans, and the goddess of Namakkal had commanded her not to stand in the way of her son fulfilling his life's purpose. This, say Ramanujan's Indian biographers, was a very agreeable surprise to all concerned.

In 1914 Ramanujan arrived in England. So far as Hardy could detect (though in this respect I should not trust his insight far) Ramanujan, despite the difficulties of breaking the caste proscriptions, did not believe much in theological doctrine, except for a vague pantheistic benevolence, any more than Hardy did himself. But he did certainly believe in ritual. When Trinity put him up in college – within four years he became a Fellow – there was no 'Alan St Aubyn' apolausticity for him at all. Hardy used to find him ritually changed into his pyjamas, cooking vegetables rather miserably in a frying-pan in his own room.

Their association was a strangely touching one. Hardy did not forget that he was in the presence of genius: but genius that was, even in mathematics, almost untrained. Ramanujan had not been able to enter Madras University because he could not matriculate in English. According to Hardy's report, he was always amiable and good-natured, but no doubt he sometimes found Hardy's conversation outside mathematics more than a little baffling. He seems to have listened with a patient smile on his good, friendly, homely face. Even inside mathematics they had to come to terms with the difference in their education. Ramanujan was self-taught; he knew nothing of the modern rigour: in a sense he didn't know what a proof was. In an uncharacteristically sloppy moment, Hardy once wrote that if he had been better educated, he would have been less Ramanujan. Coming back to his ironic senses, Hardy later corrected himself and said that the statement was nonsense. If Ramanujan had been better educated, he would have been even more wonderful than he was. In fact, Hardy was obliged to teach him some formal mathematics as though Ramanujan had been a scholarship candidate at Winchester. Hardy said that this was the most singular experience of his life: what did modern mathematics look like to someone who had the deepest insight, but who had literally never heard of most of it?

Anyway, they produced together five papers of the highest class, in which Hardy showed supreme originality of his own (more is known of the details of this collaboration than of the Hardy-Littlewood one). Generosity and imagination were, for once, rewarded in full.

This is a story of human virtue. Once people had started behaving well, they went on behaving better. It is good to remember that England gave Ramanujan such honours as were possible. The Royal Society elected him a Fellow at the age of thirty (which, even for a mathematician, is very young). Trinity also elected him a Fellow in the same year. He was the first Indian to be given either of these distinctions. He was amiably grateful. But he soon became ill.

Hardy used to visit him, as he lay dying in hospital at Putney. It was on one of those visits that there happened the incident of the taxi-cab number. Hardy had gone out to Putney by taxi, as usual his chosen method of conveyance. He went into the room where Ramanujan was lying. Hardy, always inept about introducing a conversation, said, probably without a greeting, and certainly as his first remark: 'The number of my taxi-cab was 1729. It seemed to me rather a dull number.' To which Ramanujan replied: 'No, Hardy! No, Hardy! It is a very interesting number. It is the smallest number expressible as the sum of two cubes in two different ways.'

That is the exchange as Hardy recorded it. It must be substantially accurate. He was the most honest of men; and further, no one could possibly have invented it.

90

It was said of Ramanujan, evidently justly, that every integer was his personal friend. Another story of the demise of a great mathematician concerns Abraham de Moivre, who died in 1754. When he began to fail he declared that he needed to sleep ten minutes or a quarter of an hour more each day than the previous: the day he exceeded the limit of twenty-three and three-quarter hours, he died in his sleep.

Mathematicians have strong principles about their craft. Littlewood told a story (recorded in his Mathematician's Miscellany*) about A.A. Markov, who was one of a panel of examiners for a Ph.D. The candidate was failed by unanimous agreement. 'The other examiners were in favour of leaving it at that. Markov wished to read the man a severe lecture on the enormity of his performance, but allowed himself to be overruled. On his death-bed he said he had never forgiven himself for this weakness, and it saddened his end.'*

Stanislas Ulam was a member of the great Polish mathematics school between the wars. He emigrated to America, participated in the Manhattan Project, played a key part in the development of the hydrogen bomb and wrote a pleasant volume of memoirs, Adventures of a Mathematician. *Here he recalls the pioneer of cybernetics, Norbert Wiener.*

Wiener seemed childish in many ways. Being very ambitious about his place in the history of mathematics, he needed constant reassurance about his creative ability. I was almost stunned a few weeks after our first encounter when he asked me point blank: 'Ulam! Do you think I am through in mathematics?' Mathematicians tend to worry about their diminishing power of concentration much as some men do about their sexual potency. Impudently, I felt a strong temptation to say 'yes' as a joke, but refrained; he would not have understood. Speaking of that remark, 'Am I through,' several years later at the first World Congress of Mathematicians held in Cambridge, I was walking on Massachusetts Avenue and saw Wiener in front of a bookstore. His face was glued to the window and when he saw me, he said, 'Oh! Ulam! Look! There is my book!' Then he added, 'Ulam, the work we two have done in probability theory has not been noticed much before, but see! Now, it is in the center of everything.' I found this disarmingly and blessedly naive.

Anecdotes about Wiener abound; every mathematician who knew him has his own collection. I will add my story of what happened when I came to MIT as a visiting professor in the fall of 1957. I was assigned an office across the hall from his. On the second day after my arrival, I met him in the corridor and he stopped me to say, 'Ulam! I can't tell you what I am working on now, you are in a position to put a secret stamp on it!' (This presumably because of my position in Los Alamos.) Needless to say, I could do no such thing.

91

Marc Kac, who became a founder of probability theory, travelled a parallel path from Poland to the United States. In his memoirs, Enigmas of Chance, *he tells how the brilliant Lwów school got started.*

The Lwów School owes its existence to a lucky accident. In 1916 Hugo Steinhaus, who received his doctorate in Göttingen in 1911 and was then an up-and-coming mathematician, was walking in a park in Cracow when he overhead a snatch of conversation in which the term 'Lebesgue measure' was used. Now standard fare in first-year graduate courses in Real Analysis, Lebesgue measure in 1916 was almost unknown outside of France and even there very few were familiar with the concept. Startled and curious, Steinhaus soon located three young men who were heatedly discussing a problem. One of them was Stefan Banach.

As soon as Steinhaus settled in Lwów, he helped Banach to find a job as an assistant at the Engineering School and he guided the younger man through the early stages of his academic career. Banach's doctoral dissertation, written under the sympathetic tutelage if not the actual supervision of Steinhaus, marked the beginning of the modern period in functional analysis. It turned out to be a most fruitful field which became the hallmark of the Lwów branch of the Polish School of Mathematics. In 1929 the Lwów branch founded its own journal, *Studia Mathematica*. Like its older sister, *Fundamenta*, it became internationally acclaimed and it, too, survives to this day.

Banach was the unquestioned superstar of Polish mathematics and his name is known wherever mathematics is taught. In the short fifty-three years of his life (he died in 1945) he succeeded in combining an overwhelming flow of brilliant ideas with a style of high living that few men could sustain. In an obituary of Banach, Steinhaus, who outlived him by twenty-seven years, wrote: 'He combined a spark of genius with an inner compulsion which incessantly reminded him of the words of the poet: "*Il n'y a que la gloire ardente du métier*" (Verlaine). Mathematicians know that their craft and that of the poets share the same mystery.'

92

Leopold Infeld, the third of this line of Polish-Jewish mathematicians, of whose life more will be heard later, gave the following description in his autobiography, Quest, *of one of the greatest of the patricians of theoretical physics, P.A.M. Dirac.*

The greatest theoretical physicist in Cambridge was P. A. M. Dirac, one of the outstanding scientists of our generation, then a young man about thirty. He still occupies the chair of mathematics, the genealogy of which can be traced directly to Newton.

I knew nothing of Dirac, except that he was a great mathematical physicist. His papers, appearing chiefly in the *Proceedings of the Royal Society*, were written with wonderful clarity and great imagination. His name is usually linked with those of Heisenberg and Schroedinger as the creators of quantum mechanics. Dirac's book *The Principles of Quantum Mechanics* is regarded as the bible of modern physics. It is deep, simple, lucid and original. It can only be compared in its importance and maturity to Newton's *Principia*. Admired by everyone as a genius, as a great star in the firmament of English physics, he created a legend around him. His thin figure with its long hands, walking in heat and cold without overcoat or hat, was a familiar one to Cambridge students. His loneliness and shyness were famous among physicists. Only a few men could penetrate his solitude. One of the fellows, a well-known physicist, told me:

'I still find it very difficult to talk with Dirac. If I need his advice I try to formulate my question as briefly as possible. He looks for five minutes at the ceiling, five minutes at the windows, and then says "Yes" or "No." And he is always right.'

Once – according to a story which I heard – Dirac was lecturing in the United States and the chairman called for questions after the lecture. One of the audience said:

'I did not understand this and this in your arguments.'

Dirac sat quietly, as though the man had not spoken. A disagreeable silence ensured, and the chairman turned to Dirac uncertainly:

'Would you not be kind enough, Professor Dirac, to answer this question?'

To which Dirac replied: 'It was not a question; it was a statement.'

Another story also refers to his stay in the United States. He lived in an apartment with a famous French physicist and they invariably talked English to each other. Once the French physicist, finding it difficult to explain something in English, asked Dirac, who is half English and half French:

'Do you speak French?'

'Yes. French is my mother's tongue,' answered Dirac in an unusually long sentence. The French professor burst out:

'And you say this to me now, having allowed me to speak my bad, painful English for weeks! Why did you not tell me this before?'

'You did not ask me before,' was Dirac's answer.

But a few scientists who knew Dirac better, who managed after years of acquaintance to talk to him, were full of praise of his gentle attitude toward everyone. They believed that his solitude was a result of shyness and could be broken in time by careful aggressiveness and persistence.

These idiosyncrasies made it difficult to work with Dirac. The result has been that Dirac has not created a school of personal contact. He has created a school by his papers, by his book, but not by collaboration. He is one of the very few scientists who could work even on a lonely island if he had a library and could perhaps even do without books and journals.

When I visited Dirac for the first time I did not know how difficult it was to talk to him as I did not then know anyone who could have warned me.

I went along the narrow wooden stairs in St. John's College and knocked at the door of Dirac's room. He opened it silently and with a friendly gesture indicated an armchair. I sat down and waited for Dirac to start the conversation. Complete silence. I began by warning my host that I spoke very little English. A friendly smile but again no answer. I had to go further:

'I talked with Professor Fowler. He told me that I am supposed to work with you. He suggested that I work on the internal conversion effect of positrons.'

No answer. I waited for some time and tried a direct question:

'Do you have any objection to my working on this subject?'

'No.'

At least I had got a word out of Dirac.

Then I spoke of the problem, took out my pen in order to write a formula. Without saying a word Dirac got up and brought paper. But my pen refused to write. Silently Dirac took out his pencil and handed it to me. Again I asked him a direct question to which I received an answer in five words which took me two days to digest. The conversation was finished. I made an attempt to prolong it.

'Do you mind if I bother you sometimes when I come across difficulties?'

'No.'

I left Dirac's room, surprised and depressed. He was not forbidding, and I should have had no disagreeable feeling had I known what everyone in Cambridge knew. If he seemed peculiar to Englishmen, how much more so he seemed to a Pole who had polished his smooth tongue in Lwow cafés! One of Dirac's principles is:

'One must not start a sentence before one knows how to finish it.'

Someone in Cambridge generalized this ironically:

'One must not start a life before one knows how to finish it.'

A story that encapsulates Dirac's enigmatic character is the following: a student visiting Dirac in his rooms in College discovered, to his surprise, a copy of a novel: E.M. Forster's A Passage to India. *The student, who knew Forster, was struck by the idea of bringing the two grand old men, the one from King's, the other from Trinity, together. A tea was arranged. Introductions were followed by a heavy silence. Then Dirac spoke: 'What happened in the cave?' Forster pondered. 'I don't know,' he said. The two old men lapsed into silence and after more time had passed parted without a further word.*

There are many stories of Dirac's laconic manner. Typical is an exchange, recorded by Rudolf Peierls in his memoirs, Bird of Passage. *Dirac's wife, thinking of giving a party to allow Peierls's daughter, who was staying, to meet some younger people, inquired of her husband: 'Paul, do you have any students?' 'I had one,' came the reply, 'but he died.' (This is reminiscent of the agitated remark by the President of Trinity College, Oxford, to C.N. Hinshelwood, then a young chemistry tutor, later to become Professor of Physical Chemistry and a Nobel Laureate, as he was being removed to hospital with acute appendicitis: 'Hinshelwood, this is most unfortunate, most unfortunate, our last Science Tutor died of this.')*

Here finally, is how the precocious I.I. Rabi, growing up in the poverty of an immigrant community in Brooklyn, experienced the first stirrings of scientific curiosity, as he related it to Jeremy Bernstein.

I asked Rabi if he had had any particular experience in those early days that turned his mind toward science. 'Yes,' he said. 'A very profound one. One time, I was walking along and looked down the street – looked right down the street, which faced east. The moon was just rising. And it scared hell out of me! Absolutely scared the hell out of me. Another profound experience that I had revolved around the first verses of Genesis. They were very moving to me as a kid. The whole idea of the creation – the mystery and the philosophy of it. It sank in on me, and it's something I still feel. But, as a matter of fact, I got into science in a funny way. I read all the fairy stories and other stories in the children's section of the library. I started with Alcott and worked down through Trowbridge – all those children's books, all those writers. Then I came to the end of *those* shelves, and there was a science shelf. So I started with astronomy. That was what determined my later life more than anything else – reading a little book on astronomy. That's where I first heard of the Copernican system and the explanation of the changes of the seasons, the phases of the moon, and the idea that the stars were suns, very distant suns. Ours was such a fundamentalist family that my parents hadn't heard of the Copernican system, and for me it was a tremendous revelation. I was so impressed – the beauty of it all, and the simplicity.'

Science Collides with the World

Charles Babbage, engineer and mathematician, is now remembered as the father of the computer, although the construction of his 'difference engine' was abandoned long before it was completed. He was proud and quarrelsome and died embittered. Not long before his death he spoke with rancour to a friend about his life. 'He spoke as if he hated mankind in general, Englishmen in particular, and the English Government and organ grinders most of all.' Darwin has the following passage in his Autobiography*:*

I remember a funny dinner at my brother's, where, amongst a few others, were Babbage and Lyell, both of whom liked to talk. Carlyle, however, silenced everyone by haranguing during the whole dinner on the advantages of silence. Babbage, in his grimmest manner, thanked Carlyle for his very interesting lecture on silence.

Joel Shurkin, in his history of the computer, Engines of the Mind, *offers the following story:*

Babbage was at the center of a social set that contained many of the prominent figures in the arts and sciences of the day. Darwin and Dickens might come to dinner on Saturday night. So might Thackeray, Herschel, and Longfellow. After he read a poem by Tennyson, which contains the lines 'Every minute dies a man/Every minute one is born,' Babbage wrote to the poet, an occasional guest at his famous dinner parties, and suggested a few changes.

I need hardly to tell you that this calculation would tend to keep the sum total of the world's population in a state of perpetual equipoise, whereas it is a well-known fact that the said sum total is constantly on the increase. I would therefore take the liberty of suggesting that in the next edition of your excellent poem the erroneous calculation to which I refer should be corrected as follows:

Every moment dies a man,
And one and a sixteenth is born.

I may add that the exact figures are 1.167, but something must, of course, be conceded to the laws of metre.

Tennyson changed the poem to read:

Every moment dies a man,
Every moment one is born.

95

And here is the authentic voice of Babbage himself, from his book, Passages from the Life of a Philosopher.

Amongst the various questions which have been asked respecting the Difference Engine, I will mention a few of the most remarkable: – One gentleman addressed me thus: 'Pray, Mr. Babbage, can you explain to me in two words what is the principle of this machine?' Had the querist possessed a moderate acquaintance with mathematics I might in four words have conveyed to him the required information by answering, 'The method of differences.' The question might indeed have been answered with six characters thus –

$$\Delta^7 u_x = 0.$$

but such information would have been unintelligible to such inquirers.

On two occasions I have been asked, – 'Pray, Mr. Babbage, if you put into the machine wrong figures, will the right answers come out?' In one case a member of the Upper, and in the other a member of the Lower, House put this question. I am not able rightly to apprehend the kind of confusion of ideas that could provoke such a question.

96

From Logan Pearsall Smith's Unforgotten Years, *a story that gives an indication of what Babbage and his like had to put up with from the* savants *of the day.*

[The] ideal of endowment for research was particularly shocking to Benjamin Jowett, the great inventor of the tutorial system which it threatened. I remember once, when staying with him at Malvern, inadvertently pronouncing the ill-omened word. 'Research!' the Master exclaimed. 'Research!' he said. 'A mere excuse for idleness; it has never achieved, and will never achieve any results of the slightest value.' At this sweeping statement I protested; whereupon I was peremptorily told, if I knew of any such results of value, to name them without delay. My ideas on the subject were by no means profound, and anyhow it is difficult to give instances of a general proposition at a moment's notice. The only thing that came into my head was the recent discovery, of which I had read somewhere, that on striking a patient's kneecap sharply he would give an involuntary kick, and by the vigour or lack of vigour of this 'knee jerk', as it is called, a judgement could be formed of his general state of health.

'I don't believe a word of it', Jowett replied. 'Just give my knee a tap.'

I was extremely reluctant to perform this irreverent act upon his person, but the Master angrily insisted, and the undergraduate could do nothing but obey. The little leg reacted with a vigour which almost alarmed me, and must, I think, have considerably disconcerted that elderly and eminent opponent of research.

97

Robert Frost's view of science was in the sour tradition of Blake.

WHY WAIT FOR SCIENCE

Sarcastic Science, she would like to know,
In her complacent ministry of fear,
How we propose to get away from here
When she has made things so we have to go
Or be wiped out. Will she be asked to show
Us how by rocket we may hope to steer
To some star off there, say, a half light-year
Through temperature of absolute zeró?
Why wait for Science to supply the how
When any amateur can tell it now?
The way to go away should be the same
As fifty million years ago we came –
If anyone remembers how that was.
I have a theory, but it hardly does.

98

Babbage's relations with the political establishment were turbulent. Faraday's celebrated encounter with the Chancellor of the Exchequer,

Gladstone, is remembered for the answer that Faraday gave when the Chancellor asked, having seen a demonstration of electromagnetism, 'But what use is it?' Faraday replied that he did not know, but he dared hazard that one day the Chancellor would be able to tax it. Babbage's view on the Chancellor and his kind, expressed in Passages from the Life of a Philosopher, *are less familiar.*

Who, possessing one grain of common sense, could look upon the unrivalled workmanship of the then existing portion of the Difference Engine No. 1, and doubt whether a simplified form of the same engine could be executed?

As to any doubt of its mathematical principles, this was excusable in the Chancellor of the Exchequer, who was himself too practically acquainted with the fallibility of his own figures, over which the severe duties of his office had stultified his brilliant imagination. Far other figures are dear to him – those of speech, in which it cannot be denied he is indeed pre-eminent.

Any junior clerk in his office might, however, have told him that the power of computing Tables by differences merely required a knowledge of simple addition.

As to the impossibility of ascertaining the expenditure, this merges into the first objection; but a poetical brain must be pardoned when it repeats or amplifies. I will recall to the ex-Chancellor of the Exchequer what Lord Rosse really proposed, namely, that the Government should take the opinion of the President of the Institution of Civil Engineers upon the question, whether a contract could be made for constructing the Difference Engine, and if so, for what sum.

But the very plan proposed by Lord Rosse and refused by Lord Derby, for the construction of the *English* Difference Engine, was adopted some few years after by another administration for the *Swedish* Difference Engine. Messrs Donkin, the eminent Engineers, *made an estimate*, and a *contract was* in consequence executed to construct for Government a facsimile of the *Swedish* Difference Engine, which is now in use in the department of the Registrar-General, at Somerset House. There were far greater mechanical difficulties in the production of that machine than in the one the drawings of which I had offered to the Government.

From my own experience of the cost of executing such works, I have no doubt, although it was highly creditable to the skill of the able firm who constructed it, but that it must have been commercially unprofitable. Under such circumstances, surely it was harsh on the part of the Government to refuse Messrs Donkin permission to exhibit it as a specimen of English workmanship at the Exhibition of 1862.

But the machine upon which everybody could calculate, had little chance of fair play from the man on whom nobody could calculate.

If the Chancellor of the Exchequer had read my letter to Lord Derby, he would have found the opinion of the Committee of the Royal Society expressed in these words:–

'They consider the former [the abstract mathematical principle] as not only sufficiently clear in itself, but as already admitted and acted on by the Council in their former proceedings.

'The latter [its public utility] they consider as obvious to every one who considers the immense advantage of accurate numerical tables in all matters of calculation, especially in those which relate to astronomy and navigation.' – *Report of the Royal Society,* 12th Feb., 1829.

Thus it appears:–

1st. That the Chancellor of the Exchequer presumed to set up his *own idea* of the utility of the Difference Engine in direct opposition to that of the Royal Society.

2nd. That he *refused* to take the opinion of the highest mechanical authority in the country on its probable cost, and even *to be informed* whether a contract for its construction at a definite sum might not be attainable: he then boldly pronounced the expense to be 'utterly incapable of being calculated.'

This much-abused Difference Engine is, however, like its prouder relative the Analytical, a being of sensibility, of impulse, and of power.

It can not only calculate the millions the ex-Chancellor of the Exchequer squandered, but it can deal with the smallest quantities; nay, it feels even for zeros. It is as conscious as Lord Derby himself is of the presence of a *negative quantity*, and it is not beyond the ken of either of them to foresee the existence of *impossible ones.*

Yet should any unexpected course of events ever raise the ex-Chancellor of the Exchequer to his former dignity, I am sure he will be its *friend* as soon as he is convinced that it can be made *useful* to him.

It may possibly enable him to un-muddle even his own financial accounts, and to –

But as I have no wish to crucify him, I will leave his name in obscurity.

The Herostratus of Science, if he escape oblivion, will be linked with the destroyer of the Ephesian Temple.

99

A century later, François Jacob tried it on, when confronted by General de Gaulle (in whose army he had fought a few years before). The exchange loses some of its savour in translation.

At this moment he is receiving the old guard. Some three hundred Companions from all over France. Many of my old Free French comrades. Several politicians, sprung from the Resistance. Buffet. Champagne. *Petits fours.* The General moves about the hall, passing from group to group. A word for everyone. A longer pause before the parents of vanished Companions. Suddenly I find myself in front of him. I introduce myself. He extends his hand.

Ah! Jacob. Glad to see you again. (A pause) What are you doing now?

I stutter:

– De la recherche scientifique, mon Général.
– Ah! Très intéressant. Dans quel domaine?
– Biologie, mon Général.
– Ah! Très intéressant. Quel genre?
– Génétique, mon Général.
– Ah! Très intéressant. Et ou travaillez-vous?
– A l'Institut Pasteur, mon Général.
– Ah! Très intéressant. Vous avez ce qu'il vous faut?

Après une hésitation:

– Non, mon Général.
– Au revoir!

100

A fictional scientist goes into politics: the infamous Dr Almus Pickerbaugh in Arrowsmith *was supposedly William de Kleine, medical director of the American Red Cross.*

His opponent was a snuffy little lawyer whose strength lay in his training. He had been state senator, lieutenant governor, county judge. But the Democratic slogan, 'Pickerbaugh the Pick-up Candidate,' was drowned in the admiration for the hero of the health fair. He dashed about in motors, proclaiming, 'I am not running because I want office, but because I want the chance to take to the whole nation my ideals of health.' Everywhere was plastered:

For Congress
PICKERBAUGH
The two-fisted fighting poet doc

Just elect him for a term
And all through the nation he'll swat the germ.

Enormous meetings were held. Pickerbaugh was ample and vague about his Policies. Yes, he was opposed to our entering the European War, but he assured them, he certainly did assure them, that he was for using every power of our Government to end this terrible calamity. Yes, he was for high tariff, but it must be so adjusted that the farmers in his district could buy everything cheaply. Yes, he was for high wages for each and every workman, but he stood like a rock, like a boulder, like a moraine, for protecting the prosperity of all manufacturers, merchants, and real-estate owners.

While this larger campaign thundered, there was proceeding in Nautilus a smaller and much defter campaign, to re-elect as mayor one Mr. Pugh, Pickerbaugh's loving chief. Mr. Pugh sat nicely at desks, and he was pleasant and promissory to everybody who came to see him; clergymen, gamblers, G. A. R. veterans, circus advance-agents, policemen, and ladies of reasonable virtue – everybody except perhaps socialist agitators, against whom he staunchly protected the embattled city. In his speeches Pickerbaugh commended Pugh for 'that firm integrity and ready sympathy with which His Honor had backed up every movement for the public weal,' and when Pickerbaugh (quite honestly) begged, 'Mr. Mayor, if I go to Congress you must appoint Arrowsmith in my place; he knows nothing about politics but he's incorruptible,' then Pugh gave his promise, and amity abode in that land. . . . Nobody said anything at all about Mr. F. X. Jordan.

F. X. Jordan was a contractor with a generous interest in politics. Pickerbaugh called him a grafter, and the last time Pugh had been elected – it had been on a Reform Platform, though since that time the reform had been coaxed to behave itself and be practical – both

Pugh and Pickerbaugh had denounced Jordan as a 'malign force.' But so kindly was Mayor Pugh that in the present election he said nothing that could hurt Mr. Jordan's feelings, and in return what could Mr. Jordan do but speak forgivingly about Mr. Pugh to the people in blind-pigs and houses of ill fame?

On the evening of the election, Martin and Leora were among the company awaiting the returns at the Pickerbaughs'. They were confident. Martin had never been roused by politics, but he was stirred now by Pickerbaugh's twitchy pretense of indifference, by the telephoned report from the newspaper office, 'Here's Willow Grove township – Pickerbaugh leading, two to one!' by the crowds which went past the house howling, 'Pickerbaugh, Pickerbaugh, Pickerbaugh!'

At eleven the victory was certain, and Martin, his bowels weak with unconfidence, realized that he was now Director of Public Health, with responsibility for seventy thousand lives.

He looked wistfully toward Leora and in her still smile found assurance.

Orchid had been airy and distant with Martin all evening, and dismayingly chatty and affectionate with Leora. Now she drew him into the back parlor and 'So I'm going off to Washington – and you don't care a bit!' she said, her eyes blurred and languorous and undefended. He held her, muttering, 'You darling child, I can't let you go!' As he walked home he thought less of being Director than of Orchid's eyes.

In the morning he groaned, 'Doesn't anybody ever learn anything? Must I watch myself and still be a fool, all my life? Doesn't any story ever end?'

He never saw her afterward, except on the platform of the train.

Leora surprisingly reflected, after the Pickerbaughs had gone, 'Sandy dear, I know how you feel about losing your Orchid. It's sort of Youth going. She really is a peach. Honestly, I can appreciate how you feel, and sympathize with you – I mean, of course, providin' you aren't ever going to see her again.'

Over the *Nautilus Cornfield's* announcement was the vigorous headline:

ALMUS PICKERBAUGH WINS
First Scientist Ever Elected
to Congress

Side-kick of Darwin and Pasteur
Gives New Punch to Steering
Ship of State

Pickerbaugh's resignation was to take effect at once; he was, he explained, going to Washington before his term began, to study legislative methods and start his propaganda for the creation of a national Secretaryship of Health.

101

The great Richard Feynman's encounters with authority invariably left authority reeling. Here is what happened when he reluctantly allowed himself to be roped into a committee to approve science text-books for use in Californian schools. The passage is from Surely You're Joking, Mr Feynman!

A few days later a guy from the book depository called me up and said, 'We're ready to send you the books, Mr. Feynman; there are three hundred pounds.'

I was overwhelmed.

'It's all right, Mr. Feynman; we'll get someone to help you read them.'

I couldn't figure out how you *do* that: you either read them or you don't read them. I had a special bookshelf put in my study downstairs (the books took up seventeen feet), and began reading all the books that were going to be discussed in the next meeting. We were going to start out with the elementary school-books.

It was a pretty big job, and I worked all the time at it down in the basement. My wife says that during this period it was like living over a volcano. It would be quiet for a while, but then all of a sudden, 'BLLLLLOOOOOOWWWWW!!!!' – there would be a big explosion from the 'volcano' below.

The reason was that the books were so lousy. They were false. They were hurried. They would try to be rigorous, but they would use examples (like automobiles in the street for 'sets') which were *almost* OK, but in which there were always some subtleties. The definitions weren't accurate. Everything was a little bit ambiguous

– they weren't *smart* enough to understand what was meant by 'rigor.' They were faking it. They were teaching something they didn't understand, and which was, in fact, *useless*, at that time, for the child.

I understood what they were trying to do. Many people thought we were behind the Russians after Sputnik, and some mathematicians were asked to give advice on how to teach math by using some of the rather interesting modern concepts of mathematics. The purpose was to enhance mathematics for the children who found it dull.

I'll give you an example: They would talk about different bases of numbers – five, six, and so on – to show the possibilities. That would be interesting for a kid who could understand base ten – something to entertain his mind. But what they had turned it into, in these books, was that *every* child had to learn another base! And then the usual horror would come: 'Translate these numbers, which are written in base seven, to base five.' Translating from one base to another is an *utterly useless* thing. If you *can* do it, maybe it's entertaining; if you *can't* do it, forget it. There's no *point* to it.

Anyhow, I'm looking at all these books, all these books, and none of them has said anything about using arithmetic in science. If there are any examples on the use of arithmetic at all (most of the time it's this abstract new modern nonsense), they are about things like buying stamps.

Finally I come to a book that says, 'Mathematics is used in science in many ways. We will give you an example from astronomy, which is the science of stars.' I turn the page, and it says, 'Red stars have a temperature of four thousand degrees, yellow stars have a temperature of five thousand degrees . . .' – so far, so good. It continues: 'Green stars have a temperature of seven thousand degrees, blue stars have a temperature of ten thousand degrees, and violet stars have a temperature of . . . (some big number).' There are no green or violet stars, but the figures for the others are roughly correct. It's *vaguely* right – but already, trouble! That's the way everything was: Everything was written by somebody who didn't know what the hell he was talking about, so it was a little bit wrong, always! And how we are going to teach well by using books written by people who don't *quite* understand what they're talking about. I *cannot* understand. I don't know why, but the books are lousy; UNIVERSALLY LOUSY!

Anyway, I'm *happy* with this book, because it's the first example of applying arithmetic to science. I'm a *bit* unhappy when I read about the stars' temperatures, but I'm not *very* unhappy because it's more or less right – it's just an example of error. Then comes the list of problems. It says, 'John and his father go out to look at the stars. John sees two blue stars and a red star. His father sees a green star, a violet star, and two yellow stars. What is the total temperature of the stars seen by John and his father?' – and I would explode in horror.

My wife would talk about the volcano downstairs. That's only an example: it was *perpetually* like that. Perpetual absurdity! There's no purpose whatsoever in adding the temperature of two stars. Nobody *ever* does that except, maybe, to then take the *average* temperature of the stars, but *not* to find out the *total* temperature of all the stars! It was awful! All it was was a game to get you to add, and they didn't understand what they were talking about. It was like reading sentences with a few typographical errors, and then suddenly a whole sentence is written backwards. The mathematics was like that. Just hopeless!

Then I came to my first meeting. The other members had given some kind of ratings to some of the books, and they asked me what *my* ratings were. My rating was often different from theirs, and they would ask, 'Why did you rate that book low?'

I would say the trouble with that book was this and this on page so-and-so – I had my notes.

They discovered that I was kind of a goldmine: I would tell them, in detail, what was good and bad in all the books; I had a reason for every rating.

I would ask them why they had rated this a book so high, and they would say, 'Let us hear what you thought about such and such a book.' I would never find out why they rated anything the way they did. Instead, they kept asking me what *I* thought.

We came to a certain book, part of a set of three supplementary books published by the same company, and they asked me what I thought about it.

I said, 'The book depository didn't send me that book, but the other two were nice.'

Someone tried repeating the question: 'What do you think about the book?'

'I said they didn't send me that one, so I don't have any judgment on it.'

The man from the book depository was there, and he said,

'Excuse me; I can explain that. I didn't send it to you because that book hadn't been completed yet. There's a rule that you have to have every entry in by a certain time, and the publisher was a few days late with it. So it was sent to us with just the covers, and it's blank in between. The company sent a note excusing themselves and hoping they could have their set of three books considered, even though the third one would be late.'

It turned out that the blank book had a rating by some of the other members! They couldn't believe it was blank, because they had a rating. In fact, the rating for the missing book was a little bit higher than for the two others. The fact that there was nothing in the book had nothing to do with the rating.

102

In a later collection, What do *you* care what other people think? *published posthumously, Feynman describes what happened when he was persuaded to join the presidental committee of investigation after the Challenger disaster, in which a crew of astronauts was incinerated. The episode in the extract given here received saturation coverage in the mass media, and caused the tormented NASA even more hideous embarrassment. (The 'tang', to which he refers is the engineering parlance for the male part of a joint, and the 'clevis' is the female.)*

Mr. Mulloy explains how the seals are supposed to work – in the usual NASA way: he uses funny words and acronyms, and it's hard for anybody else to understand.

In order to set things up while I'm waiting for the ice water, I start out: 'During a launch, there are vibrations which cause the rocket joints to move a little bit – is that correct?'

'That is correct, sir.'

'And inside the joints, these so-called O-rings are supposed to expand to make a seal – is that right?'

'Yes, sir. In static conditions they should be in direct contact with the tang and clevis and squeezed twenty-thousandths of an inch.'

'Why don't we take the O-rings out?'

'Because then you would have hot gas expanding through the joint . . .'

'Now, in order for the seal to work correctly, the O-rings must be made of rubber – not something like lead, which, when you squash it, it stays.'

'Yes, sir.'

'Now, if the O-ring weren't resilient for a second or two, would that be enough to be a very dangerous situation?'

'Yes, sir.'

That led us right up to the question of cold temperature and the resilience of the rubber. I wanted to prove that Mr. Mulloy must have known that temperature had an effect, although – according to Mr. MacDonald – he claimed that the evidence was 'incomplete.' But still, no ice water! So I had to stop, and somebody else started asking questions.

The model comes around to General Kutyna, and then to me. The clamp and pliers come out of my pocket, I take the model apart, I've got the O-ring pieces in my hand, but I still haven't got any ice water! I turn around again and signal the guy I've been bothering about it, and he signals back, 'Don't worry, you'll get it!'

Pretty soon I see a young woman, way down in front, bringing in a tray with glasses on it. She gives a glass of ice water to Mr. Rogers, she gives a glass of ice water to Mr. Armstrong, she works her way back and forth along the rows of the dais, giving ice water to everybody! The poor woman had gotten everything together – jug, glasses, ice, tray, the whole thing – so that everybody could have ice water.

So finally, when I get my ice water, I don't drink it! I squeeze the rubber in the C-clamp, and put them in the glass of ice water.

After a few minutes, I'm ready to show the results of my little experiment. I reach for the little button that activates my microphone.

General Kutyna, who's caught on to what I'm doing, quickly leans over to me and says, 'Co-pilot to pilot: not now.'

Pretty soon, I'm reaching for my microphone again.

'Not now!' He points in our briefing book – with all the charts and slides Mr. Mulloy is going through – and says, 'When he comes to this slide, here, that's the right time to do it.'

Finally Mr. Mulloy comes to the place, I press the button for my microphone, and I say, 'I took this rubber from the model and put it in a clamp in ice water for a while.'

I take the clamp out, hold it up in the air, and loosen it as I talk: 'I discovered that when you undo the clamp, the rubber doesn't spring back. In other words, for more than a few seconds, there is no resilience in this particular material when it is at a temperature of 32 degrees. I believe that has some significance for our problem.'

Before Mr. Mulloy could say anything, Mr. Rogers says, 'That is a matter we will consider, of course, at length in the session that we will hold on the weather, and I think it is an important point which I'm sure Mr. Mulloy acknowledges and will comment on in a further session.'

During the lunch break, reporters came up to me and asked questions like, 'Were you talking about the O-ring, or the putty?' and 'Would you explain to us what an O-ring is, exactly?' So I was rather depressed that I wasn't able to make my point. But that night, all the news shows caught on to the significance of the experiment, and the next day, the newspaper articles explained everything perfectly.

103

A.D. Godley was an Oxford don, who wrote verses, among which this paean to donnish scruples.

THE MEGALOPSYCHIAD

Great and good is the typical Don, and of evil and wrong the foe,
Good, and great, I'm a Don myself, and therefore I ought to know:
But of all the sages I ever have met, and of all the Dons I've known,
There never was one so good and great as Megalopsychus Brown.

Megalopsychus Brown was blessed with a Large and Liberal View:
Six sides he saw of a question vexed, when commonplace men saw two:
He looked at it East, and he looked at it West, and he looked at it upside down –
Such was the large and liberal mind of Megalopsychus Brown.
He held one creed which he made for himself, and he held it fast and strong –

That to act in an obvious logical cause is shallow, and base, and wrong;
And all that was said for Freedom of Trade so plausible seemed and plain,
That he nearly made up his mind to vote for Mr Chamberlain –
Yes! if any one urged that the moon was a cheese, he would always at once admit,
Though the point of view was undoubtedly new, there was much to be said for it.
But out and alas! for his charity wide had a tendency sad to see
(And it much impaired the practical use of Megalopsychus B.); –
For since, as I've said, no strange ideas could cause him the least alarm,
As he never believed that anyone else intended the smallest harm,
He became the sport and the natural prey of men both bold and bad
Who hadn't at heart the Highest Good (as Megalopsychus had);
Men with a crank, and men with a fad, and men with an axe to grind,
Men with an eye to the main main chance and an unacademical mind.
They told him of Science, they told him of Greek, they told him of verses and prose,
They led him about in the strangest ways by his highly respectable nose:–

Till the Public awoke and was pained to find that Megalopsychus' rule
Had changed what once was the Muses' seat to a kind of Technical School;
And everyone said when that learned spot was shorn of its old renown,
'Behold the large and liberal views of Megalopsychus Brown!'

104

Isaac Asimov is the most prolific science writer of all time. His oeuvre *includes science fiction, short stories, essays and books of instruction at all levels. Here is an extract from a good short story, called 'Paté de foie gras'.*

The head of my section at the Department of Agriculture is Louis P. Bronstein. (Don't bother looking him up. The 'P' stands for Pittfield if you want more misdirection.)

He and I are on good terms and I felt I could explain things without being placed under immediate observation. Even so, I took no chances. I had the egg with me and when I got to the tricky part, I just laid it on the desk between us.

Finally he touched it with his finger as though it were hot.

I said, 'Pick it up.'

It took him a long time, but he did, and I watched him take two tries at it as I had.

I said, 'It's a yellow metal and it could be brass, only it isn't because it's inert to concentrated nitric acid. I've tried that already. There's only a shell of gold because it can be bent with moderate pressure. Besides, if it were solid gold, the egg would weigh over ten pounds.'

Bronstein said, 'It's some sort of hoax. It *must* be.'

'A hoax that uses real gold? Remember, when I first saw this thing, it was covered completely with authentic unbroken eggshell. It's been easy to check a piece of the eggshell. Calcium carbonate. That's a hard thing to gimmick. And if we look inside the egg – (I didn't want to do that on my own, Chief) – and find real egg, then we've got it, because that would be impossible to gimmick. Surely this is worth an official project.'

'How can I approach the secretary with –' He stared at the egg.

But he did in the end. He made phone calls and sweated out most of the day. One or two of the department brass came to look at the egg.

Project Goose was started. That was July 20, 1955.

I was the responsible investigator to begin with, and I remained in titular charge throughout, though matters quickly got beyond me.

We began with the one egg. Its average radius was 35 millimeters (major axis, 72 millimeters; minor axis, 68 millimeters). The gold shell was 2.45 millimeters in thickness. Studying other eggs later on, we found this value to be rather high. The average thickness turned out to be 2.1 millimeters.

Inside *was* egg. It looked like egg and it smelled like egg.

Aliquots were analyzed and the organic constituents were reasonably normal. The white was 9.7 per cent albumen. The yolk had the normal complement of vitellin, cholesterol, phospholipid and carotenoid. We lacked enough material to test for trace constituents, but later on, with more eggs at our disposal, we did and nothing unusual showed up as far as the contents of vitamins, coenzymes, nucleotides, sulfhydryl groups, etc., etc., were concerned.

One important gross abnormality that showed was the egg's behavior on heating. A small portion of the yolk, heated, 'hard-boiled' almost at once. We fed a portion of the hard-boiled egg to a mouse. It survived.

I nibbled at another bit of it. Too small a quantity to taste, really, but it made me sick. Purely psychosomatic, I'm sure.

Boris W. Finley of the Department of Biochemistry of Temple University (a Department Consultant) supervised these tests.

He said, referring to the hard-boiling, 'The ease with which the egg proteins are heat-denatured indicates a partial denaturation to begin with, and, considering the nature of the shell, the obvious guilt would lie at the door of heavy-metal contamination.'

So a portion of the yolk was analyzed for inorganic constituents, and it was found to be high in chloraurate ion, which is a singly-charged ion containing an atom of gold and four of chlorine, the symbol for which is $AuCl_4$. (The 'Au' symbol for gold comes from the fact that the Latin word for gold is 'aurum.') When I say the chloraurate ion content was high, I meant it was 3.2 parts per thousand, or 0.32 percent. That's high enough to form insoluble complexes of 'gold-protein,' which would coagulate easily.

Finley said, 'It's obvious this egg cannot hatch. Nor can any other such egg. It is heavy-metal poisoned. Gold may be more glamorous than lead, but it is just as poisonous to proteins.'

I agreed gloomily, 'At least it's safe from decay too.'

'Quite right. No self-respecting bug would live in this chlorauriferous soup.'

The final spectrographic analysis of the gold of the shell came in. Virtually pure. The only detectable impurity was iron, which amounted to 0.23 percent of the whole. The iron content of the egg yolk had been twice normal also. At the moment, however, the matter of the iron was neglected.

One week after Project Goose was begun, an expedition was sent into Texas. Five biochemists went (the accent was still on biochemistry, you see) along with three truck-loads of equipment and a squadron of army personnel. I went along too, of course.

As soon as we arrived, we cut MacGregor's farm off from the world.

That was a lucky thing, you know – the security measures we took right from the start. The reasoning was wrong, at first, but the results were good.

The Department wanted Project Goose kept quiet at the start simply because there was always the thought that this might still be an elaborate hoax, and we couldn't risk the bad publicity if it were. And if it weren't a hoax, we couldn't risk the newspaper hounding that would definitely result for any goose-and-golden-egg story.

It was only well after the start of Project Goose, well after our arrival at MacGregor's farm, that the real implications of the matter became clear.

Naturally MacGregor didn't like the men and equipment settling down all about him. He didn't like being told The Goose was government property. He didn't like having his eggs impounded.

He didn't like it, but he agreed to it – if you can call it agreeing when negotiations are being carried on while a machine gun is being assembled in a man's barnyard and ten men, with bayonets fixed, are marching past while the arguing is going on.

He was compensated, of course. What's money to the government?

The Goose didn't like a few things either – like having blood samples taken. We didn't dare anesthetize it for fear of doing anything to alter its metabolism, and it took two men to hold it each time. Ever try to hold an angry goose?

The Goose was put under a twenty-four-hour guard with the threat of summary court-martial to any man who let anything happen to it. If any of those soldiers read this article, they may get a sudden glimmering of what was going on. If so, they will probably have the sense to keep quiet about it. At least if they know what's good for them, they will.

The blood of The Goose was put through every test conceivable.

It carried two parts per hundred thousand (0.002 percent) of chloraurate ion. Blood taken from the hepatic vein was richer than the rest, almost four parts per hundred thousand.

Finley grunted. 'The liver,' he said.

We took X-rays. On the X-ray negative the liver was a cloudy mass of light gray, lighter than the viscera in its neighborhood, because it stopped more of the X-rays, because it contained more gold. The blood vessels showed up lighter than the liver proper, and the ovaries were pure white. No X-rays got through the ovaries at all.

It made sense, and in an early report Finley stated it as bluntly as possible. Paraphrasing the report, it went, in part:

'The chloraurate ion is secreted by the liver into the bloodstream. The ovaries act as a trap for the ion, which is there reduced to metallic gold and deposited as a shell about the developing egg. Relatively high concentrations of unreduced chloraurate ion penetrate the contents of the developing egg.

'There is little doubt that The Goose finds this process useful as a means of getting rid of the gold atoms which, if allowed to accumulate, would undoubtedly poison it. Excretion by eggshell may be novel in the animal kingdom, even unique, but there is no denying that it is keeping The Goose alive.

'Unfortunately, however, the ovary is being locally poisoned to such an extent that few eggs are laid, probably not more than will suffice to get rid of the accumulating gold, and those few eggs are definitely unhatchable.'

That was all he said in writing, but to the rest of us he said, 'That leaves one peculiarly embarrassing question.'

I knew what it was. We all did.

Where was the gold coming from?

No answer to that for a while, except for some negative evidence. There was no perceptible gold in The Goose's feed, nor were there any gold-bearing pebbles about that it might have swallowed. There was no trace of gold anywhere in the soil of the area, and a search of the house and grounds revealed nothing. There were no gold coins, gold jewelry, gold plate, gold watches, or gold anything. No one on the farm even had as much as gold fillings in his teeth.

There was Mrs. MacGregor's wedding ring, of course, but she had only had one in her life and she was wearing that one.

So where was the gold coming from?

The beginnings of the answer came on August 16, 1955.

Albert Nevis, of Purdue, was forcing gastric tubes into The Goose (another procedure to which the bird objected strenuously) with the idea of testing the contents of its alimentary canal. It was one of our routine searches for exogenous gold.

Gold *was* found, but only in traces, and there was every reason to suppose those traces had accompanied the digestive secretions and were therefore endogenous (from within, that is) in origin.

However, something else showed up, or the lack of it, anyway.

I was there when Nevis came into Finley's office in the temporary building we had put up overnight (almost) near the goosepen.

Nevis said, 'The Goose is low in bile pigment. Duodenal contents show about none.'

Finley frowned and said, 'Liver function is probably knocked loop-the-loop because of its gold concentration. It probably isn't secreting bile at all.'

'It *is* secreting bile,' said Nevis. 'Bile acids are present in normal quantity. Near normal, anyway. It's just the bile pigments that are missing. I did a fecal analysis and that was confirmed. No bile pigments.'

Let me explain something at this point. Bile acids are steroids secreted by the liver into the bile and *via* that are poured into the upper end of the small intestine. These bile acids are detergentlike molecules which help to emulsify the fat in our diet (or The Goose's) and distribute them in the form of tiny bubbles through the watery intestinal contents. This distribution, or homogenization, if you'd rather, makes it easier for the fat to be digested.

Bile pigments, the substance that was missing in The Goose, are something entirely different. The liver makes them out of hemoglobin, the red oxygen-carrying protein of the blood. Worn-out hemoglobin is broken up in the liver, the heme part being split away. The heme is made up of a ringlike molecule (called a 'porphyrin') with an iron atom in the center. The liver takes the iron out and stores it for future use, then breaks the ringlike molecule that is left. This broken porphyrin is bile pigment. It is colored brownish or greenish (depending on further chemical changes) and is secreted into the bile.

The bile pigments are of no use to the body. They are poured into the bile as waste products. They pass through the intestines and come out with the feces. In fact, the bile pigments are responsible for the color of the feces.

Finley's eyes began to glitter.

Nevis said, 'It looks as though porphyrin catabolism isn't following the proper course in the liver. Doesn't it to you?'

It surely did. To me too.

There was tremendous excitement after that. This was the first metabolic abnormality, not directly involving gold, that had been found in The Goose!

We took a liver biopsy (which means we punched a cylindrical sliver out of The Goose, reaching down into the liver). It hurt The Goose but didn't harm it. We took more blood samples too.

This time we isolated hemoglobin from the blood and small quantities of the cytochromes from our liver samples. (The cytochromes are oxidizing enzymes that also contain heme.) We separated out the heme, and in acid solution some of it precipitated in the form of a brilliant orange substance. By August 22, 1955, we had five micrograms of the compound.

The orange compound was similar to heme, but it was not heme. The iron in heme can be in the form of a doubly charged ferrous ion (Fe^{++}) or a triply charged ferric ion (Fe^{+++}), in which latter case, the compound is called hematin. (Ferrous and ferric, by the way, come from the Latin word for iron, which is 'ferrum.')

The orange compound we separated from heme had the porphyrin portion of the molecule all right, but the metal in the center was gold – to be specific, a triply charged auric ion (Au^{+++}). We called this compound 'aureme,' which is simply short for 'auric heme.'

Aureme was the first naturally occurring, gold-containing organic compound ever discovered. Ordinarily it would rate headline news in the world of biochemistry. But now it was nothing; nothing at all in comparison to the further horizons its mere existence opened up.

The liver, it seemed, was not breaking up the heme to bile pigment. Instead, it was converting it to aureme; it was replacing iron with gold. The aureme, in equilibrium with chloraurate ion, entered the bloodstream and was carried to the ovaries, where the gold was separated out and the porphyrin portion of the molecule disposed of by some as yet unidentified mechanism.

Further analyses showed that 29 percent of the gold in the blood of The Goose was carried in the plasma in the form of chloraurate ion. The remaining 71 percent was carried in the red blood corpuscles in the form of 'auremoglobin.' An attempt was made to feed The Goose traces of radioactive gold so that we could pick up radioactivity in plasma and corpuscles and see how readily the auremoglobin molecules were handled in the ovaries. It seemed to us the auremoglobin should be much more slowly disposed of than the dissolved chloraurate ion in the plasma.

The experiment failed, however, since we detected no radioactivity. We put it down to inexperience since none of us were isotopes men, and that was too bad since the failure was highly significant, really, and by not realizing it, we lost several days.

The auremoglobin was, of course, useless as far as carrying oxygen was concerned, but it only made up about 0.1 percent of the total hemoglobin of the red blood cells so there was no interference with the respiration of The Goose.

105

Haem-pigments also enter transiently into Arthur Conan Doyle's story, 'A Physiologist's Wife'. Conan Doyle had been a doctor and had an excellent background in biology.

His sister sighed.

'You have no faith,' she said.

'I have faith in those great evolutionary forces which are leading the human race to some unknown but elevated goal.'

'You believe in nothing.'

'On the contrary, my dear Ada, I believe in the differentiation of protoplasm.'

She shook her head sadly. It was the one subject upon which she ventured to dispute her brother's infallibility.

'This is rather beside the question,' remarked the Professor, folding up his napkin. 'If I am not mistaken, there is some possibility of another matrimonial event occurring in the family. Eh, Ada? What!'

His small eyes glittered with sly facetiousness as he shot a twinkle at his sister. She sat very stiff, and traced patterns upon the cloth with the sugar-tongs.

'Dr. James M'Murdo O'Brien –' said the Professor sonorously.

'Don't, John, don't!' cried Miss Ainslie Grey.

'Dr. James M'Murdo O'Brien,' continued her brother inexorably, 'is a man who has already made his mark upon the science of the day. He is my first and my most distinguished pupil. I assure you, Ada, that his "Remarks upon the Bile-Pigments, with special reference to Urobilin," is likely to live as a classic. It is not too much to say that he has revolutionised our view about urobilin.'

He paused, but his sister sat silent, with bent head and flushed cheeks. The little ebony cross rose and fell with her hurried breathings.

'Dr. James M'Murdo O'Brien has, as you know, the offer of the physiological chair at Melbourne. He has been in Australia five years, and has a brilliant future before him. To-day he leaves us for Edinburgh, and in two months' time he goes out to take over his new duties. You know his feeling towards you. It rests with you as to whether he goes out alone. Speaking for myself, I cannot imagine any higher mission for a woman of culture than to go through life in the company of a man who is capable of such a research as that which Dr. James M'Murdo O'Brien has brought to a successful conclusion.'

'He has not spoken to me,' murmured the lady.

'Ah, there are signs which are more subtle than speech,' said her brother, wagging his head. 'You are pale. Your vasomotor system is excited. Your arterioles have contracted. Let me entreat you to compose yourself. I think I hear the carriage. I fancy that you may have a visitor this morning, Ada. You will excuse me now.'

With a quick glance at the clock he strode off into the hall, and within a few minutes he was rattling in his quiet, well-appointed brougham through the bricklined streets of Birchespool.

His lecture over, Professor Ainslie Grey paid a visit to his laboratory, where he adjusted several scientific instruments, made a note as to the progress of three separate infusions of bacteria, cut half a dozen sections with a microtome, and finally resolved the difficulties of seven different gentlemen, who were pursuing researches in as many separate lines of inquiry. Having thus conscientiously and methodically completed the routine of his duties, he returned to his carriage and ordered the coachman to drive him to The Lindens. His face as he drove was cold and impassive, but he drew his fingers from time to time down his prominent chin with a jerky, twitchy movement. . . .

'I trust that you have been successful, O'Brien,' said he. 'I should be loath to exercise any undue pressure upon my sister Ada; but I have given her to understand that there is no one whom I should prefer for a brother-in-law to my most brilliant scholar, the author of "Some Remarks upon the Bile-Pigments, with special reference to Urobilin." '

'You are very kind, Professor Grey – you have always been very kind,' said the other. 'I approached Miss Grey upon the subject; she did not say No.'

'She said Yes, then? '

'No; she proposed to leave the matter open until my return from Edinburgh. I go today, as you know, and I hope to commence my research tomorrow.'

'On the comparative anatomy of the vermiform appendix, by James M'Murdo O'Brien,' said the Professor sonorously. 'It is a glorious subject – a subject which lies at the very root of evolutionary philosophy.'

'Ah, she is the dearest girl,' cried O'Brien, with a sudden little spurt of Celtic enthusiasm – 'she is the soul of truth and of honour.'

'The vermiform appendix –' began the Professor.

'She is an angel from heaven,' interrupted the other. 'I fear that it is my advocacy of scientific freedom in religious thought which stands in my way with her.'

'You must not truckle upon that point. You must be true to your convictions; let there be no compromise there.'

'My reason is true to agnosticism, and yet I am conscious of a void – a vacuum.'

106

There follows a tale 'Von Kempelen and His Discovery', with a related theme by Edgar Allan Poe, who published a series of scientific short stories, densely written and formidably well-informed, according to the wisdom of the time. Von Kempelen has come upon a diary of Sir Humphry Davy's and there found a momentous observation.

Von Kempelen had never been even tolerably well off during his residence at Bremen; and often, it was well known, he had been put to extreme shifts, in order to raise trifling sums. When the great excitement occurred about the forgery on the house of Gutsmuth & Co., suspicion was directed towards Von Kempelen, on account of his having purchased a considerable property in Gasperitch Lane, and his refusing, when questioned, to explain how he became possessed of the purchase money. He was at length arrested, but nothing decisive appearing against him, was in the end set at liberty. The police, however, kept a strict watch upon his movements, and thus discovered that he left home frequently, taking always the same road, and invariably giving his watchers the slip in the neighborhood of that labyrinth of narrow and crooked passages known by the flash-name of the 'Dondergat'. Finally, by dint of great perseverance, they traced him to a garret in an old house of seven stories, in an alley called Flatzplatz; and, coming upon him suddenly, found him, as they imagined, in the midst of his counterfeiting operations. His agitation is represented as so excessive that the officers had not the slightest doubt of his guilt. After hand-cuffing him, they searched his room, or rather rooms; for it appears he occupied all the *mansarde*.

Opening into the garret where they caught him, was a closet, ten feet by eight, fitted up with some chemical apparatus, of which the object has not yet been ascertained. In one corner of the closet was a very small furnace, with a glowing fire in it, and on the fire a kind of duplicate crucible – two crucibles connected by a tube. One of these crucibles was nearly full of *lead* in a state of fusion, but not reaching up to the aperture of the tube, which was close to the brim. The other crucible had some liquid in it, which, as the officers entered, seemed to be furiously dissipating in vapor. They relate that, on finding himself taken, Von Kempelen seized the crucibles with both hands (which were encased in gloves that afterwards turned out to be asbetic), and threw the contents on the tiled floor. It was now that they hand-cuffed him; and, before proceeding to ransack the premises, they searched his person, but nothing unusual was found about him, excepting a paper parcel, in his coat pocket, containing what was afterwards ascertained to be a mixture of antimony and some *unknown substance*, in nearly, but not quite, equal proportions. All attempts at analyzing the unknown substance have, so far, failed, but that it will ultimately be analyzed, is not to be doubted.

Passing out of the closet with their prisoner, the officers went through a sort of ante-chamber, in which nothing material was found, to the chemist's sleeping-room. They here rummaged some drawers and boxes, but discovered only a few papers, of no importance, and some good coin, silver and gold. At length, looking under the bed, they saw *a large, common hair trunk, without hinges, hasp, or lock*, and with the top lying carelessly *across* the bottom portion. Upon attempting to draw this trunk out from under the bed, they found that, with their united strength (there were three of them, all powerful men), they 'could not stir it one inch'. Much astonished at this, one of them crawled under the bed, and looking into the trunk, said:

'No wonder we couldn't move it – why, it's full to the brim of old bits of brass!'

Putting his feet, now, against the wall, so as to get a good purchase, and pushing with all his force, while his companions pulled with all theirs, the trunk, with much difficulty, was slid out from under the bed, and its contents examined. The supposed brass with which it was filled was all in small, smooth pieces, varying from the size of a pea to that of a dollar; but the pieces were irregular in shape, although all more or less flat – looking, upon the whole, 'very much as lead looks when thrown upon the ground in a molten state, and there suffered to grow cool'. Now, not one of these officers for a moment suspected this metal to be anything *but* brass. The idea of its being *gold* never entered their brains, of course; how *could* such a wild fancy have entered it? And their astonishment may be well conceived, when next day it became known, all over Bremen, that the 'lot of brass' which they had carted so contemptuously to the police office, without putting themselves to the trouble of pocketing the smallest scrap, was not only gold – real gold – but gold far finer than any employed in coinage – gold, in fact, absolutely pure, virgin, without the slightest appreciable alloy!

I need not go over the details of Von Kempelen's confession (as far as it went) and release, for these are familiar to the public. That he has actually realized, in spirit and in effect, if not to the letter, the old chimera of the philosopher's stone, no sane person is at liberty to doubt. The opinions of Arago are, of course, entitled to the greatest consideration; but he is by no means infallible; and what he says of *bismuth*, in his report to the academy, must be taken *cum grano salis*. The simple truth is, that up to this period, *all* analysis has failed; and until Von Kempelen chooses to let us have the key to his own published enigma, it is more than probable that the matter will remain, for years, *in statu quo*. All that yet can fairly be said to be known, is, that *'pure gold can be made at will, and very readily, from lead, in connection with certain other substances, in kind and in proportions, unknown'*.

Speculation, of course, is busy as to the immediate and ultimate results of this discovery – a discovery which few thinking persons will hesitate in referring to an increased interest in the matter of gold generally, by the late developments in California; and this reflection brings us inevitably to another – the exceeding *inopportuneness* of Von Kempelen's analysis. If many were prevented from adventuring to California, by the mere apprehension that gold would so materially diminish in value, on account of its plentifulness in the mines there, as to render the speculation of going so far in search of it a doubtful one – what impression will be wrought *now*, upon the minds of those about to emigrate, and especially upon the minds of those actually in the mineral region, by the announcement of this astounding discovery of Von Kempelen? a discovery which declares, in so many words, that beyond its intrinsic worth for manufacturing purposes, (whatever that worth may be), gold now is, or at least soon will be (for it cannot be supposed that Von Kempelen can *long* retain his secret) of no greater *value* than lead, and of far inferior value to silver. It is, indeed, exceedingly difficult to speculate prospectively upon the consequences of the discovery; but one thing may be positively maintained – that the announcement of the discovery six months ago, would have had material influence in regard to the settlement of California.

In Europe, as yet, the most noticeable results have been a rise of two hundred per cent in the price of lead, and nearly twenty-five per cent in that of silver.

107

And here is Edgar Allan Poe, the poet:

TO SCIENCE

Science! true daughter of Old Time thou art!
  Who alterest all things with thy peering eyes.
Why preyest thou thus upon the poet's heart,
  Vulture, whose wings are dull realities?
How should he love thee? or how deem thee wise,
  Who wouldst not leave him in his wandering
To seek for treasure in the jewelled skies,
  Albeit he soared with an undaunted wing?
Hast thou not dragged Diana from her car?
  And driven the Hamadryad from the wood
To seek a shelter in some happier star?
  Hast thou not torn the Naiad from her flood,
The Elfin from the green grass, and from me
The summer dream beneath the tamarind tree?

'The Gold-Makers' was J.B.S. Haldane's only short story. It is full of authentic detail, including some of what he had learned in the Army during the First World War. The notion developed in Haldane's story, that gold could be economically extracted from sea-water, is by no means far-fetched, and may have been inspired by Fritz Haber's efforts in this direction, aimed at paying off at a stroke the war debt imposed on Germany in 1919 (see no. 174, introduction).

I am more shameless than my colleagues about some things. I don't believe I know French better than most of them. But I don't mind talking it at a great rate without too meticulous a regard for genders. So every now and then I take a perfectly good holiday by giving a course of lectures in alleged French in Paris. Everyone is pleased by this arrangement. My university feels it is doing something for international co-operation, I manage to tell my French colleagues some things they don't know, and I learn any number of things I don't know myself. It's no good going to Germany, because the Germans read everything that is printed anywhere, and publish all they have done, and a bit more, at immense length. The French remain beautifully oblivious to a lot of work done outside France until everyone says French science is going to the dogs. Then it turns out that some perfectly obscure French man or woman has just discovered something really original and unlikely, such as radioactivity, or wave mechanics, which makes Einstein seem as simple as Rule of Three, and incidentally landed Eugène Galois in Devil's Island, and me (I sincerely hope) in the local gaol at Ambert.

I shouldn't have put in all this preface if I was publishing this narrative in the *Chemical Gazette* or the *British Journal of Physics*, as I originally thought of doing. But readers of this magazine might wonder what I was doing in the Rue Cujas at 11 P.M. on June 28, 1930, and why the man with no front teeth should have known who I am, and that I actually understand something about the application of wave mechanics to chemistry. The streets round the Sorbonne are placarded with lecture announcements, and my portrait had appeared in *L'Illustration* with a highly misleading biography. On the date in question I had just delivered my sixth and last lecture, and subsequently consumed a considerable quantity of very light beer with some French colleagues at the Café Soufflet, which is at the corner of the Boulevard St. Michel and the Rue des Écoles. I had sat there with my eminently respectable colleague Henriot and his wife, whilst one of his most brilliant pupils played backgammon at a neighboring table with a little lady whose calling was not in doubt. I had reflected on the improbability of such a scene at Oxford or Cambridge, yet remembered a not utterly dissimilar occasion at the Cosmopolitan Club off Leith Walk in Edinburgh. But the Professor's wife had not been there.

The man with no front teeth was remarkably shabbily dressed, and looked hungry, which is unusual in Paris, where there is work for almost everybody today. He sidled up to me, and in a voice which was not rendered more intelligible by the absence of his teeth, said, 'For the love of Science read this, and if you want more, follow me.' He slunk on ahead of me and waited in the shadow of a doorway while I stood under a lamp and looked at the paper he had given me. It was the first part of the wave equation for carbon, or, rather, of the set of forty-two simultaneous differential equations which would enable one to predict the behavior of that element if one could solve them. But it was expressed in a notation new to me, and certainly unpublished. Now a beggar or a tout for some unsavory concern might conceivably have copied out some of Kultchagin's equations to act as ground bait for me, but he could not possibly have transposed them like that. Imagine a man handing you a copy of *A Shropshire Lad* translated faultlessly into Icelandic, and then written out in Egyptian hieroglyphics, and you can get an idea of the intellectual effort involved and the special knowledge needed. This was something really queer, and I am a student of the really queer in physical chemistry, but do not despise it when I meet it elsewhere. I followed him. 'Bar du Progrès, Porte de la Villette, minuit,' he whispered. He was obviously in very considerable terror, and motioned me to go on.

I had an hour before midnight. I felt that I might be in for something odd, and after making sure that I was not followed, I went into the Café d'Harcourt and over a coffee wrote a note to my friend Bertaux, giving him the facts and asking him to ring me up at my hotel at noon next day, and to inform the police if I were missing. Then I boarded the Metro for the Porte de la Villette, an exit from Paris which so far was only known to me by its proximity on the map to the municipal slaughterhouse. I was not as calm as I could have wished, for as I entered the train I found that I had just lit a threepenny cigar in oblivion of the fact that there are no smoking compartments on the Metro, a fact which I have always resented most keenly.

The Bar du Progrès is dim, but not really sinister. It is extremely like some thousands of other bars. There is the same fat lady behind the same zinc counter, the same surprising variety of bottles behind her, the same rather consumptive-looking waiter. At the back there is a table in a recess, with two chairs. One occupant of the table can be seen from the door. The other is screened. I went in just before midnight, ordered a café cognac, and sat down facing the door. The only other customer was an inoffensive-looking lorry driver who was describing in considerable detail a collision in which he had, of course, been the innocent party, but which had detained him beyond his usual bedtime. On the stroke of midnight my friend of the Rue Cujas came in and without a word sat down in the seat opposite me. I ordered him another café cognac, and repeated the latter at suitable intervals during the next hour. I observed that the missing teeth were only one effect of what must have been a thoroughly nasty wound in the face. But the scars were old; it looked like a war wound. He spoke in a low voice for the best part of an hour in rapid and not easily intelligible French. Occasionally, at critical places in the story, he put in certain key words in English. Once or twice he made a scientific point in German. He was obviously suffering from extreme terror, but it was not the terror of the raw recruit during his first heavy shelling. It was the much grimmer emotion of the old soldier who realizes that there is a definite limit to human endurance, the terror of 1918. This is roughly what he told me. I don't think my recollection contains any serious errors on matters of fact.

'You have heard what happened to Eugène Galois?'

'I know he was found guilty of murder and sent to Devil's Island. But I can't believe he murdered a colleague for money. He's as big a mathematician as his namesake was a century ago. He might have committed a *crime passionnel*. Anyone with guts might do that. But you can't murder for gain unless your mind is obsessed by money, and his mind was too full of loxodromic groups to leave room for that sort of obsession. I hear they're trying to get his case retried. If I can do anything in reason to help, I will.'

'I'm glad you feel like that about Galois,' said my neighbor, 'but it's too late. He died last month of parrot disease. The convicts were allowed to keep pets, and there was an epidemic. He was a martyr. I am only talking to you because he is dead. He was on to the biggest thing since the invention of the steam engine. He was murdered because he knew too much. If you listen to me, you may make world history. You may quite possibly become the richest man on the planet. But you are also likely to be murdered. Indeed, if you have been seen with me you probably will be. But if you're afraid, you'd better clear out at once.'

I don't mind admitting that I was afraid. But since November 11, 1918, my adventures had been intellectual and emotional only. Moreover, I am ambitious. I fell a victim to my really lamentable propensity for quotation, and reminded myself,

> He either fears his fate too much, or his deserts are small,
> Who dares not put it to the touch, to gain or lose it all.

'Go on,' I said. I also repeated under my breath a line from a less reputable poem, which I had found consolatory on unpleasant occasions during the war, to the effect that, whatever happened, I should be 'damnable mouldy a hundred years hence.' He went on.

'Galois was a man of genius. You know that. But you don't perhaps realize how broad his interests were. He felt very deeply that the evils of the present day were due to the application of science by unscientific men. "We have given humanity a large degree of control over matter, and they have given us modern war and modern industry," he used to say. So he determined to apply his science according to his own ideas, not those of financiers. He had a special down on financiers. He realized that wave mechanics meant a new era in chemistry. When he heard that Eucken and Bonhöfer had proved hydrogen to be a mixture, he said it was only the beginning. He had some private means, and after his last published paper he went off to the country, and worked out the wave equations for the gold atom. You will realize the stupendous nature of that. A man with the mathematical ability to do it could have determined the orbit of the new planet Pluto in one evening in a café with the band playing. He bought a cottage in the country and had the walls papered white. He went round the different rooms with a stepladder, covering the paper with calculations. Of course he filled masses of notebooks too. But he said he needed the walls for the main results, and by writing them up in that way he knew how to find what he wanted. He worked eight hours a day for a year and a half, and at the end of that time he had his principal results in a single notebook. I have seen it, but you will soon hear why I haven't got it. In another half-year he had worked out that gold must have an enormous and quite unsuspected affinity for a certain group of organic compounds. Then he got hold of Riquier, an organic chemist who had been with him at the École Normale, and Riquier made one of the compounds in question. They showed that their scheme would work on a laboratory scale. Then they approached me.

'My name is Martin. That is irrelevant. I do not think that I shall live long. I was a works chemist at Nanterre and a friend of Riquier's. We went down together to Ste. Leocadie, a little village on the sea coast, in Bouches du Rhône, near Aigues Mortes, where there is a large lagoon. We started a salt pan. I don't know that the salt was particularly good, but we managed to sell it, anyway. That, however, wasn't what we were after. You know there is gold in sea water. Not very much, about one part in twenty million. When you evaporate the water in a salt pan, most of the salt crystallizes out, and you are left with a sticky solution full of Epsom salts and what not. Almost all the gold is in there, so it is easy to concentrate it a hundred times in the sunlight of southern France. You can crystallize out most of the rest of the salts without losing much gold. The brine left behind has about one part in 200,000 of gold. That's a lot. Gravel with only one part in a million has been worked profitably, and even on the Rand the quartz only averages twelve parts in a million of gold. You take your residual solution and add about eight parts in a million of Riquier's compound, which we call auron. I don't know what it is, but it is bright blue, and it is made from a saponine, and I believe has two pyrrole rings in its molecule. You leave the mixture for an hour, and then bubble air through it. The blue stuff has combined with the gold to make a red compound. This is a surface active substance, and it all comes into the froth which you blow off the top. You dry the froth, add a little acid, and out comes the gold. You can use the blue stuff again, but we used to lose about 5 per cent at each operation.

'My job was to run the bubbling tanks. Riquier made the stuff, and Galois saw to sales and purchases. There were some local men, chosen for their stupidity, who looked after the salt pans, and I acted as foreman there. We started in January 1929, but it wasn't till May that we got the process working perfectly, and from May to September we got out about four million francs' worth of gold, £30,000, or a little more. Most of that went in paying off our debts, but we had a million or so of clear profit. Of course before the show started we had decided what to do with the money. We were all somewhat idealistic. You have to be idealistic to go in for science in France today, when a professor gets £300 a year or so as the reward of a distinguished career. Our immediate idea was to go straight ahead until we had a thousand million francs, and then to start endowing science as it ought to be endowed, so that a good scientific worker was paid as well as a good engineer or surgeon, and a reasonable amount was available for apparatus. Naturally, we thought most of France, Belgium, and Italy, where scientific workers are worst paid. But we hadn't forgotten Germany, and we had a few schemes even for England and America. Well, that's all over! If you succeed where we failed, don't forget French science.'

'I'm not likely to,' I said.

'We reckoned to make some hundred millions of francs without exciting much notice, but obviously the thing couldn't go on indefinitely. But here Galois had his plans. He believed that the world was not producing gold quickly enough. If humanity increases its stock of gold more slowly than its other material wealth, prices will fall and you will get unemployment. That is what is happening now. If we made gold too quickly, say thirty thousand million francs a year, prices would rise, and all the world would be like France and Germany after the war. Galois's idea was to produce gold just fast enough to keep prices steady on an average. "The thing will be too big for any one man," said Galois, "and if I gave it to the French Government the country would be flooded with gold, and our agriculture and manufactures would die like those of Spain did after the conquest of Mexico and Peru. No, we'll give France enough to pay her foreign debts, but the secret, and the control of the thing, must go to the League of Nations, and the day they get it America and the Soviet will join up."

'Well, everything went swimmingly till the end of last August. Then I got a typewritten document from Paris. There was no address, but it was headed Association Internationale pour la Défense des Interêts Rentiers. It ran roughly like this:

DEAR SIR – As it is possible that in future the operations on which you are engaged will incommode us, I have the honor to offer you an income of two hundred thousand francs per year should you abandon them. Your colleages have also been approached. In the event of your resigning your occupation, you will receive your first quarter's salary within one week, the notes being despatched to your mother's house. In earnest of our good intentions we enclose 10,000 francs.

Should the offer not be accepted within one week from today, we shall be compelled to take steps to eliminate the concern in which you are a partner.

A.I.D.I.R.

'I was impressed by the ten thousand francs, but still more by the fact that when I looked at the letter three days later the paper had crumbled to powder. As a chemist I can imagine how this might be done, but it would take some working out. For that showed that our enemies had skill and knowledge behind them, as well as money. The fact that they had destroyed this evidence meant that there was probably something in their threats.

'I talked it over with my colleagues. They had had similar letters. Unfortunately they were against warning the police, as they didn't want to give away what we were doing. Galois thought the A.I.D.I.R. might be what it purported to be, a representative of a financial group interested in fixed-interest-bearing securities, which would of course fall in value if we flooded the world with gold, while ordinary shares and equities would rise. Riquier and I, rightly as it turned out, believed that they really stood for a gold-mining group.

'I never found out how they discovered our secret. Galois and I used to take our gold by car to a bank at Cette. Someone connected with that may have got suspicious and tracked us. Possibly Riquier may have talked too much to a lady who I think was his mistress. But I doubt it.

'Next week I was rung up on the telephone. The voice said, "A.I.D.I.R. speaking. Our offer is and remains open. We are even prepared to raise it if you state your terms in the advertisement column of the *Petit Nimois*. If you do not accept, you will all be killed. This is our last communication."

'I found later that the call had been made at a public call office in Nîmes. We agreed to take no notice, but started a scheme of defences. We all had automatic pistols, and Riquier made us a supply of lachrymatory gas bombs. The factory was easily defensible, and we had burglar alarms, and a couple of fairly excitable dogs. The other two were enthusiasts, and I am not much afraid of death. As you see, my face got fairly smashed up in the war.'

He lifted his rather long and dirty hair, and I noticed that, besides the damage to his mouth, he had no left ear.

'At that time I had some false teeth, which I have just pawned in order to live. But I have been in constant pain since 1916, so I do not find life immensely attractive, even when I am not being hunted.

'Towards the end of September I developed a boil on my neck, and had to go into a clinic for two days to have it lanced. While I was away, the blow fell. Riquier was found shot outside the factory door. Two Swiss tourists swore that they had seen Galois shoot him after a quarrel. The bullets fitted a pistol which a Marseilles gunsmith swore Galois had bought from him, and which was found near the body. Several other witnesses turned up later, and swore to the most incredible lies, which hung together to make a pretty damning story. Almost simultaneously an alleged Chilian millionaire called Fernandez sued Galois for six million francs which he claimed to have lent him for a scheme for extracting gold from sea water. Apparently he had a large outfit of forged documents. As you know, the jury found that Galois was a swindler and a murderer. I got hold of his lawyer, and offered to give evidence, but he thought it would be useless, and I lay low. But the A.I.D.I.R. people found me. As I was coming home one night I was attacked by three men. I didn't want to shoot and get jailed, as I probably would have been. I managed to burst a lachrymatory bomb among them, and left them weeping. But I have been on the run ever since. Meanwhile Fernandez was able to seize the factory for debt, and presumably got our documents and about a kilogram of auron.

'Even after his condemnation Galois's lawyer believed in him. He found out some odd facts. He established that both Fernandez and one of the supposed Swiss were connected with the same gold-mining group.'

That was the end of our conversation, because at this point I noticed a bomb coming towards me through the air. Up till that moment I had refused to decide between two alternatives – that M. Martin's story was true, or that he was a very good liar. I had a strong suspicion that he would shortly ask me to lend him a hundred francs. I felt that I had had my money's worth, and proposed to lend him a hundred and fifty, for good lying is a rare gift which should be encouraged.

The bomb, however, convinced me that he had been speaking the truth. But its immediate effect was to jerk me back for thirteen years into the past. It was very unfortunate indeed for the throwers of this bomb that hand grenades had been my special line during the

Great War. I am one of the few people who ran a bombing school for nine months without casualties. Among the things which we occasionally did as demonstrations was to catch lighted bombs and throw them back, or more accurately, sideways, out of the trench. I had a one-eyed and rarely quite sober corporal who used to do this, but I sometimes did it myself. I admit that we used to lengthen the time fuse beforehand. Provided you are a good judge of time, it is no more dangerous than crossing the road among motor traffic, but it is more impressive to onlookers. Some idiot asked questions about it in Parliament, and got an army order issued forbidding the practice.

This bomb was a 'stick bomb' with a long wooden handle. I think it was a German type with a five-seconds time fuse. Looking past the bomb through the door of the café, I saw two men in a car, one at the wheel, and one who had clearly just thrown it. The car was moving slowly. I reckoned that the charge would explode in another three seconds. As the bomb, which was thrown with a very good aim, landed in my coffee cup, I caught it by the handle, and ran towards the door, swinging it as I did so. The man standing in the car was expecting me to run out, so he fired at me. But he was not expecting me to return the bomb, so he fired very erratically. One bullet went through my raincoat. Another, as I afterwards learnt, hit the lady behind the counter, but not fatally. As I reached the door, I pitched the bomb neatly into the car, which was now accelerating, and threw myself flat on my face with more speed than elegance, as I had been accustomed to do when a machine gun opened fire on me.

The bomb burst as I reached the ground. The man with the revolver was jumping from the car as it did so. A piece wounded him, and he fell on the pavement. The driver could not escape, and the explosion lifted him into the air. His body, oddly twisted, fell back into the car as it struck a lamp post and burst into flames. I got up and ran past the blazing car. As I passed the man on the pavement, I kicked his head as hard as I could. Some bone in it broke with a crack. I ran my fastest down a side street, dodged round several corners, and was violently sick. The burning car lit up the sky behind me as I walked with deliberate slowness on to the Rue de Flandres. I saw no sign of Martin, and nobody followed me. I was fortunate enough to catch, within a very few minutes, an omnibus going to the Châtelet, one of those which run at hourly intervals throughout the night. I got off just before the terminus, and walked to my hotel by a round-about route.

I am fairly sure that I was not followed, for on several occasions I turned corners when no one was in sight. I reflected for a short time on my adventures, and on the fact that I had not paid for the coffee. I regretted this, for I am rather scrupulous in small matters. I also hoped that I had killed the man on the pavement, or at least given him severe enough concussion to blot out his memory of recent events. In that case there would probably be no one who had seen me with Martin, and I stood a chance of becoming the modern equivalent of Midas. Meanwhile, however, there was nothing to be done, and rather unexpectedly I fell asleep within half an hour of getting to bed.

I did not wake up next morning until about ten. I came to the conclusion that I had better clear out for England at once. As soon as I had dressed I telephoned to Bertaux that I was all right, but had to leave Paris. I told him to keep his mouth shut, and to lunch with me at one o'clock. I then took a bus to my bank in the Place de la Concorde, and drew out two thousand francs for my hotel bill and railway ticket. There had been no account of last night's affair in the morning paper, but when I left the bank I bought a *Paris-Midi,* which devoted a column to the outrage of the Bar du Progrès. The man in the car was dead, although a baker had burned his hands badly in an attempt to rescue him. Another man, presumably the one with the revolver, was in the hospital with a wound in the shoulder and a fractured jaw. So was the proprietress. The waiter had seen me throwing a bomb, but apparently no one had seen the bomb coming in! Fortunately he gave a very vague description of me, and had overheard Martin and me talking in German, but not in English. He also said that I spoke French with a German accent, in which he was not so far out. Being a Scotsman, I do not talk it with an English accent. If I go into a café full of young ladies eager to make my acquaintance, it is often simplest to keep off the rest by standing one of them a drink. In such a case I always ask her to guess my nationality. She then suggests Dutch, Danish, Polish, and Czechoslovak. I assume that I am really taken for a German, but that the young lady, being too polite to make such a suggestion, names the various neighbouring states.

I went to the Gare du Nord for a ticket, and for a reservation on the four o'clock train. As I left the station I noticed a man with a black moustache in a bowler hat looking at me rather intently. I got into a taxi and ordered the driver to go to the restaurant where

Bertaux was expecting me. I gave the directions in a fairly low voice, so I hardly think the watcher could have overheard me. He got into another taxi, and I noticed it following us. I did not wish to involve Bertaux in my little troubles. I also wanted to be perfectly sure that I was being followed, and, if possible, to shake the man off. So instead of going to my restaurant near the Sorbonne, I told the driver to go to the Gare du Luxembourg.

I can just remember the time when the London Metropolitan was worked by steam-engines. Those of my readers who regret those romantic days are advised to travel by the Chemin de Fer de Sceaux et Limours, a suburban railway line which starts from the Gare du Luxembourg and leaves Paris by an extremely long tunnel which is always full of a particularly suffocating smoke. I bought a ticket for Massy-Verrières, an undistinguished station on that undistinguished line. My pursuer followed me. I went down and got into the fullest compartment I could find. He got in close behind me. At the first stop, the Gare de Port Royal, still in Paris, I dashed out through the smoke and up the stairs. My unknown friend followed. But here I had a stroke of luck. Just outside the station there was one – and only one – taxi. I boarded it and told the man to go to the Institut Pasteur. I crouched to offer as small a target as possible if my unknown friend opened fire. Fortunately he did not. As we turned the first corner I looked back and saw him running after me. I do not know when he finally lost sight of me, but, as I went down to the Metro station in the Boulevard Pasteur, to which I had diverted my vehicle, I saw a taxi driving up at a rather dangerous speed. Four different lines, however, leave that station, and if he was still chasing me, he must have taken the wrong train.

After several changes on the Metro, I arrived rather late at my restaurant. Bertaux was waiting, but had nearly finished lunch. I told him I was being followed by would-be murderers, and was going to run for it. As I refused to take his advice and go to the police, I think that he suspected my pursuer of being an irate husband. That is the worst of these romantics. I had no intention of going back to my hotel, for if my identity was known it might be watched. I asked Bertaux to go for a small suitcase which I had already packed with my more essential clothes and shaving things. He was to say I would call round for my trunk later. I had just finished lunch when he came back. I determined to run in the opposite direction from England, in the hope that my enemies were only watching the west-bound trains, so I went straight to the Gare de l'Est and bought a ticket for Berne, hoping to work round from there to one of the German airports and fly home.

Even as I bought my ticket I noticed, or seemed to notice, a man who looked at me closely and then went to a telephone booth. I did not see him again after this, and my journey was uneventful as far as Belfort. But in the *Soir*, which I had bought, I found two interesting items. The man with the broken jaw had given his name. It was presumably false, but this meant that he could speak. On the same page was a not uncommon headline, *'Un Inconnu se suicide'* – 'Suicide of Unknown Man.' The man in question, who had been found hanged from the scaffolding of a suburban cinema under construction, had no left ear, and his front teeth were missing. Nothing was found on his person but a crude scrawl stating that he was fed up. So they had got Martin. Possibly they had forced my name out of him before his murder. In any case, I proposed to do my best to escape his fate.

It was night when I got to Belfort. As the train ran into the station I saw on the platform my friend who had chased me earlier in the day, with two other men. Presumably he at least had flown to Belfort.

I may well have done him an injustice, but the events of the past twenty hours had somewhat prejudiced me against him, and, although I was not certain whether he intended to murder me or merely to hand me over to the police on a charge of bombing his friends, I did not feel called upon to put his intentions to an experimental test. I am sure he had not seen me when I rapidly left the compartment, and before the train had drawn up, bolted myself into a lavatory, where I remained. I had taken my handbag, as there was still a chance that my name was not known to my hunters.

Five minutes after leaving Belfort I had some luck. If I had not, I should not be alive and writing this account. The train drew up, as I had hoped it would, and clearly not in a station, as there were no lights. I dashed out of my retreat, opened a door, and jumped out of the train on the left-hand side. As I did so, the train began to move. Someone fired two or three shots at me. Presumably they missed because the train was moving and I was already fifteen yards away. Then a train going towards Paris, for which we had no doubt been waiting, cut in between me and the gunman. I did not stop running, and soon reached a road, where my luck still held. A lorry was waiting at a level crossing for the trains to pass. I asked for, and got, a lift. It appeared that the lorry was going from Strasbourg to Lons-le-Saunier with the household effects of a subprefect. For a hundred francs the driver was willing to take me to Besançon, where he was stopping for the night. I suspected that I might be chased, so I took him into my confidence, or, more accurately, halfway in. Having received a sound classical education, I remembered Odysseus, and the advantage of economy of truth when in tight corners. I judged the driver to be romantic in the worst sense of the word. Sailors, it is said, have a wife in every port, but ports are confined to sea coasts and large rivers and canals, whereas lorries can visit any town in a civilized country. Lorry drivers of a polygamous disposition are thus peculiarly favoured by their professional duties.

I therefore informed him that I had formed a romantic attachment for the *poule* of a millionaire, and persuaded her to flee to Switzerland. The irate lover had pursued me while I was attempting to join her, and had tried to have me murdered in the train. As it was likely that he, or one of his myrmidons, would chase me in a car, I desired to hide among the furniture of the subprefect. Another fifty francs secured me a place on the top of a bale of carpets under a table, the tarpaulin was drawn over me, and we started off. I had sufficient experience of lorry-jumping in World War I to guess that I had not chosen a bed of roses. And this was a particularly ancient lorry. By bracing my arms against the table-top I managed on the whole to avoid hitting the under side of the table when the lorry kicked its tail into the air, but the strain was considerable, and I was already aching, and had hit the table once, when, after ten minutes, the lorry stopped with a jarring of the brakes and an explosion of language.

The French tongue is peculiarly ill-adapted for abuse. Theological invective is useless among a nation who are now mostly rationalists, and even in their religious days had a sneaking regard for the Devil. And anatomical and physiological terms which horrify the Anglo-Saxon do not shock the Latin. So the special vocabulary of abuse is largely confined to the monosyllable which Marshal Cambronne used at Waterloo. This word circulated freely while my saviour denied having seen me, much less given me a lift. He even applied it to an offer of fifty francs to look under the tarpaulin. He also mentioned the police. I heard the voices of two other men in discussion, and drew out my only weapon, a large penknife, not so much in the hope of saving my life, as on the principle enunciated by Macbeth, that, 'Whiles I see lives, the gashes do better upon them.' I am a Scotsman, like Macbeth. I proposed to close with one of my assailants and aim at his jugular vein, for the bullets from a small-bore automatic pistol, though quite efficient killers, have little immediate stopping power. The discussion continued, the details being inaudible. Finally my saviour produced his last verbal card. *'Tristes individus,'* he began a sentence, but before he had finished it I heard the noise of a large car accelerating. They had thought better of it. After all, one cannot hold up all the recalcitrant lorry drivers of a Department at the pistol's point.

At this moment I made the bravest decision of my life. I refused an offer to come out from my hiding-place. The next two hours were the most unpleasant I have ever passed, and I have been through three intense bombardments and had one septic wound. When we arrived at Besançon, I was bruised all over, and bleeding in a number of places. As the lorry driver stood me a stiff cognac for which he insisted on paying, he informed me that the men in the car had passed him again on their way back to Belfort, and looked at him closely. So at least I had saved my life.

I was extremely tired as well as sore, but I lost no time about my next step. With their clearly efficient organization I assumed that the A.I.D.I.R. would discover my identity in a day or so, if they did not know it already. I wrote to a friend in London (for obvious reasons I do not mention his name) to say that I desired the story to appear in the press that I had disappeared suddenly from Paris, leaving my luggage behind. I also wrote a letter dated from Munich to my cousin Polly, better known as Meg o' Mayfair, the lady who meets a duchess a day in the gossip columns of the *Daily Excess*. I told her that I was feeling very run down after my lectures, and had been so absent-minded as to leave most of my luggage in Paris. I was going off for a walking tour in the Böhmer Wald, and hoped to see her when I returned in a month or two. I did not feel that Polly would be able to avoid contributing the gist of this note to the news columns. Another letter from the same address to my Parisian landlady stated that I would return to call for my luggage and pay my bill, although I guessed that the A.I.D.I.R. would probably do both on my behalf. I enclosed these letters to a friend who was working at the Bavarian Academy of Sciences, with an urgent request to post them, burn the covering letter, and keep his mouth shut.

I reckoned that although the A.I.D.I.R. would probably not take this too seriously, they would be bound to divert a little energy to Germany, and if they failed to find me within a fortnight or so, they might begin to transfer their energies across the frontier. Meanwhile I slept, although it was clear from the state of the sheets that the spot on which I laid my head had supported many boots in the past.

In the grey dawn my chauffeur woke me. I suppressed a desire to see Goya's 'Scenes of Cannibalism' in the local picture gallery, though this would have accorded well with my mood, and continued my journey. I again refused to sit outside, but this time I was packed more scientifically and rattled rather little. By request the lorry halted on a deserted part of the road near Poligny, about thirty miles on. I kept up my romance, and said that I proposed to walk to the Swiss frontier, and cross by the Col de St. Georges, a good day's walk up through the Jura. But my plans were different. I walked in the opposite direction, making for the centre of France by unfrequented roads.

As I walked I tried to consider my situation as a purely intellectual problem. The A.I.D.I.R. probably believe that Martin knew how to make auron, and has told me the secret. But in any case I know enough to make it worth their while to murder me. If the bomb had been thrown five minutes later I should be in a better position to judge whether they intend to flood the world with gold, or merely to suppress the secret and guard their profits. I suspect the latter, for they would not be likely to keep the secret for long. While they believe that by murdering me they can hush the thing up, they will try to do so. I am therefore taking what I think to be the safest, though not the noblest course. I am publishing all I know on the subject of gold extraction from sea water. Until publication they will try to murder me. But I hope that my account will appear simultaneously in England and the United States in magazines which generally contain fiction. Many of their readers may suppose this story to be fiction. But if I am murdered, or imprisoned on a false charge, like Galois, this will constitute an advertisement to the whole world that I have written the truth.

Even the A.I.D.I.R. cannot comb out the whole of France for me, though they may possibly put the French police on my track. But they are doubtless watching the French frontiers, the British ports, and my university. They may even contrive to open letters to my friends, but they will hardly suspect me of publishing the most practically important discovery of the century in popular magazines. I am writing this manuscript in a little walled town in Auvergne, where I arrived a week after the rather hectic day which I have here chronicled. On the way I have done my best to sink my identity. I have removed all names and marks from my clothes, burnt my passport, and exchanged my suitcase for a knapsack. I have even gone so far as to remove the buttons from my clothes, and substitute French buttons. Sewing is not one of my accomplishments, and not only did I prick my fingers, but several of my buttons are insecure. My beard is growing nicely. In fact I think of wearing one permanently in future. Also, I got ten francs for my razor. In Clermont-Ferrand I bought a French-Danish dictionary. As soon as the manuscript is accepted for publication, I am asking the firm who will act as my agents to put an advertisement in the *Petit Parisien* on three consecutive days, giving the date when the news will be published. The advertisement will be a request to the creditors of a mythical American lately deceased, to write to a lawyer (also I hope mythical) in Baltimore, before a certain date. I, and I alone, shall know that that date is the date of publication, and until then I must conceal my identity. I shall wander about France as long as my money lasts, and even try to earn a little more. I got five francs yesterday for assisting a motorist, whose engine had broken down, to carry a suitcase to the nearest village. I dare not write to England for money, as this would give my address away if letters to my acquaintances are opened. Also it might be difficult to get a letter without producing evidence of my identity.

I am living as cheaply as I can, but my expenses are much increased by the law of the land. In England anyone may sleep in the open, provided he has a shilling on his person to rebut the charge of being without visible means of support. But in France, to quote Anatole France from memory, 'The law with majestic impartiality forbids the rich, no less than the poor, to sleep in ditches or under haystacks.' That is the worst of equality. Unless I can get a job, my money will run out in a month at most. Shortly before it is all spent, I propose to seek the shelter of the only available free lodging, namely prison. I may, of course, be run in anywhere for having no identity papers, but my plan is to go to Ambert, get rather drunk, and be arrested. I shall then say that I am a native of Iceland, and have lost my passport. The local Danish vice-consul will presumably appear. I worked for three months in Copenhagen at the Institute of Theoretical Physics, and know enough Danish to be fairly rude, and a smattering of the modern Icelandic dialect. Besides, I am spending two hours a day with my dictionary. So I am going to be Mr Thorgrim Magnusson, an ardent Icelandic home-ruler, who objects to all Danish officials as representatives of a foreign domination. I hope that I shall be so unpleasant that the Danes will refuse to accept me, and the French will keep me in jail. I do not particularly relish the idea of hard labour in a French prison, but it is preferable to the fate of Galois, Riquier, and Martin. Also I have a theory, which I devoutly hope is correct, that prisoners are allowed to smoke in France.

The moment my story is published, I appeal to my friends and relations to take all possible steps to get me released. I shall also give my real name. But if no prisoner of the name of Thorgrim Magnusson is to be found in the jails of Auvergne, it may be presumed that I have met the fate of Martin. Only yesterday I saw an individual who appeared to be following me. I went up the nearest hill and outdistanced him. But I cannot believe that the A.I.D.I.R. have agents everywhere, and I suspect he was merely struck by my rather unkempt appearance. I believe that I shall get away with it.

What will happen after this story is published I do not know. I do not propose to emulate Galois and shall not try to make gold. I will not even take holidays at the seaside. Obviously if I knew the formula for auron and intended to work the process, I should not have given away half the secret. I shall certainly be watched, but I credit the A.I.D.I.R. with sufficient intelligence not to assassinate me. They will hope that this story will be taken as the ingenious attempt of a professor to explain his otherwise discreditable arrest for drunkenness. I take it, however, that someone will have the wits to see that it is a perfectly true story, and that Galois' process will be working somewhere within the next ten years. I hope so, because I should like to see the men who organized the murder of Galois and his friends picking crusts out of the gutter. A team of fairly good mathematicians could do the requisite calculations in four years or so. So about six years hence I recommend my readers to sell out shares in gold-mines and fixed-interest-bearing securities, and to buy industrials. But there are some very good mathematical physicists in Russia, and if the Bolsheviks get hold of the process first there will be about £1,000,000,000 per year available for the purposes of the world revolution. In that case the purchase of securities of any kind will be pointless.

Edgar Allan Poe once more, to show the breadth of his erudition, in a modern version of the tales of Scheherazade.

' "Among this nation of necromancers there was also one who had in his veins the blood of the salamanders; for he made no scruple of sitting down to smoke his chibouc in a red-hot oven until his dinner was thoroughly roasted upon its floor.[1] Another had the faculty of converting the common metals into gold, without even looking at them during the process.[2] Another had such delicacy of touch that he made a wire so fine as to be invisible.[3] Another had such quickness of perception that he counted all the separate motions of an elastic body, while it was springing backwards and forwards at the rate of nine hundred million of times in a second." '[4]

'Absurd!' said the king.

' "Another of these magicians, by means of a fluid that nobody ever yet saw, could make the corpses of his friends brandish their arms, kick out their legs, fight, or even get up and dance at his will.[5] Another had cultivated his voice to so great an extent that he could have made himself heard from one end of the earth to the other.[6] Another had so long an arm that he could sit down in Damascus and indite a letter at Bagdad – or indeed at any distance whatsoever.[7] Another commanded the lightning to come down to him out of the heavens, and it came at his call; and served him for a plaything when it came. Another took two loud sounds and out of them made a silence. Another constructed a deep darkness out of two brilliant lights.[8] Another made ice in a red-hot

[1]*Chabert*, and since him, a hundred others.

[2]The Electrotype.

[3]*Wollaston* made of platinum for the field of views in a telescope a wire one eighteen-thousandth part of an inch in thickness. It could be seen only by means of the microscope.

[4]Newton demonstrated that the retina beneath the influence of the violet ray of the spectrum, vibrated 900,000,000 of times in a second.

[5]The Voltaic pile.

[6]The Electro Telegraph transmits intelligence instantaneously – at least so far as regards any distance upon the earth.

[7]The Electro Telegraph Printing Apparatus.

[8]Common experiments in Natural Philosophy. If two red rays from two luminous points be admitted into a dark chamber so as to fall on a white surface, and differ in their length by 0.0000258 of an inch, their intensity is doubled. So also if the difference in length by any whole-number multiple of that fraction. A multiple

furnace.[1] Another directed the sun to paint his portrait, and the sun did.[2] Another took this luminary with the moon and the planets, and having first weighed them with scrupulous accuracy, probed into their depths and found out the solidity of the substance of which they are made. But the whole nation is, indeed, of so surprising a necromantic ability, that not even their infants, nor their commonest cats and dogs have any difficulty in seeing objects that do not exist at all, or that for twenty thousand years before the birth of the nation itself, had been blotted out from the face of creation." '[3]

'Preposterous!' said the king.

' "The wives and daughters of these incomparably great and wise magi," ' continued Scheherazade, without being in any manner disturbed by these frequent and most ungentlemanly interruptions on the part of her husband – ' "the wives and daughters of these eminent conjurors are everything that is accomplished and refined; and would be everything that is interesting and beautiful, but for an unhappy fatality that besets them, and from which not even the miraculous powers of their husbands and fathers has, hitherto, been

by 2¼, 3¼, &c., gives an intensity equal to one ray only; but a multiple by 2½, 3½, &c., gives the result of total darkness. In violet rays similar effects arise when the difference in length is 0.000157 of an inch; and with all other rays the results are the same – the difference varying with a uniform increase from the violet to the red.

Analogous experiments in respect to sound produce analogous results.

[1]Place a platina crucible over a spirit lamp, and keep it a red heat; pour in some sulphuric acid, which, though the most volatile of bodies at a common temperature, will be found to become completely fixed in a hot crucible, and not a drop evaporates – being surrounded by an atmosphere of its own, it does not, in fact touch the sides. A few drops of water are now introduced, when the acid immediately coming in contact with the heated sides of the crucible, flies off in sulphurous acid vapor, and so rapid is its progress, that the caloric of the water passes off with it, which falls a lump of ice to the bottom; by taking advantage of the moment before it is allowed to re-melt, it may be turned out a lump of ice from a red-hot vessel.

[2]The Daguerreotype.

[3]Although light travels 200,000 miles in a second, the distance of what we suppose to be the nearest fixed star (Sireus) is so inconceivably great, that its rays would require *at least* three years to reach the earth. For stars beyond this 20 – or even 1000 years – would be a moderate estimate. Thus, if they had been annihilated 20 or 1000 years ago, we might still see them to-day, by the light which *started* from their surfaces, 20 or 1000 years in the past time. That many which we see daily are really extinct, is not impossible – not even improbable [BROADWAY JOURNAL NOTE.]

The elder Herschel maintains that the light of the faintest nebulae seen through his great telescope, must have taken 3,000,000 years in reaching the earth. Some, made visible by Lord Ross' instrument must, then, have required at least 20,000,000. [GRISWOLD NOTE.]

adequate to save. Some fatalities come in certain shapes, and some in others – but this of which I speak, has come in the shape of a crotchet." '

'A what?' said the king.

' "A crotchet," ' said Scheherazade. ' "One of the evil genii who are perpetually upon the watch to inflict ill, has put it into the heads of these accomplished ladies that the thing which we describe as personal beauty, consists altogether in the protuberance of the region which lies not very far below the small of the back. – Perfection of loveliness, they say, is in the direct ratio of the extent of this hump. Having been long possessed of this idea, and bolsters being cheap in that country, the days have long gone by since it was possible to distinguish a woman from a dromedary –" '

'Stop!' said the king, – 'I can't stand that, and I won't. You have already given me a dreadful headache with your lies. The day, too, I perceive, is beginning to break. How long have we been married? – my conscience is getting to be troublesome again. And then that dromedary touch – do you take me for a fool? Upon the whole you might as well get up and be throttled.'

These words, as I learn from the 'Isitsöornot,' both grieved and astonished Scheherazade; but, as she knew the king to be a man of scrupulous integrity, and quite unlikely to forfeit his word, she submitted to her fate with a good grace. She derived, however, great consolation, (during the tightening of the bowstring,) from the reflection that much of the history remained still untold, and that the petulance of her brute of a husband had reaped for him a most righteous reward, in depriving him of many inconceivable adventures.

110

Julian Huxley, grandson of T.H. and brother of Aldous, was a serious populariser of science, an essayist, poet and man of letters, public sage, Director of the London Zoo and the first Director General of UNESCO. Here is part of his story, 'The Tissue-Culture King', an early parable of the responsibility of the scientist for the consequences of his discoveries.

That was how it all started. Perhaps the best way of giving some idea of how it had developed will be for me to tell my own impressions when Hascombe took me round his laboratories. One whole quarter of the town was devoted entirely to religion – it struck me as excessive, but Hascombe reminded me that Tibet spends one-fifth of its revenues on melted butter to burn before its shrines. Facing the main square was the chief temple, built impressively enough of solid mud. On either side were the apartments where dwelt the servants of the gods and administrators of the sacred rites. Behind were Hascombe's laboratories, some built of mud, others, under his later guidance, of wood. They were guarded night and day by patrols of giants, and were arranged in a series of quadrangles. Within one quadrangle was a pool which served as an aquarium; in another, aviaries and great hen houses; in yet another, cages with various animals; in the fourth a little botanic garden. Behind were stables with dozens of cattle and sheep, and a sort of experimental ward for human beings.

He took me into the nearest of the buildings. 'This,' he said, 'is known to the people as the Factory (it is difficult to give the exact sense of the word, but it literally means producing-place), the Factory of Kingship or Majesty, and the Well-spring of Ancestral Immortality.' I looked round, and saw platoons of buxom and shining African women, becomingly but unusually dressed in tight-fitting dresses and caps, and wearing rubber gloves. Microscopes were much in evidence, also various receptacles from which steam was emerging. The back of the room was screened off by a wooden screen in which were a series of glass doors; and these doors opened into partitions, each labelled with a name in that unknown tongue, and each containing a number of objects like the one I had seen taken out of the basket by the giant before we were captured. Pipes surrounded this chamber, and appeared to be distributing heat from a fire in one corner.

'Factory of Majesty!' I exclaimed. 'Wellspring of Immortality! What the dickens do you mean?'

'If you prefer a more prosaic name,' said Hascombe, 'I should call this the Institute of Religious Tissue Culture.' My mind went back to a day in 1918 when I had been taken by a biological friend in New York to see the famous Rockefeller Institute; and at the word tissue culture I saw again before me Dr. Alexis Carrel and troops of white-garbed American girls making cultures, sterilizing, microscopizing, incubating, and the rest of it. The Hascombe Institute was, it is true, not so well equipped, but it had an even larger, if differently colored, personnel.

Hascombe began his explanations. 'As you probably know,

Frazer's "Golden Bough"★ introduced us to the idea of a sacred priest-king, and showed how fundamental it was in primitive societies. The welfare of the tribe is regarded as inextricably bound up with that of the King, and extraordinary precautions are taken to preserve him from harm. In this kingdom, in the old days, the King was hardly allowed to set his foot to the ground in case he should lose divinity; his cut hair and nail-parings were entrusted to one of the most important officials of state, whose duty it was to bury them secretly, in case some enemy should compass the King's illness or death by using them in black magic rites. If anyone of base blood trod on the King's shadow, he paid the penalty with his life. Each year a slave was made mock-king for a week, allowed to enjoy all the king's privileges, and was decapitated at the close of his brief glory; and by this means it was supposed that the illnesses and misfortunes that might befall the King were vicariously got rid of.

'I first of all rigged up my apparatus, and with the aid of Aggers, succeeded in getting good cultures, first of chick tissues and later, by the aid of embryo-extract, of various and adult mammalian tissues. I then went to Bugala, and told him that I could increase the safety, if not of the King as an individual, at least of the life which was in him, and that I presumed that this would be equally satisfactory from a theological point of view. I pointed out that if he chose to be made guardian of the King's subsidiary lives, he would be in a much more important position than the chamberlain or the burier of the sacred nail-parings, and might make the post the most influential in the realm.

'Eventually I was allowed (under threats of death if anything untoward occurred) to remove small portions of His Majesty's subcutaneous connective tissue under a local anaesthetic. In the presence of the assembled nobility I put fragments of this into culture medium, and showed it to them under the microscope. The cultures were then put away in the incubator, under a guard – relieved every eight hours – of half a dozen warriors. After three days, to my joy they had all taken and showed abundant growth. I could see that the Council was impressed, and reeled off a magnificent speech, pointing out that this growth constituted an actual increase in the quantity of the divine principle inherent in royalty; and what was more, that I could increase it indefinitely. With that I cut each of my cultures into eight, and sub-cultured all the pieces. They were

★A very elaborate treatise on a division of Roman mythology, especially on the cult of Diana.

again put under guard, and again examined after three days. Not all of them had taken this time, and there were some murmurings and angry looks, on the ground that I had killed some of the King; but I pointed out that the King was still the King, that his little wound had completely healed, and that any successful cultures represented so much extra sacredness and protection to the state. I must say that they were very reasonable, and had good theological acumen, for they at once took the hint.

'I pointed out to Bugala, and he persuaded the rest without much difficulty, that they could now disregard some of the older implications of the doctrines of kingship. The most important new idea which I was able to introduce was *mass production*. Our aim was to multiply the King's tissues indefinitely, to ensure that some of their protecting power should reside everywhere in the country. Thus by concentrating upon quantity, we could afford to remove some of the restrictions upon the King's mode of life. This was of course agreeable to the King; and also to Bugala, who saw himself wielding undreamt-of power. One might have supposed that such an innovation would have met with great resistance simply on account of its being an innovation; but I must admit that these people compared very favorably with the average business man in their lack of prejudice.

'Having thus settled the principle, I had many debates with Bugala as to the best methods for enlisting the mass of the population in our scheme. What an opportunity for scientific advertising! But, unfortunately, the population could not read. However, war propaganda worked very well in more or less illiterate countries – why not here?'

111

George Bernard Shaw was not deterred by deep ignorance of science from putting forward strong views. He was clearly in some sense interested, and The Doctor's Dilemma *was the most tangible result. Colonel Sir Almroth Wright was Professor of Bacteriology at St Mary's Hospital Medical School in London, and was known to Shaw. Fleming worked in the Department, and in his admirable biography,* Alexander Fleming, *Gwyn Macfarlane (who also wrote an equally fine study of H.W. Flory) gives the following description of the genesis of* The Doctor's Dilemma:

Fleming could meet great men of literature and learning not only in his new club, but in the unlikely setting of the Inoculation Department. Almroth Wright had a reputation as an original thinker, talker, writer and polemicist, that went far beyond the bounds of medicine and science, and he attracted equally radical thinkers sometimes of opposite views. One of these was Bernard Shaw, who used to visit Wright at St Mary's for a good argument. On one of these visits, probably in 1905, Wright showed him the work on phagocytes. Shaw was fascinated – though he refused to believe that there were hundreds of millions of cells in every drop of blood. This was, he said, no more credible than the 'yarn that the sun was 98 million miles from the earth'. While he was talking, John Freeman came in to ask Wright if he could take on the treatment of another patient. After a discussion, they decided that they could not – they already had as much work as they could handle. Turning to Shaw, Wright said, 'The time is coming when we shall have to decide whether this man or that is most worth saving.' 'Ha!' said Shaw, 'there I smell drama!'

A week or so later, Wright heard that Shaw was writing a play in which he would be portrayed as one of the main characters. Wright protested to Shaw, arguing that he was already unpopular with the medical pundits, and to be staged as a hero would increase their dislike. Shaw replied that he need have no fears. In the play, the hero made no medical mistakes and was loathed in consequence, whereas the king's physician made nothing but mistakes and was loved by all. So *The Doctor's Dilemma* was produced in 1906 at the Court Theatre, a brilliant satire on the medical profession. Granville Barker played Sir Colenso Rigeon, a somewhat exaggerated picture of Wright. The dilemma that confronted Rigeon was whether to treat an honest, decent 'sixpenny doctor' or a scoundrelly blackguard of an artist with real genius – he could not treat both. After much verbal agonizing, Rigeon treats the doctor and refers the artist to the eminent, stupid physician, Sir Ralph Bonnington, who kills his patient by failing to recognize the importance of determining the opsonic index when stimulating the phagocytes by vaccination. The fact that the artist left a charming widow who had become an admirer of Sir Colenso Rigeon adds another dimension to the ethics of the dilemma.

When Wright went to see the play he walked out during the performance at the point when the fashionable physician kills his patient by injecting vaccine at the wrong time. He was indignant that Shaw had contrived a situation for which there was no scientific justification. 'Shaw is not interested in truth – only in circus tricks,' he said afterwards.

112

The passage below from the preface of the play shows that Shaw did not think much of biological research.

The smattering of science that all – even doctors – pick up from the ordinary newspapers nowadays only makes the doctor more dangerous than he used to be. Wise men used to take care to consult doctors qualified before 1860, who were usually contemptuous of or indifferent to the germ theory and bacteriological therapeutics; but now that these veterans have mostly retired or died, we are left in the hands of the generations which, having heard of microbes much as St Thomas Aquinas heard of angels, suddenly concluded that the whole art of healing could be summed up in the formula: Find the microbe and kill it. And even that they did not know how to do. The simplest way to kill most microbes is to throw them into an open street or river and let the sun shine on them, which explains the fact that when great cities have recklessly thrown all their sewage into the open river the water has sometimes been cleaner twenty miles below the city than thirty miles above it. But doctors instinctively avoid all facts that are reassuring, and eagerly swallow those that make it a marvel that anyone could possibly survive three days in an atmosphere consisting mainly of countless pathogenic germs. They conceive microbes as immortal until slain by a germicide administered by a duly qualified medical man. All through Europe people are adjured, by public notices and even under legal penalties, not to throw their microbes into the sunshine, but to collect them carefully in a handkerchief; shield the handkerchief from the sun in the darkness and warmth of the pocket; and send it to a laundry to be mixed up with everybody else's handkerchiefs, with results only too familiar to local health authorities.

In the first frenzy of microbe killing, surgical instruments were dipped in carbolic oil, which was a great improvement on not dipping them in anything at all and simply using them dirty; but as microbes are so fond of carbolic oil that they swarm in it, it was not a success from the anti-microbe point of view. Formalin was squirted into the circulation of consumptives until it was discovered that formalin nourishes the tubercle bacillus handsomely and kills men. The popular theory of disease is the common medical theory: namely, that every disease had its microbe duly created in the garden of Eden, and has been steadily propagating itself and producing widening circles of malignant disease ever since. It was plain from the first that if this had been even approximately true, the whole human race would have been wiped out by the plague long ago, and that every epidemic, instead of fading out as mysteriously as it rushed in, would spread over the whole world. It was also evident that the characteristic microbe of a disease might be a symptom instead of a cause. An unpunctual man is always in a hurry; but it does not follow that hurry is the cause of unpunctuality: on the contrary, what is the matter with the patient is sloth. When Florence Nightingale said bluntly that if you overcrowded your soldiers in dirty quarters there would be an outbreak of smallpox among them, she was snubbed as an ignorant female who did not know that smallpox can be produced only by the importation of its specific microbe.

If this was the line taken about smallpox, the microbe of which has never yet been run down and exposed under the microscope by the bacteriologist, what must have been the ardor of conviction as to tuberculosis, tetanus, enteric fever, Maltese fever, diphtheria, and the rest of the diseases in which the characteristic bacillus had been identified! When there was no bacillus it was assumed that, since no disease could exist without a bacillus, it was simply eluding observation. When the bacillus was found, as it frequently was, in persons who were not suffering from the disease, the theory was saved by simply calling the bacillus an impostor, or pseudo-bacillus. The same boundless credulity which the public exhibit as to a doctor's power of diagnosis was shewn by the doctors themselves as to the analytic microbe hunters. These witch finders would give you a certificate of the ultimate constitution of anything from a sample of the water from your well to a scrap of your lungs, for seven-and-sixpence. I do not suggest that the analysts were dishonest. No doubt they carried the analysis as far as they could afford to carry it for the money. No doubt also they could afford to carry it far enough to be of some use. But the fact remains that just as doctors perform for half-a-crown, without the least misgiving, operations which could not be thoroughly and safely performed with due scientific rigor and the requisite apparatus by an unaided private practitioner for less than some thousands of pounds, so did they proceed on the assumption that they could get the last word of science as to the constituents of their pathological samples for a two-hours cab fare.

113

So to the play itself, in which Sir Colenso Ridgeon is Almroth Wright. Of the others, Cutler Walpole, the surgeon, has been identified as Sir William Arbuthnot Lane. (Walpole's King Charles's Head was a useless organ, called the nuciform sac, in which, he asserted, resided all agents of infection and decay, and which had to be removed to extirpate the chronic blood-poisoning that afflicted nearly all the population. Sir William's obsession in real life was with the contents of the colon, for he believed that constipation lay behind most human ills. He invented castor oil as a radical cure, and it was related that barrels of the fluid used to be delivered at Guy's Hospital like so much beer at a pub. Only years later did many of his patients start to die of liver failure. It was said that at autopsy oil could be wrung from their bloated livers like water out of sponge (for an account, see Alex Comfort: The Anxiety Makers*). Lane died, laden with years and honours, in 1945.) Who the lethal B.B. was is not recalled. As to the treatment – stimulate the phagocytes – this also passed into the language. In H.G. Wells's* Tono Bungay, *the wonder patent medicine of that name is advertised as 'Worster sauce for the Phagocyte. It gives an appetite'.*

SIR PATRICK. What did you find out from Jane's case?

RIDGEON. I found out that the inoculation that ought to cure sometimes kills.

SIR PATRICK. I could have told you that. Ive tried these modern inoculations a bit myself. Ive killed people with them; and Ive cured people with them; but I gave them up because I never could tell which I was going to do.

RIDGEON [*taking a pamphlet from a drawer in the writing-table and handing it to him*] Read that the next time you have an hour to spare; and you'll find out why.

SIR PATRICK [*grumbling and fumbling for his spectacles*] Oh, bother your pamphlets. Whats the practice of it? [*Looking at the pamphlet*] Opsonin? What the devil is opsonin?

RIDGEON. Opsonin is what you butter the disease germs with

to make your white blood corpuscles eat them. [*He sits down again on the couch*].

SIR PATRICK. Thats not new. Ive heard this notion that the white corpuscles – what is it that whats his name? – Metchnikoff – calls them?

RIDGEON. Phagocytes.

SIR PATRICK. Aye, phagocytes: yes, yes, yes. Well, I heard this theory that the phagocytes eat up the disease germs years ago: long before you came into fashion. Besides, they dont always eat them.

RIDGEON. They do when you butter them with opsonin.

SIR PATRICK. Gammon.

RIDGEON. No: it's not gammon. What it comes to in practice is this. The phagocytes wont eat the microbes unless the microbes are nicely buttered for them. Well, the patient manufactures the butter for himself all right; but my discovery is that the manufacture of that butter, which I call opsonin, goes on in the system by ups and downs – Nature being always rhythmical, you know – and that what the inoculation does is to stimulate the ups or downs, as the case may be. If we had inoculated Jane Marsh when her butter factory was on the up-grade, we should have cured her arm. But we got in on the down-grade and lost her arm for her. I call the up-grade the positive phase and the down-grade the negative phase. Everything depends on your inoculating at the right moment. Inoculate when the patient is in the negative phase and you kill: inoculate when the patient is in the positive phase and you cure.

SIR PATRICK. And pray how are you to know whether the patient is in the positive or the negative phase?

RIDGEON. Send a drop of the patient's blood to the laboratory at St Anne's; and in fifteen minutes I'll give you his opsonin index in figures. If the figure is one, inoculate and cure: if it's under point eight, inoculate and kill. Thats my discovery: the most important that has been made since Harvey discovered the circulation of the blood. My tuberculosis patients dont die now.

SIR PATRICK. And mine do when my inoculation catches them in the negative phase, as you call it. Eh?

RIDGEON. Precisely. To inject a vaccine into a patient without first testing his opsonin is as near murder as a respectable practitioner can get. If I wanted to kill a man I should kill him that way.

Leo Szilard, the rootless theoretical physicist who lived all his life in hotel rooms, with only as many personal possessions as could fit comfortably into a medium-sized suitcase, published a volume of ingenious stories with the title, The Voice of the Dolphins. *'Report on Grand Central Terminal' imagines that anthropologists from another planet come to explore the ruins of a terrestrial civilisation, after all life has been destroyed by a nuclear war in the twentieth century.*

The cars stored in this station were labeled – we discovered – either 'Smokers' or 'Nonsmokers,' clearly indicating some sort of segregation of passengers. It occurred to me right away that there may have lived in this city two strains of earth-dwellers, a more pigmented variety having a dark or 'smoky' complexion, and a less pigmented variety (though not necessarily albino) having a fair or 'nonsmoky' complexion.

All remains of earth-dwellers were found as skeletons, and no information as to pigmentation can be derived from them. So at first it seemed that it would be difficult to obtain confirmation of this theory. In the meantime, however, a few rather spacious buildings were discovered in the city which must have served as some unknown and rather mysterious purposes. These buildings had painted canvases in frames fastened to the walls of their interior – both landscapes and images of earth-dwellers. And we see now that the earth-dwellers fall indeed into *two* classes – those whose complexion shows strong pigmentation (giving them a smoky look) and those whose complexion shows only weak pigmentation (the nonsmoky variety). This is exactly as expected.

I should perhaps mention at this point that a certain percentage of the images disclose the existence of a third strain of earth-dwellers. This strain has, in addition to a pair of hands and legs, a pair of wings, and apparently *all* of them belonged to the less pigmented variety. None of the numerous skeletons so far examined seems to have belonged to this winged strain, and I concluded therefore that we have to deal here with images of an extinct variety. That this view is indeed correct can no longer be doubted, since we have determined that the winged forms are *much more frequently found among the older paintings than among the more recent paintings.*

I cannot of course describe to you here *all* the puzzling discoveries which we made within the confines of the 'Grand

Central Terminal,' but I want to tell you at least about the most puzzling one, particularly since Xram is basing his war theory on it.

This discovery arose out of the investigation of an insignificant detail. In the vast expanse of the 'Grand Central Terminal' we came upon two smaller halls located in a rather hidden position. Each of these two halls (labeled 'Men' or 'Women') contains a number of small cubicles which served as temporary shelter for earth-dwellers while they were depositing their excrements. The first question was, How did the earth-dwellers locate these hidden depositories within the confines of 'Grand Central Terminal'?

An earth-dweller moving about at random within this large building would have taken about one hour (on the average) to stumble upon one of them. It is, however, possible that the earth-dwellers located the depositories with the aid of olfactory guidance, and we have determined that if their sense of smell had been about thirty to forty times more sensitive than the rudimentary sense of smell of our own species, the average time required would be reduced from one hour to about five or ten minutes. This shows there is no real difficulty connected with this problem.

Another point, however, was much harder to understand. This problem arose because we found that the door of each and every cubicle in the depository was locked by a rather complicated gadget. Upon investigation of these gadgets it was found that they contained a number of round metal disks. By now we know that these ingenious gadgets barred entrance to the cubicle until an additional disk was introduced into them through a slot; at that very moment the door became unlocked, permitting access to the cubicle.

These disks bear different images and also different inscriptions which, however, all have in common the word 'Liberty.' What is the significance of these gadgets, the disks in the gadgets and the word 'Liberty' on the disks?

Though a number of hypotheses have been put forward in explanation, consensus seems to veer toward the view that we have to deal here with a ceremonial act accompanying the act of deposition, similar perhaps to some of the curious ceremonial acts reported from the planets Sigma 25 and Sigma 43. According to this view, the word 'Liberty' must designate some virtue which was held in high esteem by the earth-dwellers or else their ancestors. In this manner we arrive at a quite satisfactory explanation for the sacrificing of disks immediately preceding the act of deposition.

But why was it necessary to make sure (or, as Xram says, to enforce), by means of a special gadget, that such a disk was in fact sacrificed in each and every case? This too can be explained if we assume that the earth-dwellers who approached the cubicles were perhaps driven by a certain sense of urgency, that in the absence of the gadgets they might have occasionally forgotten to make the disk sacrifice and would have consequently suffered pangs of remorse afterward. Such pangs of remorse are not unknown as a consequence of omissions of prescribed ceremonial performances among the inhabitants of the planets Sigma 25 and Sigma 43.

I think that this is on the whole as good an explanation as can be given at the present, and it is likely that further research will confirm this view. Xram, as I mentioned before, has a theory of his own which he thinks can explain everything, the disks in the gadgets as well as the uranium explosions which extinguished life.

He believes that these disks were given out to earth-dwellers as rewards for services. He says that the earth-dwellers were not rational beings and that they would not have collaborated in co-operative enterprises without some special incentive.

He says that, by barring earth-dwellers from depositing their excrements unless they sacrificed a disk on each occasion, they were made eager to acquire such disks, and that the desire to acquire such disks made it possible for them to collaborate in co-operative efforts which were necessary for the functioning of their society.

He thinks that the disks found in the depositories represent only a special case of a more general principle and that the earth-dwellers probably had to deliver such disks not only prior to being given access to the depository but also prior to being given access to food, etc.

He came to talk to me about all this a couple of days ago; I am not sure that I understood all that he said, for he talked very fast, as he often does when he gets excited about one of his theories. I got the general gist of it, though, and what he says makes very little sense to me.

Apparently, he had made some elaborate calculations which show that a system of production and distribution of goods based on a system of exchanging disks cannot be stable, but is necessarily subject to great fluctuations vaguely reminiscent of the manic-depressive cycles of the insane. He goes so far as to say that in such a depressive phase war becomes psychologically possible even within the same species.

No one is more ready than I to admit that Xram is brilliant. His theories have invariably been proved to be wrong, but so far all of them had contained at least a grain of truth. In the case of his present theory the grain must be a very small grain indeed, and, moreover, this once I can *prove* that he is wrong.

In the last few days we made a spot check on ten different lodging houses of the city, selected at random. We found a number of depositories but not a single one that was equipped with a gadget containing disks – not in any of the houses which we checked so far. In view of this evidence, Xram's theory collapses.

It seems now certain that the disks found in the depositories at 'Grand Central Terminal' had been placed there as a ceremonial act. Apparently such ceremonial acts were connected with the act of deposition *in public places* and in public places only.

I am glad that we were able to clear this up in time, for I should have been sorry to see Xram make a fool of himself by including his theory in the report. He is a gifted young man, and in spite of all the nonsensical ideas he can put forward at the drop of a hat, I am quite fond of him.

115

Szilard it was who conceived the idea of a nuclear chain reaction, while in London after escaping from Germany. He was temporarily working at Bart's Hospital and living in the Strand Palace Hotel; the brainwave hit him as he was walking to work one morning, waiting at the traffic lights in Kingsway to cross into High Holborn. He took out a patent in the hope of limiting the practical investigation of chain reactions. The experimental discovery of nuclear fission was made, though not at once recognised, shortly afterwards in Germany. The story of the realization of what it portended is nowhere better told than in Robert Jungk's Brighter than a Thousand Suns.

On one occasion, in 1935, Fräulein Meitner directed her pupil von Droste to repeat in Dahlem certain experiments which had been carried out in Paris with the bombardment of thorium. Irène Joliot-Curie had declared that the thorium isotope sent out alpha rays under radiation. Droste did not find any such rays. Once more Lise Meitner believed she had convicted her rival of inaccuracy. And once more she was mistaken.

Droste experimented not only with thorium, but also with uranium. If he had not, in the latter case, introduced a filter to avert particles with a range of under three centimetres, he would not only on that occasion have realised that Madame Joliot's results were as she had stated but would also have necessarily found, there and then, fission products from uranium. So near had experiment come, even at that date, to the discovery of uranium fission.

Irène Joliot-Curie's next paper on transuranic elements appeared in the summer of 1938, with the Jugoslav Savitch as co-author. In this article a substance was mentioned which did not fit in with the pattern which had been meanwhile worked out by Hahn and his collaborators for these elements.

It was said in Dahlem that 'Madame Joliot-Curie is still relying on the chemical knowledge she received from her famous mother and that knowledge is just a bit out of date today.' Hahn considered it necessary to be tactful. He thought it desirable not to reveal his French colleague's 'negligence' to the entire world in a scientific periodical. 'There is quite enough vexatious strain just now in the relations between Germany and France,' he said. 'Let us not help to increase it.' He accordingly wrote a private note to the laboratory in the rue d'Ulm suggesting that the experiments should be repeated somewhat more carefully.

But no reply to Hahn's note came from Paris. On the contrary, Madame Joliot commited a further 'sin'. She published a second article, based on the data of the first. Hahn refused to read it, in spite of being urged to do so by his assistant Strassmann. He was thoroughly disgusted with the 'unteachability' of his Parisian colleague. He was being worried, that same summer of 1938, by a problem which had nothing to do with physics. Attempts were being made to deprive him of his *alter ego*, Lise Meitner. For over a quarter of a century the Viennese physicist and her 'cockerel' [*Translator's note:* Hahn means 'cock' in German] had worked side by side. Their identities were so closely fused, even in their own minds, that Fräulein Meitner once absent-mindedly replied to a colleague who spoke to her at a congress: 'I think you've mistaken me for Professor Hahn.'

As an Austrian, Lise Meitner, despite the sudden discovery that she was 'not Aryan', had been permitted to go on working at the Kaiser Wilhelm Institute after 1932. But the *Anschluss* of March 1938 made the racial legislation of the Third Reich applicable to her case. Hahn and Max Planck intervened to no purpose, though the latter actually approached Hitler himself. Hahn's Jewish colleague could not be saved for the Institute. She had to go. It was not even certain whether the Government would at least allow her to leave Germany. So she was obliged to slip across the Dutch frontier, disguised as a 'tourist', without saying goodbye to her former collaborators. Apart from Hahn, only two or three people in the Dahlem Institute knew that Lise Meitner would not come back from her summer holiday.

That autumn Madame Joliot-Curie published a third article which summarised and enlarged her last two. Since Fräulein Meitner's departure Strassmann had become Hahn's closest associate in the field of radium chemistry. He saw at a glance that no mistake had been made in the Curie laboratory but that on the contrary a remarkable new avenue of approach to the problem had probably been opened up. He rushed excitedly upstairs to Hahn on the first floor and exclaimed with the greatest emphasis: 'You simply must read this report!'

But Hahn was adamant.

'I'm not interested in our lady friend's latest effusion,' he answered, puffing quietly at his cigar.

Strassmann, however, would not give in to this rebuff. Before Hahn could repeat it he rapidly gave his chief a succinct account of the most important points in this new performance. 'It struck Hahn like a thunderbolt,' he recollected later. 'He never finished that cigar. He laid it, still glowing on his desk, and ran downstairs with me to the laboratory.'

It proved very difficult to persuade Hahn that he, in common with investigators all over the world, had been following a false trail for years. But as soon as he had recognised the fact he instantly turned in his tracks and made every effort to get at the truth. It did not come easy to confess to a series of failures. But he owed to that admission, shortly afterwards, the greatest success of his career.

In almost incessant work, for weeks on end, the experiments of Madame Joliot and Savitch were thoroughly tested with the most exact methods of radium chemistry. The process showed that the bombardment of uranium with neutrons did in fact produce a substance which, as the Paris 'team' had stated, closely resembled lanthanum. The more precise analyses of Hahn and Strassmann, however, led to the chemically incontrovertible but physically inexplicable result that the real element concerned was barium, which occupies a position in the centre of the list and only weighs a little more than half as much as uranium.

The discovery was only made later that this at first incomprehensible presence of barium could be explained by the 'bursting', as Hahn called it, of the nucleus. But at the time what Hahn and Strassmann had found in their chemical probings seemed so incredible to them that they jotted down the following sceptical sentences, which have since become famous.

'We come to this conclusion. Our "radium" istopes have the properties of barium. As chemists, we are in fact bound to affirm that the new bodies are not radium but barium. For there is no question of elements other than radium or barium being present. . . . As nuclear chemists we cannot decide to take this step in contradiction to all previous experience in nuclear physics.'

The two German atomic scientists perceived that they had made a notable discovery, even though it might not yet be explicable in terms of physics. The date was just before Christmas 1938 and it seemed important to Hahn to publish an account of his work as soon as possible. He therefore took the unusual step of telephoning to the Director of the Springer Verlag, Dr Paul Rosbaud, who was a personal friend of his, and asking whether he could find room for some urgent 'information' in the forthcoming number of *Naturwissenschaften*. Rosbaud agreed to do so. The paper, dated the 22nd December 1938, accordingly left Hahn's desk.

Nearly twenty years later Hahn told the author: 'After the manuscript had been posted, the whole thing once more seemed so extraordinary to me that I wished I could get the document back out of the letter-box.'

It was with such hesitations and doubts that the age of atomic fission began.

It was by way of a demonstration against the racial legislation of the Third Reich, and yet at the same time no more than plain proof of a confidence that had now lasted for decades, when Otto Hahn immediately despatched his new data to his former collaborator Lise Meitner, now an emigrant in Stockholm. The fateful letter to Sweden was already on its way before any other member of Hahn's department in the Kaiser Wilhelm Institute heard anything about the new and still, for the present, quite inexplicable discovery. Hahn waited tensely to hear what the partner with whom he had begun the experiments in question in 1934, and with whom he had so often discussed them, would say to these astonishing new results, which contradicted all previous experience. He was a little afraid of what her answer might be. She had always been a stern critic of his work. Probably, he thought, she would tear these latest data to pieces.

Lise Meitner received his letter in the small township of Kungelv, near Göteborg. She had gone to this seaside resort, which was almost wholly lifeless in winter, to spend her first Christmas in exile alone, in a small family boarding-house, far from the wide world. Her young nephew, the physicist O. R. Frisch, since 1934 himself a refugee working at the Institute of Niels Bohr in Copenhagen, felt sorry for his aunt's loneliness. He determined to pay her a visit. He happened, therefore, to be present when Hahn's letter arrived in the quiet little provincial town. As might have been expected, the information from Dahlem excited its recipient to an unusually high degree. If the radium chemistry analyses carried out by Hahn and Strassmann were accurate – and Fräulein Meitner, who knew the precision of Hahn's work, could hardly doubt it – then certain ideas in nuclear physics, hitherto taken to be unassailable, could not be true. She perceived even more clearly than had her former partner Hahn that something tremendous had here unexpectedly come to light.

Fräulein Meitner naturally longed to discuss, with some duly qualified person, the multitude of questions and conjectures that arose in her mind. It seemed a bit of luck that her nephew, who was regarded as one of the leading lights in Bohr's circle, happened to be with her just then. But after all Frisch had not come to Kungelv to 'talk shop' to his aunt. He had come for the sake of a holiday. 'It took her some time before she could make me listen,' he reported later. To be precise, Frisch actually tried to run away, in the literal sense, from Lise Meitner's expositions. In fact, he buckled on a pair of skis and would certainly soon have been beyond the reach of his aunt if the ground had not been so hopelessly flat all round the town. For this reason she could keep pace with him and maintain a continuous flow of talk while they stamped together through the snow. At last she succeeded, by dint of a bombardment of words, in piercing the blank wall of his indifference and releasing a veritable chain-reaction of ideas in his brain.

The consequences of the inspiring debates that then took place during evenings in the old-fashioned lounge of the boarding-house are described by Frisch as follows:

'Very gradually we realised that the break-up of a uranium nucleus into two almost equal parts . . . had to be pictured in quite a different way. The picture is one . . . of the gradual deformation of the original uranium nucleus, its elongation, formation of a waist and finally separation of the two halves. The striking similarity of that picture with the process of fission by which bacteria multiply caused us to use the phrase 'nuclear fission' in our first publication.

'That publication was somewhat laboriously composed by long-distance telephone (Professor Meitner had gone to Stockholm, and I had returned to Copenhagen) and eventually appeared in *Nature* in February 1939. . . . The most striking feature of this novel form of nuclear reaction was the large energy liberated. . . . But the really important question was whether neutrons were liberated in the process, and that was a point which I, for one, completely missed.'

At that time Frisch was somewhat uneasy about his discovery. He wrote to his mother: 'I felt as if I had caught an elephant by its tail, without meaning to, while walking through a jungle. And now I don't know what to do with it.'

116

Transmutation of the elements is well known to poetry – as in this snatch from John Updike's 'To Crystallization'.

When, on those anvils at the center of stars
and those even more furious anvils
of the exploding supernovae,
the heavy elements were beaten together
to the atomic number of 94
and the crystalline metals with their easily lost
valence electrons arose,
their malleability and conductivity
made Assyrian goldsmithing possible,
and most of New York City.

117

As high-energy physics developed into an increasingly expensive pursuit, inextricably linked to special interests and national prestige, so scientists broke through the barrier that had excluded them from political and administrative power. The dictum in Whitehall had been that scientists must be on tap and not on top. The rise of the scientific politicians is well portrayed in Dan Greenberg's scholarly study, The Politics of Pure Science. *In the mid-1950s when the Cold War was thickening, physicists played the Soviet card shamelessly.*

Representative Melvin Price relates the following conversation with a Soviet physicist at the Dubna laboratory: 'The Dubna Laboratory asked our group when we were there two years ago how we got the money to build our accelerators. We told him the legislative process of getting money on our program. He said, "That is not the way I understand." He said, "I understand you get it by saying the Russians have a 10-million electron volt synchroton and we need a 20-billion electron synchroton and that is how you get your money." I said, "There may be something to it." I said, "How do you get your money?" He said, "The same way." ' Price related this tale to John Williams, director of the AEC research division, who commented, 'That is certainly a very true story.'

118

C.P. Snow was fascinated by science and power. William Cooper, who, as head of the Civil Service Commission, had an insider's view, caught what seems to be the authentically heady atmosphere, notably in his novel Memoirs of a New Man.

Poor old Herbert! Norman, Bill Taylor, Amos Wilson – they'd all got K.s. Plunging still further into trouble, I tried to be funny, I said:

'With the Power Board on one hand and the University on the other, I must be a classic example of *rising* between two stools.'

Bert said in a low voice: 'I expect you're content, anyway.'

I looked at him and said cheerfully: 'On the Power Board's quota it must be your turn next, mustn't it?'

He was apparently occupied with extracting some bones from

the fish on his plate. He didn't answer. I couldn't help seeing two large tears forming in his eyes.

I doubt if Emily Post could have told me what to say then. I knew he shed tears readily, but that was little help.

After a dreadful pause he said: 'Do you know what happened?'

I had to look at him. The tears had rolled down his heavy cheeks into, I supposed, his *sole véronique*. What a fate for *sole véronique*! What a fate for tears!

'What?' I said.

'It *was* my turn next. I was at the top of the Power Board's list when Amos Wilson got it.' He paused. 'I saw it, Jack, with my own eyes – *Norman Standsfield personally switched our names!*'

'What on earth for?' His tears made me believe.

'Just because he meant to. Amos and I are about the same age and seniority. The Chairman made a case for Amos but it wouldn't deceive a cat. Amos got it.' He paused and his breath came in a loud sort of sob. 'It *should* have been *mine*!'

Oh dear, oh dear. . . . I've heard some cries from the heart in my time.

I said nothing. After all, it might well not be true.

Bert Hobbs became steady again.

'Divide and Rule,' he said. 'This is Norman Standsfield's principle, Jack.'

For the moment I was thinking about myself – having said that when Norman was reputed not to have a friend in the world I should have liked to be his friend. I suddenly saw myself as grossly impetuous. However, I said:

'That could be true at Board level, I agree. But actually it isn't true lower down, is it?'

'I'm glad you've noticed this, Jack! Lower down, where it comes to the lads out in the Divisions, Norman Standsfield has done a wonderful job. Really wonderful. Though he may have all us members at each other's throats, he's pulled the lads out in the Divisions together nothing short of marvellously. I know my own lads think the world of him. They think he's really put the National Power Board on the map. And for this they'd go through thick and thin for him, this they would.'

I bowed my head at the thought of those lads, out there, going through thick and thin. Bert went on.

'If any man has succeeded in pulling the National Power Board together, this man is Norman Standsfield.' He paused. 'And *not* the

Great Scientist who held the job prior to him, Lord Forbes!'

I was afraid he was right about that; but it was not, as it happened, because Lord Forbes had indeed been a distinguished scientist.

'You know, Jack,' – he became confidential – 'the Power Board is really too big to be properly pulled together by any man. It's just an impossibility. A sheer impossibility.'

Of course he was now entirely right. The original concept of a National Power Board might have been justified on the grounds of its being grandiose; but it couldn't have been justified on the grounds of its being practicable. The general consensus of opinion nowadays was that the Prime Minister at the time favoured the grandiose.

'The Great Scientist, Lord Forbes,' said Bert, 'bit off more than he could chew. This is the top and bottom of it, Jack. He wanted the Power Board formed, because it would make a nice job for him. And then when he got it he was a wash-out in it.' He paused. He spoke in a friendly tone. 'I know chaps like you and Taylor think we oughtn't to have a non-technical man as Chairman. And I agree with you wholeheartedly. But you *have* got a Scientist to blame. . . . With the best will in the world I would have to say that my thought is the old Treasury was right when they put in one of their own chaps, an Administrator, to replace Forbes, the Scientist. People with administrative expertise make a better job of pulling a big organization together than long-haired scientists.'

I kept my mouth shut.

'My friend, I don't count you as a long-haired scientist – you know this. You've knocked about too much.' He leaned forward and amiably put his hand on my forearm. 'You're for all practical purposes,' he said, 'an Engineer.'

Then he called the waiter. 'Those *crêpes suzette* looked good. Bring some for me and my guest!'

He sat back fatly, pleased with his order – and I was not displeased with it, either. Then suddenly he looked at me with an expression I hadn't seen before, as relaxed as ever, yet wily and watchful.

'While we're talking about pulling things together in the Board, Jack,' he said, 'there's one job ahead of the Chairman that really wants doing urgently, in my opinion. I've had it in mind for a long time now. . . . These recent happenings, that we were talking about at today's meeting, have made me think we oughtn't to put off doing something about it any longer.'

He hadn't waited till we got to the brandy. As he spoke he waved his hand with a reassuring, take-it-easy sort of gesture.

'My thought is,' he said, 'about pulling the Scientists together.'

I have to admit that I didn't catch his drift.

'My thinking goes like this,' he explained. 'As things stand, each Division does its own research and development. Therefore each Division has its own group of scientists. So each Division's group of scientists can make its own technical mistakes independently.' He paused. 'All that the rest of us can do is help pick up the bits afterwards. You see what I mean, don't you, Jack? And I expect you see the way my mind is working. Just in a practical way . . .'

I looked at him steadily.

'Now the way I put it is this – and I want *your* opinion on it. Do you think we should get on more efficiently if we united all the scientists, right across the board, in *one* organization? We'd just have one Research and Development organization for the whole Power Board. With its own Member – the entire responsibility for it would be his. And as a member, he'd be one of *us*.'

The trolly with the chafing-dishes on it was standing beside us and the light of the spirit lamp was flickering in our faces. I was thinking: This is it! I'd told Alice that when the repercussions of the row got going they'd be *something*, but she was to be prepared for the rumble of them to stay subliminal for quite a while. Then one day it would cross the threshold of audibility. . . . Well, if I knew anything about it, the rumble was crossing the threshold now.

The Scientists were to be hived off from the working Divisions – leaving the working Divisions to be run by the Engineers. (I couldn't believe the Administrators were meant to get much of a look-in.)

I looked at him.

'Well,' he said, 'Jack. That's my scenario for the future.'

And then he turned to start giving instructions to the *chef de restaurant* about the *crêpes suzette*.

I must say I admired the beauty and simplicity of his plan – or 'scenario', as the case might be. I hadn't the slightest doubt it would be taken seriously in the Power Board. In fact enough factions might line up behind him to put it into force. The Scientists, seen as arrogant, confident, new men, had not made themselves universally loved. Lots of people would like to see them put in the corner. Whether the plan had any actual merits or not was beside the point.

Herbert Hobbs turned back from the *crêpes suzette,* a smile of anticipation shining round his lips and dispersing into his wobbly chins. As if he were half-surprised to find my thoughts still engaged by his proposal, he said:

'Well, what do you think of my scenario? Get all those scientists off our necks, eh?'

Quickly he patted my forearm again. 'Just my little joke – I don't mean a word of it!' His small hazel eyes flickered brightly. 'You know what a lot I think of the Scientists. The Power Board couldn't run without them, and this is the truth!'

119

In After Many a Summer *by Aldous Huxley, it is a private patron who underwrites the odious Dr Obispo. The aim is to extend man's (in particular the patron's) life-span. Dr Obispo bears a certain resemblance to Serge Voronoff, the leader of the monkey testis transplanters, but Huxley may also have had in mind Casimir Funk, the man to whom we owe the word vitamin. He ran a private laboratory in the South of France, and caused a scandal, so it was reported, when during the Abyssinian war, he read that the tribesmen were in the habit of castrating their Italian prisoners. Being anxious to procure the ensuing material, which was not otherwise easily come by, with a view to isolating the male sex hormone, testosterone, Funk attempted to negotiate with the Abyssinians for samples. This was construed by the Italian government as an insult to the flag, and a diplomatic incident is said to have resulted. Dr Obispo's endeavours are based not on the transplantation of glandular tissue but on the theory that carp – those Methuselahs among the fishes – secrete in their gut a substance that retards the processes of ageing.*

Dr. Obispo halted at last. 'Here we are,' he said, opening a door. A smell of mice and absolute alcohol floated out into the corridor. 'Come on in,' he said cordially.

Jeremy entered. There were the mice all right – cage upon cage of them, in tiers along the wall directly in front of him. To the left, three windows, hewn in the rock, gave on to the tennis court and a distant panorama of orange trees and mountains. Seated at a table in front of one of these windows, a man was looking through a microscope. He raised his fair, tousled head as they approached and turned towards them a face of almost child-like candour and openness. 'Hullo, doc,' he said with a charming smile.

'My assistant,' Dr. Obispo explained, 'Peter Boone. Pete, this is Mr. Pordage.' Pete rose and revealed himself an athletic young giant.

'Call me Pete,' he said, when Jeremy had called him Mr. Boone. 'Everyone calls me Pete.'

Jeremy wondered whether he ought to invite the young man to call him Jeremy – but wondered, as usual, so long that the appropriate moment for doing so passed, irrevocably.

'Pete's a bright boy,' Dr. Obispo began again in a tone that was affectionate in intention, but a little patronizing in fact. 'Knows his physiology. Good with his hands, too. Best mouse surgeon I ever saw.' He patted the young man on the shoulder.

Pete smiled – a little uncomfortably, it seemed to Jeremy, as though he found it rather difficult to make the right response to the other's cordiality.

'Takes his politics a bit too seriously,' Dr. Obispo went on. 'That's his only defect. I'm trying to cure him of that. Not very successfully so far, I'm afraid. Eh, Pete?'

The young man smiled again, more confidently; this time he knew exactly where he stood and what to do.

'*Not* very successfully,' he echoed. Then, turning to Jeremy, 'Did you see the Spanish news this morning?' he asked. The expression on his large, fair, open face changed to one of concern.

Jeremy shook his head.

'It's something awful,' said Pete gloomily. 'When I think of those poor devils without planes or artillery or . . .'

'Well, don't think of them,' Dr. Obispo cheerfully advised. 'You'll feel better.'

The young man looked at him, then looked away again without saying anything. After a moment of silence he pulled out his watch. 'I think I'll go and have a swim before lunch,' he said and walked towards the door.

Dr. Obispo picked up a cage of mice and held it within a few inches of Jeremy's nose. 'These are the sex-hormone boys,' he said with a jocularity that the other found curiously offensive. The animals squeaked as he shook the cage. 'Lively enough while the effect lasts. The trouble is that the effects are only temporary.'

Not that temporary effects were to be despised, he added, as he replaced the cage. It was always better to feel temporarily good than temporarily bad. That was why he was giving old Jo a course of that testosterone stuff. Not that the old bastard had any great need of it with that Maunciple girl around . . .

Dr. Obispo suddenly put his hand over his mouth and looked around towards the window. 'Thank God,' he said, 'he's out of the room. Poor old Pete!' A derisive smile appeared on his face. 'Is he in love?' He tapped his forehead. 'Thinks she's like something in the works of Tennyson. You know, chemically pure. Last month he nearly killed a man for suggesting that she and the old boy . . . Well, you know. God knows what he figures the girl is doing here. Telling Uncle Jo about the spiral nebulae, I suppose. Well, if it makes him happy to think that way, I'm not the one that's going to spoil his fun.' Dr. Obispo laughed indulgently. 'But to come back to what I was saying about Uncle Jo . . .'

Just having that girl around the house was the equivalent of a hormone treatment. But it wouldn't last. It never did. Brown-Sequard and Voronoff and all the rest of them – they'd been on the wrong track. They'd thought that the decay of sexual power was the cause of senility. Whereas it was only one of the symptoms. Senescence started somewhere else and involved the sex mechanism along with the rest of the body. Hormone treatments were just palliatives and pick-me-ups. Helped you for a time, but didn't prevent your growing old.

Jeremy stifled a yawn.

For example, Dr. Obispo went on, why should some animals live much longer than human beings and yet show no signs of old age? Somehow, somewhere we had made a biological mistake. Crocodiles had avoided that mistake; so had tortoises. The same was true of certain species of fish.

'Look at this,' he said; and, crossing the room, he drew back a rubber curtain, revealing as he did so the glass front of a large aquarium recessed into the wall. Jeremy approached and looked in.

In the green and shadowy translucence, two huge fish hung suspended, their snouts almost touching, motionless except for the occasional ripple of a fin and the rhythmic panting of their gills. A few inches from their staring eyes, a rosary of bubbles streamed ceaselessly up towards the light and all around them the water was spasmodically silver with the dartings of smaller fish. Sunk in their mindless ecstasy, the monsters paid no attention.

Carp, Dr. Obispo explained; carp from the fishponds of a castle in Franconia – he had forgotten the name; but it was somewhere near Bamberg. The family was impoverished; but the fish were heirlooms, unpurchasable. Jo Stoyte had had to spend a lot of money to have these two stolen and smuggled out of the country in a specially constructed automobile with a tank under the back seats. Sixty pounders they were; over four feet long; and those rings in their tails were dated 1761.

'The beginning of my period,' Jeremy murmured in a sudden access of interest. 1761 was the year of 'Fingal.' He smiled to himself; the juxtaposition of carp and Ossian, carp and Napoleon's favourite poet, carp and the first premonitions of the Celtic Twilight, gave him a peculiar pleasure. What a delightful subject for one of his little essays! Twenty pages of erudition and absurdity – of sacrilege in lavender – of a scholar's delicately *canaille* irreverence for the illustrious or unillustrious dead.

But Dr. Obispo would not allow him to think his thoughts in peace. Indefatigably riding his own hobby, he began again. There they were, he said, pointing at the huge fish; nearly two hundred years old; perfectly healthy; no symptoms of senility; no apparent reason why they shouldn't go on for another three or four centuries. There they were; and here were you. He turned back accusingly towards Jeremy. Here were you! no more than middle-aged, but already bald, already long-sighted and short-winded; already more or less edentate; incapable of prolonged physical exertion; chronically constipated (could you deny it?); your memory already not so good as it was; your digestion capricious; your potency falling off, if it hadn't, indeed, already disappeared for good.

Jeremy forced himself to smile and, at every fresh item, nodded his head in what was meant to look like an amused assent. Inwardly, he was writhing with a mixture of distress at this all too truthful diagnosis and anger against the diagnostician for the ruthlessness of his scientific detachment. Talking with a humorous self-deprecation about one's own advancing senility was very different from being bluntly told about it by someone who took no interest in you except as an animal that happened to be unlike a fish. Nevertheless, he continued to nod and smile.

Here you were, Dr. Obispo repeated at the end of his diagnosis, and there were the carp. How was it that you didn't manage your physiological affairs as well as they did? Just where and how and why did you make the mistake that had already robbed you of your teeth and hair and would bring you in a very few years to the grave?

Old Metchnikoff had asked those questions and made a bold attempt to answer. Everything he said happened to be wrong; phagocytosis didn't occur; intestinal auto-intoxication wasn't the sole cause of senility; neuronophages were mythological monsters; drinking sour milk didn't materially prolong life; whereas the removal of the large gut *did* materially shorten it. Chuckling, he recalled those operations that were so fashionable just before the War – old ladies and gentlemen with their colons cut out, and in consequence being forced to evacuate every few minutes like canaries! All to no purpose, needless to say; because of course the operation that was meant to make them live to a hundred killed them all off within a year or two. Dr. Obispo threw back his glossy head and uttered one of those peals of brazen laughter which were his regular response to any tale of human stupidity resulting in misfortune. Poor old Metchnikoff, he went on, wiping the tears of merriment from his eyes. Consistently wrong. And yet almost certainly not nearly so wrong as people had thought. Wrong, yes, in supposing that it was all a matter of intestinal statis and auto-intoxication. But probably right, in thinking that the secret was somewhere down there, in the gut. Somewhere in the gut, Dr. Obispo repeated; and, what was more, he believed that he was on its track.

He paused and stood for a moment in silence, drumming with his fingers on the glass of the aquarium. Poised between mud and air, the two obese and aged carps hung in their greenish twilight, serenely unaware of him. Dr. Obispo shook his head at them. The worst experimental animals in the world, he said in a tone of resentment mingled with a certain gloomy pride. Nobody had a right to talk about technical difficulties who hadn't tried to work with fish. Take the simplest operation; it was a nightmare. Had you ever tried to keep its gills properly wet while it was anaesthetized on the operating table? Or, alternatively, to do your surgery under water? Had you ever set out to determine a fish's basal metabolism, or take an electro-cardiograph of its heart action, or measure its blood pressure? Had you ever wanted to analyse its excreta? And, if so, did you know how hard it was even to collect them? Had you ever attempted to study the chemistry of a fish's digestion and assimilation? To determine its blood picture under different conditions? To measure the speed of its nervous reactions?

No, you had not, said Dr. Obispo contemptuously. And until you had, you had no right to complain about anything. He drew the curtain on his fish, took Jeremy by the arm and led him back to the mice.

'Look at those,' he said, pointing to a batch of cages on an upper shelf.

Jeremy looked. The mice in question were exactly like all other mice. 'What's wrong with them?' he asked.

Dr. Obispo laughed. 'If those animals were human beings,' he said dramatically, 'they'd all be over a hundred years old.'

And he began to talk, very rapidly and excitedly, about fatty alcohols and the intestinal flora of carp. For the secret was there, the key to the whole problem of senility and longevity. There, between the sterols and the peculiar flora of the carp's intestine.

Those sterols! (Dr. Obispo frowned and shook his head over them.) Always linked up with senility. The most obvious case, of course, was cholesterol. A senile animal might be defined as one with an accumulation of cholesterol in the walls of its arteries. Potassium thiocyanate seemed to dissolve those accumulations. Senile rabbits would show signs of rejuvenation under a treatment with potassium thiocyanate. So would senile humans. But again, not for very long. Cholesterol in the arteries was evidently only one of the troubles. But then cholesterol was only one of the sterols. They were a closely related group, those fatty alcohols. It didn't take much to transform one into another. But if you'd read old Schneeglock's work and the stuff they'd been publishing at Upsala, you'd know that some of the sterols were definitely poisonous – much more than cholesterol, even in large accumulations. Longbotham had even suggested a connexion between fatty alcohols and neoplasms. In other words, cancer might be regarded, in a final analysis, as a symptom of sterol-poisoning. He himself would go even further and say that such sterol-poisoning was responsible for the entire degenerative process of senescence in man and the other mammals. What nobody had done hitherto was to look into the part played by fatty alcohols in the life of such animals as carp. That was the work he had been doing for the last year. His researches had convinced him of three things: first, that the fatty alcohols in carp did not accumulate in excessive quantity; second, that they did not undergo transformation into the more poisonous sterols; and third, that both of these immunities were due to the peculiar nature of the carp's intestinal flora. What a flora! Dr. Obispo cried enthusiastically. So rich, so wonderfully varied! He had not yet succeeded in isolating the organism responsible for the carp's immunity to old age, nor did he fully understand the nature of the chemical mechanisms involved. Nevertheless, the main fact was certain. In one way or another, in combination or in isolation, these organisms contrived to keep the fish's sterols from turning into poisons. That was why a carp could live a couple of hundred years and show no signs of senility.

Could the intestinal flora of a carp be transferred to the gut of a mammal? And if transferable, would it achieve the same chemical and biological results? That was what he had been trying, for the past few months, to discover. With no success, to begin with. Recently, however, they had experimented with a new technique – a technique that protected the flora from the processes of digestion, gave it time to adapt itself to the unfamiliar conditions. It had taken root. The effect on the mice had been immediate and significant. Senescence had been halted, even reversed. Physiologically, the animals were younger than they had been for at least eighteen months – younger at the equivalent of a hundred years than they had been at the equivalent of sixty.

Outside in the corridor an electric bell began to ring. It was lunch time. The two men left the room and walked towards the elevator. Dr. Obispo went on talking. Mice, he said were apt to be a bit deceptive. He had now begun to try the thing out on larger animals. If it worked all right on dogs and baboons, it ought to work on Uncle Jo.

120

'Ere You Were Queen of Sheba' is Sir Arthur Shipley's serene vision of primordial sex life.

ERE YOU WERE QUEEN OF SHEBA

When we were a soft amoeba, in ages past and gone,
Ere you were Queen of Sheba, or I King Solomon,
Alone and undivided, we lived a life of sloth,
Whatever you did, I did; one dinner served for both.
Anon came separation, by fission and divorce,
A lonely pseudopodium I wandered on my course.

One of the most successful of H.G. Wells's scientific romances was The Food of the Gods, *which was about the consequences of the release of a growth-factor (now discovered) on the flora and fauna of the English Home Counties. If a sour note seems to creep into Wells's descriptions of Mr Bensington, Professor Redmond and their friends of the Royal Society, it may be the result of the author's lifelong, and ultimately frustrated craving for a Fellowship of that body.*

In the middle years of the nineteenth century there first became abundant in this strange world of ours a class of men, men tending for the most part to become elderly, who are called, and who, though they dislike it extremely, are very properly called 'Scientists.' They dislike that word so much that from the columns of *Nature*, which was from the first their distinctive and characteristic paper, it is as carefully excluded as if it were – that other word which is the basis of all really bad language in this country. But the Great Public and its Press know better, and 'Scientists' they are, and when they emerge to any sort of publicity, 'distinguished scientists' and 'eminent scientists' and 'well-known scientists' is the very least we call them.

Certainly both Mr. Bensington and Professor Redwood quite merited any of these terms long before they came upon the marvellous discovery of which this story tells. Mr. Bensington was a Fellow of the Royal Society and a former president of the Chemical Society, and Professor Redwood was Professor of Physiology in the Bond Street College of the London University and had been grossly libelled by the anti-vivisectionists time after time. And both had led lives of academic distinction from their very earliest youth.

They were of course quite undistinguished-looking men, as indeed all true Scientists are. There is more personal distinction about the mildest-mannered actor alive than there is about the entire Royal Society. Mr. Bensington was short and very, very bald, and he stooped slightly; he wore gold-rimmed spectacles and cloth boots that were abundantly cut open because of his numerous corns, and Professor Redwood was entirely ordinary in his appearance. Until they happened upon the Food of the Gods (as I must insist upon calling it) they led lives of such eminent and studious obscurity that it is hard to find anything whatever to tell the reader about them.

Mr. Bensington won his spurs (if one may use such an

expression of a gentleman in boots of slashed cloth) by his splendid researches upon the More Toxic Alkaloids, and Professor Redwood rose to eminence – I do not clearly remember how he rose to eminence. I know he was very eminent, and that's all. But I fancy it was a voluminous work on Reaction Times with numerous plates of sphygmograph tracings (I write subject to correction) and an admirable new terminology that did the thing for him.

The general public saw little or nothing of either of these gentlemen. Sometimes at such places as the Royal Institution and the Society of Arts it did in a sort of way see Mr. Bensington, or at least his blushing baldness and something of his collar and coat, and hear fragments of a lecture or paper that he imagined himself to be reading audibly; and once I remember – one midday in the vanished past – when the British Association was at Dover, coming on Section C or D or some such letter, which had taken up its quarters in a public-house, and following out of mere curiosity, two serious-looking ladies with paper parcels through a door labelled 'Billiards' and 'Pool' into a scandalous darkness, broken only by a magic-lantern circle of Redwood's tracings.

I watched the lantern slides come and go, and listened to a voice (I forget what it was saying) which I believe was thc voice of Professor Redwood, and there was a sizzling from the lantern and another sound that kept me there, still out of curiosity, until the lights were unexpectedly turned up. And then I perceived that this sound was the sound of the munching of buns and sandwiches and things that the assembled British Associates had come there to eat under cover of the magic-lantern darkness.

And Redwood I remember went on talking all the time the lights were up and dabbing at the place where his diagram ought to have been visible on the screen – and so it was again so soon as the darkness was restored. I remember him then as a most ordinary, slightly nervous-looking dark man, with an air of being preoccupied with something else and doing what he was doing just then under an unaccountable sense of duty.

I heard Bensington also once – in the old days – at an educational conference in Bloomsbury. Like most eminent chemists and botanists, Mr. Bensington was very authoritative upon teaching – though I am certain he would have been scared out of his wits by an average Board School class in half-an-hour – and so far as I can remember now, he was propounding an improvement of Professor Armstrong's Heuristic method, whereby at the cost of three or four hundred pounds' worth of apparatus, a total neglect of all other studies and the undivided attention of a teacher of exceptional gifts, an average child might with a peculiar sort of thumby thoroughness acquire in the course of ten or twelve years almost as much chemistry as one could learn from one of those objectionable shilling text-books that were then so common at that date. . . .

Quite ordinary persons you perceive, both of them, outside their science. Or if anything on the unpractical side of ordinary. And that you will find is the case with 'scientists' as a class all the world over. What there is great about them is an annoyance to their fellow scientists and a mystery to the general public, and what is not is evident.

There is no doubt about what is not great, no race of men have such obvious littlenesses. They live so far as their human intercourse goes, in a narrow world; their researches involve infinite attention and an almost monastic seclusion; and what is left over is not very much. To witness some queer, shy, misshapen, grey-headed, self-important little discoverer of great discoveries, ridiculously adorned with the wide ribbon of an order of chivalry and holding a reception of his fellow men, or to read the anguish of *Nature* at the 'neglect of science' when the angel of the birthday honours passes the Royal Society by, or to listen to one indefatigable lichenologist commenting on the work of another indefatigable lichenologist, such things force one to realise the unfaltering littleness of men.

And withal the reef of science that these little 'scientists' built and are yet building is so wonderful, so portentous, so full of mysterious half-shapen promises for the mighty future of man! They do not seem to realise the things they are doing. No doubt long ago even Mr. Bensington, when he chose this calling, when he consecrated his life to the alkaloids and their kindred compounds had some inkling of the vision – more than an inkling. Without some great inspiration, for such glories and positions only as a 'scientist' may expect, what young man would have given his life to this work, as young men do? No, they *must* have seen the glory, they must have had the vision, but so near that it has blinded them. The splendour has blinded them, mercifully, so that for the rest of their lives they can hold the light of knowledge in comfort – that we may see.

And perhaps it accounts for Redwood's touch of preoccupation, that – there can be no doubt of it now – he among his fellows was different; he was different inasmuch as something of the vision still lingered in his eyes.

The Food of the Gods I call it, this substance that Mr. Bensington and Professor Redwood made between them; and having regard now to what it has already done and all that it is certainly going to do, there is surely no exaggeration in the name. But Mr. Bensington would no more have called it by that name in cold blood than he would have gone out from his flat in Sloane Street clad in regal scarlet and a wreath of laurel. The phrase was a mere first cry of astonishment from him. He called it the Food of the Gods in his enthusiasm, and for an hour or so at the most altogether. After that he decided he was being absurd. When he first thought of the thing he saw, as it were, a vista of enormous possibilities – literally enormous possibilities, but upon this dazzling vista, after one stare of amazement, he resolutely shut his eyes even as a conscientious 'scientist' should. After that, the Food of the Gods sounded blatant to the pitch of indecency. He was surprised he had used the expression. Yet for all that something of that clear-eyed moment hung about him and broke out ever and again. . . .

'Really, you know,' he said, rubbing his hands together and laughing nervously, 'it has more than a theoretical interest.

'For example,' he confided, bringing his face close to the Professor's and dropping to an undertone, 'it would perhaps, if suitably handled, sell. . . .

'Precisely,' he said, walking away – 'as a Food. Or at least a food ingredient.

'Assuming of course that it is palatable. A thing we cannot know till we have prepared it.'

He turned upon the hearthrug, and studied the carefully designed slits upon his cloth shoes.

'Name?' he said, looking up in response to an inquiry. 'For my part I incline to the good old classical allusion. It – it makes Science res– Gives it a touch of old-fashioned dignity. I have been thinking. . . . I don't know if you will think it absurd of me. . . . A little fancy is surely occasionally permissible. . . . Herakleophorbia. Eh? The nutrition of a possible Hercules? You know it *might*. . . .

'Of course if you think *not* –'

Redwood reflected with his eyes on the fire and made no objection.

'You think it would do?'

Redwood moved his head gravely.

'It might be Titanophorbia, you know. Food of Titans. . . . You prefer the former?

'You're quite sure you don't think it a little *too* –'

'No.'

'Ah! I'm glad.'

And so they called it Herakleophorbia throughout their investigations, and in their report – the report that was never published, because of the unexpected developments that upset all their arrangements, it is invariably written in that way. There were three kindred substances prepared before they hit on the one their speculations had foretold, and these they spoke of as Herakleophorbia I, Herakleophorbia II, and Herakleophorbia III. It is Herakleophorbia IV which I – insisting upon Bensington's original name – call here the Food of the Gods.

122

Karel Čapek was a Czech writer (famous for coining the word robot, which he introduced into his novel, RUR, *or Rossum's Universal Robots). In* War with the Newts, *his best-known novel, he explores what might happen were a new intelligent species to arise and challenge man's supremacy.*

It certainly would be an exaggeration to state that at that time people did not converse about anything else but the talking newts. They also argued and wrote about the next war, the economic crisis, league matches, vitamins, and fashions; all the same they wrote a great deal about the talking newts, and particularly without expert knowledge. Because of that Professor Dr. Vladimír Uher (University of Brno) wrote in the *Lidové Noviny* an article in which he pointed out that the presumed ability of Andrias Scheuchzeri to articulate words, which meant in fact to mimic what it had heard, from the scientific point of view was not nearly so interesting as some other problems with regard to this unique amphibian. The scientific problem of Andrias Scheuchzeri was something quite different: for instance, where it had come from; what was its first place of origin, in which it must have outlived complete geological epochs; why had it remained so long unknown, although it was now recorded in numbers almost everywhere in the equatorial belt of the Pacific Ocean. It appeared as if in recent years it has multiplied abnormally quickly; whence has come this tremendous vitality in an ancient tertiary creature, which until recently had led a completely unknown, and therefore in all probability an extremely sporadic if not geographically isolated existence? Was it, perhaps, that in some way environmental conditions changed for this fossil newt in a direction biologically favourable, so that for a rare Miocene survival a new and strangely successful period of development had set in? In such case it was not impossible that Andrias would not only increase quantitatively but also develop in a qualitative sense, and that their biological studies would have a unique opportunity of witnessing in at least one animal species a striking mutation in actual progress. The fact that Andrias Scheuchzeri croaked a few words, and that it had acquired a few tricks which to a layman appeared to be signs of intelligence, was not in a scientific sense any miracle; but what was miraculous was that powerful vital élan which had so suddenly and to such an extent revived the archaic existence of a primitive creature, and of one almost already extinct. The circumstances were in some directions unique: Andrias Scheuchzeri was the *only* salamander living in the sea and – still more striking – the *only* salamander occurring in the Ethiopian-Australian zone, the mythical Lemuria. Might one not almost suggest that Nature was making abnormal and almost precipitous efforts to make up one of those biological potentialities and forms which in *that* zone it had missed or could not fully elaborate? Moreover, it would have been strange if in the oceanic region which lay between the Japanese Giant Salamanders on one hand and the Alleghanian ones on the other there was no connecting link of any kind. If there had been no Andrias, they would in fact have been driven to ASSUME one in the very localities in which it had appeared; it seemed almost as if it simply filled the habitat which on geographical and evolutionary grounds it *ought* to have occupied from immemorial times. Let that be as it may, the article of the learned professor wound up, on this evolutionary resurrection of a Miocene salamander we observe with respect and amazement that the genius of evolution on our planet has not by any means brought to an end its creative work.

This article appeared in print despite a silent but firmly held opinion of the editorial staff that such a learned discussion is not really suitable for a newspaper.

123

If a species with rival analytical intelligence exists, it is perhaps the dolphins. This was the premise of Leo Szilard's story, 'The Voice of the Dolphins'.

In retrospect, it would appear that among the various recommendations made by the President's Science Advisory Committee there was only one which has borne fruit. At some point or other, the Committee had recommended that there be set up, at the opportune time, a major joint Russian-American research project having no relevance to the national defense, or to any politically controversial issues. The setting up in 1963 of the Biological Research Institute in Vienna under a contract between the Russian and American governments was in line with this general recommendation of the Committee.

When the Vienna Institute came to be established, both the American and the Russian molecular biologists manifested a curious predilection for it. Because most of those who applied for a staff position were distinguished scientists, even though comparatively young, practically all of those who applied were accepted.

This was generally regarded at that time as a major setback for this young branch of science, in Russia as well as in America, and there were those who accused Sergei Dressler of having played the role of the Pied Piper. There may have been a grain of truth in this accusation, inasmuch as a conference on molecular biology held in Leningrad in 1962 was due to his initiative. Dressler spent a few months in America in 1960 surveying the advances in molecular biology. He was so impressed by what he saw that he decided to do something to stimulate this new branch of science in his native Russia. The Leningrad Conference was attended by many Americans; it was the first time that American and Russian molecular biologists came into contact with each other, and the friendships formed on this occasion were to last a lifetime.

When the first scientific communications came out of the Vienna Institute, it came as a surprise to everyone that they were not in the field of molecular biology, but concerned themselves with the intellectual capacity of the dolphins.

That the organization of the brain of the dolphin has a complexity comparable to that of man had been known for a long time. In 1960, Dr. John C. Lilly reported that the dolphins might have a language of their own, that they were capable of imitating human speech and that the intelligence of the dolphins might be equal to that of humans, or possibly even superior to it. This report made enough of a stir, at that time, to hit the front pages of the newspapers. Subsequent attempts to learn the language of the dolphins, to communicate with them and to teach them, appeared to be discouraging, however, and it was generally assumed that Dr. Lilly had overrated their intelligence.

In contrast to this view, the very first bulletin from the Vienna Institute took the position that previous failures to communicate with the dolphins might not have been due to the dolphins' lack of intellectual capacity but rather to their lack of motivation. In a second communiqué the Vienna Institute disclosed that the dolphins proved to be extraordinarily fond of Sell's liver paste, that they became quickly addicted to it and that the expectation of being rewarded by being fed this particular brand of liver paste could motivate them to perform intellectually strenuous tasks.

A number of subsequent communiqués from the Institute concerned themselves with objectively determining the exact limit of the intellectual capacity of the dolphins. These communiqués gradually revealed that their intelligence far surpassed that of man. However, on account of their submerged mode of life, the dolphins were ignorant of facts, and thus they had not been able to put their intelligence to good use in the past.

Having learned the language of the dolphins and established communication with them, the staff of the Institute began to teach them first mathematics, next chemistry and physics, and subsequently biology. The dolphins acquired knowledge in all of these fields with extraordinary rapidity. Because of their lack of manual dexterity the dolphins were not able to perform experiments. In time, however, they began to suggest to the staff experiments in the biological field, and soon thereafter it became apparent that the staff of the Institute might be relegated to performing experiments thought up by the dolphins.

During the first three years of the operation of the Institute all of its publications related to the intellectual capacity of the dolphins. The communiqués issued in the fourth year, five in number, were, however, all in the field of molecular biology. Each one of these communiqués reported a major advance in this field and was issued not in the name of the staff members who had actually performed the experiment, but in the name of the dolphins who had suggested it. (At the time when they were brought into the Institute the dolphins were each designated by a Greek syllable, and they retained these designations for life.)

Each of the next five Nobel Prizes for physiology and medicine was awarded for one or another of these advances. Since it was legally impossible, however, to award the Nobel Prize to a dolphin, all the awards were made to the Institute as a whole. Still, the credit went, of course, to the dolphins, who derived much prestige from these awards, and their prestige was to increase further in the years to come, until it reached almost fabulous proportions.

In the fifth year of its operation, the Institute isolated a mutant form of a strain of commonly occurring algae, which excreted a broad-spectrum antibiotic and was able to fix nitrogen. Because of these two characteristics, these algae could be grown in the open, in improvised ditches filled with water, and they did not require the addition of any nitrates as fertilizer. The protein extracted from them had excellent nutritive qualities and a very pleasant taste.

The algae, the process of growing them and the process of extracting their protein content, as well as the protein product itself, were patented by the Institute, and when the product was marketed – under the trade name Amruss – the Institute collected royalties.

If taken as a protein substitute in adequate quantities, Amruss markedly depresses the fertility of women, but it has no effect on the fertility of men. Amruss seemed to be the answer to the prayer of countries like India. India had a severe immediate problem of food shortage; and she had an equally severe long-term problem, because her population had been increasing at the rate of five million a year.

Amruss sold at about one tenth of the price of soybean protein, and in the first few years of its production the demand greatly exceeded the supply. It also raised a major problem for the Catholic Church. At first Rome took no official position on the consumption of Amruss by Catholics, but left it to each individual bishop to issue such ruling for his diocese as he deemed advisable. In Puerto Rico the Catholic Church simply chose to close an eye. In a number of South American countries, however, the bishops took the position that partaking of Amruss was a mortal sin, no different from other forms of contraception.

In time, this attitude of the bishops threatened to have serious consequences for the Church, because it tended to undermine the institution of the confession. In countries such as El Salvador, Ecuador, Nicaragua and Peru, women gradually got tired of confessing again and again to having committed a mortal sin, and of being told again and again to do penance; in the end they simply stopped going to confession.

When the decline in the numbers of those who went to confession became conspicuous, it came to the attention of the Pope. As is generally known, in the end the issue was settled by the papal bull 'Food Being Essential for Maintaining Life,' which stressed that Catholics ought not to be expected to starve when food was available. Thereafter, bishops uniformly took the position that Amruss was primarily a food, rather than a contraceptive.

The income of the Institute, from the royalties collected, rapidly increased from year to year, and within a few years it came to exceed the subsidies from the American and Russian governments. Because the Institute had internationally recognized tax-free status, the royalties were not subject to tax.

The first investment made by the Vienna Institute was the purchase of television stations in a number of cities all over the world. Thereafter, the television programs of these stations carried no advertising. Since they no longer had to aim their programs at the largest possible audience, there was no longer any need for them to cater to the taste of morons. This freedom from the need of maximizing their audience led to a rapid evolution of the art of television, the potential of which had been frequently surmised but never actually realized.

One of the major television programs carried by the Amruss stations was devoted to the discussion of political problems. The function of *The Voice of the Dolphins*, as this program was called, was to clarify what the real issues were. In taking up an issue, *The Voice* would discuss what the several possible solutions were and would indicate in each case what the price of that particular solution might be. A booklet circulated by *The Voice of the Dolphins* explained why the program set itself this particular task, as follows:

Political issues were often complex, but they were rarely anywhere as deep as the scientific problems which had been solved in the first half of the century. These scientific problems had been solved with amazing rapidity because they had been constantly exposed to discussion among scientists, and thus it appeared reasonable to expect that the solution of political problems could be greatly speeded up also if they were subjected to the same kind of discussion. The discussions of political problems by politicians were much less productive, because they differed in one important respect from the discussions of scientific problems by scientists: When a scientist says something, his colleagues must ask themselves only whether it is true. When a politician says something, his colleagues must first of all ask, 'Why does he say it?'; later on they may or may not get around to asking whether it happens to be true. A politician is a man who thinks he is in possession of the truth and knows what needs to be done; thus his only problem is to persuade people to do what needs to be done. Scientists rarely think that they are in full possession of the truth, and a scientist's aim in a discussion with his colleagues is not to persuade but to clarify. It was clarification rather than persuasion that was needed in the past to arrive at the solution of the great scientific problems.

124

When Vicki Baum wrote Helene, *Voronoff, Steinach and their testis transplants were still to some extent taken seriously.* Helene *is about the development of a rejuvenating drug. One might conjecture that Professor Köbellin is based on Emil Behring, the founder of the pharmaceutical company, Behringwerke. (He did indeed have a Japanese assistant.)*

Director Botstiber looked at her. With his experienced eye he took in the determination and also the exhausted, overworked and strained look of her face. Quickly he recognised the redness of her eyelids as an inflammatory condition due to too much work by artificial light. He summarised his impressions and decided that this Willfüer was an incalculably precious worker, who must be won for the business at all costs. And having made this decision he threw one leg over the other, put away his cigar and picked up a glass paperweight, into which he had a habit of peering during conferences – it was a trick, a means of concentration.

'We have asked you to come here,' he began, 'to talk over the details of the contract which we propose to make with you. You will permit me first of all to tell these gentlemen briefly the gist of the preliminary negotiations which we have already conducted. They concern, as you gentlemen already know, the manufacture of a new medium, "Testinucleose," the patent for which we have obtained from Professor Köbellin. The medium has been tested with amazing success, and has an important future. Here – if you gentlemen would care to glance through this literature,' he said, and took a pile of printed papers from the writing-desk and handed them to the stern, bearded men who were sitting in their armchairs. They had, however, already made themselves acquainted with it, and were familiar with the opinions and the reports of the scientific journals, and the printed lecture delivered at the last Physiological Congress, so that the handing round of these pamphlets was nothing more than a business gesture. Moreover, this new medium had made quite enough sensation in scientific circles. . . .

'You see,' the Director continued, burying himself again in his glass prism, 'that "Testinucleose" has a somewhat similar action to that of Steinach's much-discussed operative experiment of not so long ago. But its action is not limited so completely to the reproductive sphere, but gets to the whole root of existence, if I may be allowed to express myself so popularly and unscientifically. However, we have a draft of a small advertising pamphlet here, and it would perhaps be simplest to read you a few phrases from it. Dr. Sandhagen, if you please. . . .'

Dr. Sandhagen, holding a paper well away from his far-sighted eyes, began to read aloud in a heavy tone of voice.

'Testinucleose is a lipoid which Professor Köbellin, assisted by Dr. Sei Mitsuro and Dr. Helene Willfüer, was able in the first instance to isolate from a certain group of cells in the hypophysis. Its analysis and synthetic preparation has at last been achieved after fourteen years of experimental work.

'It is a specific remedy, which not only possesses a rejuvenating power, but has a general tonic effect. It excites the formation of new cells in the system and raises the power and enjoyment of life. Its effects are not limited to the stimulation of the sexual functions, but a general, quite surprising stimulation and resuscitation of worn-out organisms appear after a course of Testinucleose. Moreover, this effect has not only been achieved in ageing people, but also in people who have needed strengthening after illness or an operation, as well as those who are by nature and disposition lacking in energy and vitality. During the past few years science has revealed the connection between the glands and mental conditions. The dependence of mental conditions and processes on the progress of the endocrine activity of the glands has been shown and proved by the action of Testinucleose. Testinucleose is the synthetic preparation of a hormone which is only produced by certain cells of the hypophysis, and to a certain extent stimulates the action of the other glands, especially the sexual gland. It thus avoids the complex sexual action and retains only those specific influences on the formation of new cells and general condition which one may popularly describe as keeping young or rejuvenation. Testinucleose has a most reliable and astonishing action and is entirely free from poisonous or harmful ingredients. It can be prepared in tablets or ampoules for commercial use. Briefly, one may say that the age-old problem of rejuvenation has been satisfactorily solved.'

125

There follows another fragment from Arthur Hailey's tale of research and skulduggery in the pharmaceutical industry, Strong Medicine. *This must surely be a* roman à clef.

The news, when it broke, came quietly. Deceptively casual, even then it did not reveal itself entirely, and afterwards it seemed to Celia as if fate had tiptoed in, at first unheeded and wearing a prosaic scabbard from which, later, emerged a fiery sword.

It began with a telephone call when Celia was away from her office. When she returned, a message – one of several – informed her that Mr Alexander Stowe, of Exeter & Stowe Laboratories, had phoned and would like her to call him. There was nothing to indicate the request was urgent, and she dealt with several other matters first.

An hour or so later, Celia asked for a call to be placed to Stowe, and soon after was informed by a secretary that he was on the line.

She pressed a button and said into a speakerphone, 'Hello, Alex. I was thinking about you this morning, wondering how your Arthrigo-Hexin W programme is going.'

There was a moment's silence, then a surprised voice, 'We cancelled our contract with you four days ago, Celia. Didn't you know?'

Now the surprise was hers. 'No, I didn't. If you told someone at your place to cancel, are you sure they followed through?'

'I handled it myself,' Stowe said, obviously still puzzled. 'I talked directly with Vince Lord. Then today, realizing I hadn't spoken with you, thought I should, as a courtesy. It's why I called.'

Annoyed at being told something she should have known sooner, Celia answered, 'I'll have something to say to Vince.' She stopped. 'What was your reason for cancelling?'

'Well . . . frankly, we're worried about those deaths from

infections. We've had two ourselves in patients we were monitoring, and while it doesn't look as if either drug – Arthrigo or Hexin W – was directly responsible, there are still unanswered questions. We're uneasy about them, so we decided not to go on, particularly in view of those other deaths elsewhere.'

Celia was startled. For the first time since the conversation began, a shiver of chill ran through her. She had a sudden premonition there was more to come and she would not like hearing it.

'What other deaths?'

This time the silence was longer. 'You mean you don't know about those either?'

She said impatiently, 'If I did, Alex, I wouldn't be asking.'

'There are four we actually know about here, though without details, except that all the deceased were taking Hexin W and died from differing types of infection.' Stowe stopped, and when he resumed his voice was measured and serious. 'Celia, I'm going to make a suggestion, and please don't think this presumptuous since it concerns your own company. But I think you need to have a talk with Dr. Lord.'

'Yes,' Celia agreed. 'So do I.'

'Vince knows about the deaths – the other ones and ours – because we discussed them. Also, he'll have had details, so as to inform the FDA.' Another hesitation. 'I truly hope, for everyone's sake in your shop, that FDA *has* been informed.'

'Alex,' Celia said, 'there appear to be some gaps in my knowledge and I intend to fill them right away. I'm obliged to you for what you've told me. Meanwhile, there doesn't seem much point in our continuing this conversation.'

'I agree with you,' Stowe said. 'But do please call me if there's any other information you need, or any way I can be of service. Oh, and the real purpose of my calling was to say I'm genuinely sorry we had to cancel. I hope, some other time, we can work together.'

Celia answered automatically, her mind already on what must be done next. 'Thank you, Alex. I hope so too.'

She terminated the call by touching a button. She was about to press another which would have connected her with Vincent Lord, then changed her mind. She would go to see him personally. Now.

The first report of death where a patient had been taking Hexin W arrived at Felding-Roth headquarters two months after the drug's introduction. It had come, as was usual, to Dr. Lord. Moments after reading it, he dismissed it entirely.

The report was from a physician in Tampa, Florida. It revealed that while the deceased had been taking Hexin W in conjunction with another drug, the cause of death was a fever and infection. Lord reasoned that the death could have had no relation to Hexin W, therefore he tossed the report aside. However, later that day, instead of sending it for routine filing, he placed the report in a folder in a locked drawer of his desk.

The second report came two weeks later. It was from a Felding-Roth detail man and was mailed after a conversation with a doctor in Southfield, Michigan. The salesman had been conscientious in recording all the information he could find.

Reports about side effects of drugs, including adverse effects, came to pharmaceutical companies from several sources. Sometimes physicians wrote directly. At other times, hospitals did so as routine procedure. Responsible pharmacists passed on what they learned. Occasionally, word came from patients themselves. As well, the companies' detail men and women had instructions to report anything they were told about a product's effect, no matter how trivial it seemed.

Within any pharmaceutical company, reports of side effects of drugs were accumulated and, in quarterly reports, passed to the FDA. That was required by law.

Also required by law was that any serious reaction, particularly with a new drug, must be passed to FDA, and flagged as 'urgent', within fifteen days of the company's learning of it. The rule applied whether the company believed its drug to be responsible or not.

The detail man's report from Southfield, again read by Lord, revealed that the patient, while taking Hexin W and another anti-arthritic drug, died from a massive liver infection. This was confirmed at autopsy.

Again, Lord decided that Hexin W could not possibly have been the cause of death. He put the report in the folder with the first.

A month went by, then two reports came in, separately but at the same time. They recorded deaths of a man and a woman. In both cases they had been taking Hexin W with another drug. The woman, elderly, developed a serious bacterial infection in a foot after it was cut in a home accident. As an emergency measure the foot was amputated, but the infection spread quickly, causing death. The man, who had been in poor health, died from an overwhelming infection of the brain.

Lord's reaction was one of annoyance with the two dead people. *Why* had their *damned diseases*, from which they would have expired anyway, had to involve Hexin W, even though the drug was clearly not responsible in either case? Just the same, the accumulating reports were becoming an embarrassment. Also a worry.

By this time Lord was aware of his failure to comply with federal law by not reporting the earlier incidents immediately to FDA. Now, he was in an impossible position.

If he sent the latest reports to FDA, he could not omit the earlier ones. Yet those were long overdue under the fifteen-day reporting rule, and if he sent them, both Felding-Roth and he personally would be shown as guilty of a law violation. Anything could happen. He was uncomfortably conscious of Dr. Gideon Mace probably waiting at FDA to pounce on such an opportunity.

Lord put the two latest reports in his folder with the others. After all, he reminded himself, he was the only one with knowledge of the total number. Each had arrived separately. None of the individuals making a report was aware of the others.

By the time Alexander Stowe telephoned, cancelling Exeter & Stowe's contract for the use of Hexin W, Lord had accumulated twelve reports and was living in fear. He also learned – increasing his anxiety – that Stowe had somehow heard about four of those Hexin W-related deaths. Lord did not tell Stowe the actual number was twelve, *plus* the two Stowe knew about directly, which Lord learned of for the first time.

Since, legally, Lord could not ignore what Stowe had told him, the total of known deaths was now fourteen.

A fifteenth report came in on the day that Stowe telephoned Celia. By then, reluctantly but unable to avoid the scientific truth, Lord had gained an idea of what was causing the deaths – most of them, if not all.

Several months earlier in Celia's office, during that sales planning meeting where afterwards his words had been applauded, he had described the effect of Hexin W. . . . *stops free radical production, so that leukocytes – white blood cells – are not attracted to a disease site. . . Result: no inflammation . . . pain disappears.*

All of that was true.

What was also becoming clear, by deduction and some hasty

new experiments, was that banishment of leukocytes opened up a weakness, a vulnerability. In the ordinary way leukocytes, though causing pain, were also a protection. But in their absence – an absence caused by the quenching of free radicals – bacteria and other organisms flourished, creating massive infections in various body locales.

And death.

Though it had yet to be proved, Vincent Lord was sure that Hexin W was, after all, the cause of at least a dozen deaths, perhaps more.

He also realized, too late to be of use, that there had been a weakness in the Hexin W clinical testing programme. Most of the patients observed had been in hospitals under controlled conditions where infections were less apt to flourish. All of the deaths recorded in his folder had occurred *away from hospitals*, in homes or other non-controlled environments where bacteria could live and breed . . .

Lord reached the conclusion – acknowledging his failure, shattering his dreams, reinforcing his present, desperate fears – only a few minutes before Celia arrived.

He knew now that Hexin W would have to be withdrawn. He knew, with despair, that he was guilty of concealment – a concealment causing deaths that could have been prevented. As a result he faced disgrace, prosecution, and perhaps imprisonment.

Strangely, his mind went back to twenty-seven years before . . . Champaign-Urbana, the University of Illinois, and the day in the dean's office when he had asked for accelerated promotion, which had been refused.

He had sensed then that the dean believed he, Vincent Lord, was flawed by some defect of character. Now, for the first time, peeling the layers from his soul, Lord asked himself: Had the dean been right?

126

We have already met Gregory Benford's Timescape. *Here is a further extract, in which the strategy of communication through time is debated.*

Markham sighed. 'Until tachyons were discovered, everybody thought communication with the past was impossible. The incredible thing is that the physics of time communication had been worked out earlier, almost by accident, as far back as the 1940s. Two physicists named John Wheeler and Richard Feynman worked out the correct description of light itself, and showed that there were *two* waves launched whenever you tried to make a radio wave, say.'

'Two?'

'Right. One of them we receive on our radio sets. The other travels backward in time – the "advanced wave," as Wheeler and Feynman called it.'

'But we don't receive any message before it's sent.'

Markham nodded. 'True – but the advanced wave is *there*, in the mathematics. There's no way around it. The equations of physics are all time-symmetric. That's one of the riddles of modern physics. How is it that we perceive time passing, and yet all the equations of physics say that time can run either way, forward or backward?'

'The equations are wrong, then?'

'No, they're not. They can predict anything we can measure – but *only* as long as we use the "retarded wave," as Wheeler and Feynman called it. That's the one that you hear through your radio set.'

'Well, look, surely there's a way to change the equation round until you get only the retarded part.'

'No, there isn't. If you do that to the equations, there's no way to keep the retarded wave the same. You *must* have the advanced wave.'

'All right, where *are* those backward-in-time radio shows? How come I can't tune into the news from the next century?'

'Wheeler and Feynman showed that it can't get here.'

'Can't get into this year? I mean, into our present time?'

'Right. See, the advanced wave can interact with the whole universe – it's moving back, into our past, so it eventually hits all the matter that's ever been. Thing is, the advanced wave strikes all that matter before the signal was sent.'

'Yes, surely.' Peterson reflected on the fact that he was now, for the sake of argument, accepting the 'advanced wave' he would have rejected only a few moments before.

'So the wave hits all that matter, and the electrons inside it jiggle around in *anticipation* of what the radio station will send.'

'Effect preceding a cause?'

'Exactly. Seems contrary to experience, doesn't it?'

'Definitely.'

'But the vibration of those electrons in the whole rest of the universe has to be taken into account. *They* in turn send out both advanced and retarded waves. It's like dropping two rocks into a pond. They both send out waves. But the two waves don't just add up in a simple way.'

'They don't? Why not?'

'They interfere with each other. They make a criss-cross network of local peaks and troughs. Where the peaks and troughs from the separate patterns coincide, they reinforce each other. But where the peaks of the first stone meet the troughs of the second, they cancel. The water doesn't move.'

'Oh. All right, then.'

'What Wheeler and Feynman showed was that the rest of the universe, when it's hit by an advanced wave, acts like a whole lot of rocks dropped into that pond. The advanced wave goes back in time, makes all these other waves. They interfere with each other and the result is zero. Nothing.'

'Ah. In the end the advanced wave cancels itself out.'

Suddenly music blared over the Whim's stereo: 'An' de Devil, he do de dance *whump whump* with Joan de Arc –'

Peterson shouted, 'Turn that down, will you?'

The music faded. He leaned forward. 'Very well. You've shown me why the advanced wave doesn't work. Time communication is impossible.'

Markham grinned. 'Every theory has a hidden assumption. The trouble with the Wheeler and Feynman model was that all those jiggling electrons in the universe in the past might *not* send back just the right waves. For radio signals, they do. For tachyons they don't. Wheeler and Feynman didn't know about tachyons; they weren't even thought of until the middle '60s. Tachyons aren't absorbed the right way. They don't interact with matter the way radio waves do.'

'Why not?'

'They're different kinds of particles. Some guys named Feinberg and Sudarshan imagined tachyons decades ago, but nobody could find them. Seemed too unlikely. They have imaginary mass, for one thing.'

'*Imaginary* mass?'

'Yes, but don't take it too seriously.'

'Seems a serious difficulty.'

'Not really. The mass of these particles isn't what we'd call an observable. That means we can't bring a tachyon to rest, since it must always travel faster than light. So, if we can't bring it to a stop in our lab, we can't measure its mass at rest. The only definition of mass is what you can put on the scales and weigh – which you can't do, if it's moving. With tachyons, all you can measure is momentum – that is, impact.' . . .

'You still haven't got round the grandfather thing. If tachyons can carry a message back to the past, how do you avoid paradoxes?' Peterson did not mention that he had gone through a discussion with Paul Davies at King's about this, but understood none of it. He was by no means assured that the ideas made any sense.

Markham grimaced. 'It's not easy to explain. The key was suspected decades ago, but nobody worked it out into a concrete physical theory. There's even a sentence in the original Wheeler-Feynman paper – "It is only required that the description should be logically self-consistent." By that they meant our sense of the flow of time, always going in one direction, is a bias. The equations of physics don't share our prejudice – they're time-symmetric. The only standard we can impose on an experiment is whether it's *logically* consistent.'

'But it's certainly illogical that you can be alive even after you've knocked off your own grandfather. Killed him before he produced your father, I mean.'

'The problem is, we're used to thinking of these things as though there was some sort of switch involved, that only had two settings. I mean, that your grandfather is either dead or he isn't.'

'Well, *that's* certainly true.'

Markham shook his head. 'Not really. What if he's wounded, but recovers? Then if he gets out of the hospital in time, he can meet your grandmother. It depends on your aim.'

'I don't see –'

'Think about sending messages, instead of shotgunning grandfathers. Everybody assumes the receiver – back there in the past – can be attached to a switch, say. If a signal from the future comes in, the switch is programmed to turn off the transmitter – *before* the signal was sent. There's the paradox.'

'Right.' Peterson leaned forward, finding himself engrossed despite his doubts. There was something he liked about the way scientists had of setting up problems as neat little thought experiments, making a clean and sure world. Social issues were always messier and less satisfying. Perhaps that was why they were seldom solved.

'Trouble is, there's no switch that has two settings – on and off – with nothing in between.'

'Come now. What about the toggle I flip to turn on the lights?'

'Okay, so you flip it. There's a time when that switch is hanging in between, neither off nor on.'

'I can make that a very short time.'

'Sure, but you can't reduce it to zero. And also, there's a certain impulse you have to give that switch to make it jump from off to on. In fact, it's possible to hit the switch just hard enough to make it go halfway – try it. That must've happened to you sometime. The switch sticks, balanced halfway between.'

'All right, granted,' Peterson said impatiently. 'But what's the connection to tachyons? I mean, what's *new* about all this?'

'What's new is thinking of these events – sending and receiving – as related in a chain, a *loop*. Say, we send back an instruction saying, "Turn off the transmitter." Think of the switch moving over to "off." This event is like a wave moving from the past to the future. The transmitter is changing from "on" to "off." Now, that – well, let's call it a wave of information – moves forward in time. So the original signal doesn't get sent.'

'Right. Paradox.'

Markham smiled and held up a finger. He was enjoying this. 'But wait! Think of all these times being in a kind of loop. Cause and effect mean nothing in this loop. There are only *events*. Now as the switch moves towards "off," information propagates forward into the future. Think of it as the transmitter getting weaker and weaker as that switch nears the "off" position. Then the tachyon beam that transmitter is sending out gets weaker.'

'Ah!' Peterson suddenly saw it. 'So the receiver in turn gets a weaker signal from the future. The switch isn't hit so hard because the backward-in-time signal is weaker. So it doesn't move so quickly toward the "off" mark.'

'That's it. The closer it gets to "off," the slower it goes. There's an information wave traveling forward into the future, and – like a reflection – the tachyon beam comes back into the past.'

'What does the experiment *do* then?'

'Well, say the switch gets near "off," and then the tachyon beam gets weak. The switch doesn't make it all the way to "off" and – like that toggle controlling the lights – it starts to fall back toward "on." But the nearer it gets to "on," the stronger the transmitter gets in the future.'

"So the *tachyon* beam gets stronger,' Peterson finished for him. 'That in turn drives the switch away from "on" and back towards "off." The switch is hung up in the middle.'

Markham leaned back and drained his stout. His tan, weakened by the dim Cambridge winter, crinkled with the lines of his wry smile. 'It flutters around there in the middle.'

'No paradox.'

'Well . . .' Markham shrugged imperceptibly. 'No logical contradictions, yes. But we still don't actually know what that intermediate, hung-up state means. It *does* avoid the paradoxes, though. There's a lot of quantum-mechanical formalism you can apply to it, but I'm not sure what a genuine experiment will give.'

'Why not?'

Markham shrugged again. 'No experiments. Renfrew hasn't had the time to do them, or the money.'

Peterson ignored the implied criticism; or was that his imagination? It was obvious that work in these fields had been cut back for years now. Markham was simply stating a fact. He had to remember that a scientist might be more prone simply to state things as they were, without calculating a statement's impact. To change the subject Peterson asked, 'Won't that stuck-in-the-middle effect prevent your sending information back to 1963?'

'Look, the point here is that our distinctions between cause and effect are an illusion. This little experiment we've been discussing is a causal *loop* – no beginning, no end. That's what Wheeler and Feynman meant by requiring only that our description be logically consistent. *Logic* rules in physics, not the myth of cause and effect. Imposing an order to events is *our* point of view. A quaintly human view, I suppose. The laws of physics don't care. That's the new concept of time we have now – as a set of completely interrelated events, linked self-consistently. *We* think we're moving along in time, but that's just a bias.'

'But we know things happen *now*, not in the past or future.'

'When is "now"? Saying that "now" is "this instant" is going around in circles. Every instant is "now" when it "happens." The point is, how do you measure the rate of moving from one instant to the next? And the answer is, you can't. What's the rate of the passage of time?'

'Well, it's –' Peterson stopped, thinking.

'How can time move? The rate is one second of movement per second! There's no conceivable coordinate system in physics from which we can measure time passing. So there isn't any. Time is frozen, as far as the universe is concerned.'

'Then . . .' Peterson raised a finger to cover his confusion, frowning. The manager appeared as though out of nowhere.

'Yes sir?' the main said with extreme politeness.

'Ah, another round.'

'Yes *sir*.' He hustled off to fill the order himself. Peterson took a small pleasure in this little play. To get such a response with a minimum display of power was an old game to him, but still satisfying.

'But you *still* believe,' Peterson said, turning back to Markham, 'that Renfrew's experiment makes sense? All this talk of loops and not being able to close switches . . .'

'Sure it'll work.' Markham accepted a glass dark with the thick stout. The manager placed Peterson's ale carefully before him and began, 'Sir, I want to apol–'

Peterson waved him into silence, impatient to hear Markham. 'Perfectly all right,' he said quickly.

Markham eyed the manager's retreating back. 'Very effective. Do they teach that in the best schools?'

Peterson smiled. 'Of course. There's lecture, then field trips to representative restaurants. You have to get the wrist action just right.'

Markham saluted with the stout. After this silent toast he said, 'Oh yes, Renfrew. What Wheeler and Feynman didn't notice was that if you send a message back which has nothing to do with shutting off the transmitter, there's no problem. Say I want to place a bet on a horse race. I've resolved that I'll send the results of the race back in time to a friend. I do. In the past, my friend places a bet and makes money. That doesn't change the outcome of the race. Afterward, my friend gives me some of the winnings. His handing over the money won't stop me from sending the information – in fact, I can easily arrange it so I only get the money *after* I've sent the message.'

'No paradox.'

'Right. So you *can* change the past, but only if you *don't* try to make a paradox. If you try, the experiment hangs up in that stuck-in-between state.'

Peterson frowned. 'But what's it like? I mean, what does the world seem like if you can change it round?'

Markham said lightly, 'Nobody knows. Nobody's ever tried it before.'

'There were no tachyon transmitters until now.'

'And no reason to try to reach the past, either.'

'Let me get this straight. How's Renfrew going to avoid creating a paradox? If he gives them a lot of information, they'll solve the problem and there'll be no reason for him to send the message.'

'That's the trick. Avoid the paradox, or you'll get a stuck switch. So Renfrew will send a *piece* of the vital information – enough to get research started, but not enough to solve the problem utterly.'

'But what'll it be like for us? The world will change round us?'

Markham chewed at his lower lip. 'I think so. We'll be in a different state. The problem will be reduced, the oceans not so badly off.'

'But what is *this* state? I mean, us sitting here? We know the oceans are in trouble.'

'Do we? How do we know this *isn't* the result of the experiment we're about to do? That is, if Renfrew hadn't existed and thought of this idea, maybe we'd be *worse* off. The problem with causal loops is that our notion of time doesn't accept them. But think of that stuck switch again.'

Peterson shook his head as though to clear it. 'It's hard to think about.'

'Like tying time in knots,' Markham conceded. 'What I've given you is an interpretation of the mathematics. We *know* tachyons are real; what we don't know is what they imply.'

127

The concept of tachyons was perhaps prefigured by Flann O'Brien, whose philosopher de Selby, in The Third Policeman, *knows all about the velocity of light:*

His theory as I understand it is as follows:

If a man stands before a mirror and sees in it his reflection, what he sees is not a true reproduction of himself but a picture of himself when he was a younger man. De Selby's explanation of this phenomenon is quite simple. Light, as he points out truly enough, has an ascertained and finite rate of travel. Hence before the reflection of any object in a mirror can be said to be accomplished, it is necessary that rays of light should first strike the object and subsequently impinge on the glass, to be thrown back again to the object – to the eyes of a man, for instance. There is therefore an appreciable and calculable interval of time between the throwing by a man of a glance at his own face in a mirror and the registration of the reflected image in his eye.

So far, one may say, so good. Whether this idea is right or wrong, the amount of time involved is so negligible that few reasonable people would argue the point. But de Selby ever loath to leave well enough alone, insists on reflecting the first reflection in a further mirror and professing to detect minute changes in this second image. Ultimately he constructed the familiar arrangement of parallel mirrors, each reflecting diminishing images of an interposed object indefinitely. The interposed object in this case was de Selby's own face and this he claims to have studied backwards through an infinity of reflections by means of 'a powerful glass'. What he states to have seen through his glass is astonishing. He claims to have noticed a growing youthfulness in the reflections of his face according as they receded, the most distant of them – too tiny to be visible to the naked eye – being the face of a beardless boy of twelve, and, to use his own words, 'a countenance of singular beauty and nobility'. He did not succeed in pursuing the matter back to the cradle 'owing to the curvature of the earth and the limitations of the telescope.'

128

P.B. Medawar, in his autobiography, Memoirs of a Thinking Radish, *describes how the world reacts when you get a Nobel Prize.*

Immediately upon his designation as such, a Nobel Laureate becomes beneficiary – or, as most of them believe, the victim – of a variant of the kind of notoriety enjoyed by pop stars and anyone reputed by the media to be a 'personality'. Overnight a Laureate is deemed to have become an authority on all the problems that plague society. His opinion is sought upon the efficacy and propriety of fertilizing human ova inside the body, on the desirability of nuclear weapons, the fitness of women for holy orders, and much else besides. The contents of his mailbag change accordingly, for in addition to the professional man's usual ration of letters from persons offering to make his fortune, there are letters from people suggesting various means by which he may make theirs. One favourite proposal is to ask the Laureate to 'enter on the attached form' answers to an enclosed sheet which contains questions such as 'Do you believe in God?' or, 'Do you believe, if there is no personal God, that God is represented by a diffuse benevolence that somehow pervades the entire universe?' A Laureate foolish enough to answer this questionnaire may find that in a month's time he features prominently in a book entitled *Does God Exist? – The Scientist Speaks*. Over and above this there are innumerable requests for autographs, signed cabinet-size photographs, pages of manuscript, personal letters wishing the applicant good cheer and good fortune, all to be answered in the Laureate's own hand. Autograph signatures, it may be explained, have a certain monetary value – as with all other things thought worth collecting, they are valued inversely to their degree of abundance. I only sign autographs when they are accompanied by a self-addressed envelope; I rather suspect that Francis Crick signs none, and that *his* autographs are accordingly worth ten or twenty Medawars. The more obliging Laureates doubtless sign everything and don't realize that they are undermining their market value. Of greater importance, perhaps, are the manifestos and declarations that the Laureate is asked to give his authority to. If these are temperately worded pleas intended to persuade foreign despots to abstain from persecuting their fellow-men, well and good, but more often Laureates are asked to sign uncontroversial manifestos such as that which I quote in my book *Advice to a Young Scientist*: The nations of the world must henceforward live together in amity and concord and abjure the use of warfare as a means of settling political disputes.

A Laureate really desperate for time to get on with research or other proper business may well have recourse to a checklist such as that used by Francis Crick to cope with importunities.*

*I quote from Harriet Zuckerman's *Scientific Élite* (New York, 1977): 'Dr Crick thanks you for your letter but regrets that he is unable to accept your kind invitation to:

send an autograph	speak after dinner	attend a conference
provide a photograph	give a testimonial	act as chairman
cure your disease	help you in your project	become an editor
be interviewed	read your manuscript	write a book
talk on the radio	deliver a lecture	accept an honorary degree.'
appear on TV		

Crick cannot reasonably be accused of arrogance or callousness for using this checklist, because he devoted the time thus saved to research in collaboration with Dr Sidney Brenner which elucidated the nucleotide alphabet of the genetic code – work in my opinion worthy of a second Nobel Prize.

129

The Theory of Relativity, as well as the person of Einstein himself, caught the imagination of the world as perhaps no science or scientist had ever done before. The joke about Einstein's asking the porter at Paddington Station 'Does Oxford stop at this train?' needed no explaining. Here from Ronald Clark's Einstein *is a description of a clash of cultures and of disciplines all at the same time. The account comes from a Japanese paper, reprinted in English in the* Japan Weekly Chronicle. *It describes a discussion 'of quite unusual nature' in the Cabinet Council.*

One of the Ministers asked whether ordinary people would understand Professor Einstein's lectures on the theory of relativity. Mr Kamada, Minister of Education, rather rashly said of course they would. Dr Okano, Minister of Justice, contradicted Mr Kamada, saying that they would never understand. Mr Arai, Minister of Agriculture and Commerce, was rather sorry for Mr Kamada, so he said that they would perhaps understand vaguely. The headstrong Minister of Justice insisted that there could be no midway between understanding and not understanding. If they understood, they understood clearly. If they did not understand, they did not understand at all. A chill fell on the company. Mr Baba, the tactful director of the Legislation Bureau, said that they could understand if they made efforts. Their efforts would be useless, persisted the Minister of Justice. He had himself ordered a book on the theory of relativity when the theory was first introduced into Japan last year and tried to study it. On the first page he found higher mathematics, and he had to shut the book for the present. When the members of the Imperial Academy were invited to dinner at the Hama detached palace, he had mentioned the problem to Dr Tanakadate Aikitsu, who was seated next to him. Dr Fujisawa Rikitaio (an authority on mathematics) overhearing their discussion, said that in America they were collecting popular explanations of the theory, offering an enormous prize. Such being the case, Dr Fujisawa said, it was wiser not to begin the study at once. He supported Dr Okano's opinion. Hearing this elaborate explanation, Mr Baba, director of the Legislation Bureau, decided to eschew Einstein for the time being.

130

Another example of the interest, mingled with unease, that the Theory of Relativity excited in the unlikeliest quarters comes from Recollections and Reflections *by J.J. Thomson, Cavendish Professor of Experimental Physics in Cambridge, Master of Trinity and the discoverer of the electron and of isotopes.*

. . . it excited the interest and admiration of mathematicians and also of the general public to an extent which had never been approached by any other mathematical question. Lectures on it attracted large audiences, books about it became 'best sellers', and one was continually being asked by one's neighbours at a dinner party to explain in simple words what it was about. Once at the Athenaeum Club, Lord Sanderson, who was for long Permanent Secretary at the Foreign Office, came to me and asked if I could help him. He said, 'Lord Haldane has been to the Archbishop (Randall Davidson) and told him that relativity was going to have a great effect upon theology, and that it was his duty as Head of the English Church to make himself acquainted with it'; he went on, 'The Archbishop, who is the most conscientious of men, has procured several books on the subject and has been trying to read them, and they have driven him to what it is not too much to say is a state of *intellectual desperation*. I have read several of them myself and have drawn up a memorandum which I thought might be of service to him. I should be glad if you would read it and let me know what you think about it, before I send it to him.' I said I did not think relativity had anything to do with religion but I would read his memorandum. I see from the biography of the Archbishop by the Bishop of Chichester that about this time the Archbishop met Einstein at Lord Haldane's house, and asked him what effect he thought relativity would have on religion. Einstein said, 'None. Relativity is a purely scientific matter and has nothing to do with religion.'

131

The meeting at the house of Lord Haldane (at different times Secretary of State for War and Lord Chancellor, as well as brother of J.S. Haldane and uncle of J.B.S. – see nos. 43, 44, 47, 48 – and perhaps the first politician to give coherent thought to Government scientific policy), may have been the occasion referred to by H.G. Wells in his Experiment in Autobiography:

When Einstein came to England and was lionized after the war, he was entertained by Haldane. Einstein I know and can converse with very interestingly, in a sort of Ollendorffian French, about politics, philosophy and what not, and it is one of the lost good things in my life, that I was never able to participate in the mutual exploration of these two stupendously incongruous minds. Einstein must have been like a gentle bright kitten trying to make friends with a child's balloon, very large and unaccountably unpuncturable.

132

The attempt to assimilate scientific knowledge into a view of life and into twentieth-century religion is a theme that runs through one of John Updike's most consummate novels, Roger's Version. *Myron, the loudmouthed scientist speaks:*

Kriegman's constant benign smile widens into an audible chuckle. His lips are curious in being the exact same shade of swarthiness as his face, like muscles in a sepia anatomy print. As he raises his glass to these exemplary lips Dale intervenes with 'I think, sir –'

'Fuck the "sir" stuff. Name's Myron. Not Ron, mind you. *My*ron.'

'I think it's a little more than that, what I'm trying to say; the puddle analogy is as if the anthropic principle were being argued from the Earth as opposed to the other planets, which of course we can now see, if we ever doubted it, aren't suitable for life. In that sense, yes, we're here because we're here. But in the case of the universe, where you have only one, why should, say, the observed recessional velocity so exactly equal the necessary escape velocity?'

'How do you know there's only one universe? There might

be zillions. There's no logical reason to say the universe we can observe is the only one.'

'I know there's no *logical* reason –'

'Are we talking logic or not? Don't start getting all intuitive and subjective on me, my pal, because I'm pretty much a pragmatist myself on some scores. If it helps you through the night to believe the moon is green cheese –'

'I don't –'

'Don't believe it is? Good for you. I don't either. Those rocks they brought back didn't test out as green cheese. But my daughter Florence does; some zonked-out punk with purple hair tells her it is when she's as stoned as he is. She thinks she's a Tibetan Buddhist, except on weekends. Her sister Miriam talks about joining some Sufi commune over in New York State. I don't let it get to me, it's their lives. But you, if I size you up right, young fella, you're pulling my leg.'

'I –'

'You really give a damn about cosmology, I'll tell you where the interesting work is being done right now: it's the explanation of how things popped up out of nothing. The picture's filling in from a number of directions, as clear as the hand in front of your face.' He tipped his head back to see Dale better and his eyes seemed to multiply in the trifocals. 'As you know,' he said, 'inside the Planck length and the Planck duration you have this space-time foam where the quantum fluctuations from matter to non-matter really have very little meaning, mathematically speaking. You have a Higgs field tunnelling in a quantum fluctuation through the energy barrier in a false-vacuum state, and you get this bubble of broken symmetry that by negative pressure expands exponentially, and in a couple of microseconds you can have something go from next to nothing to the size and mass of the observable present universe. How about a drink? You look pretty dry, standing there.'

Kriegman takes another plastic glass of white wine from the tray one of the Irish girls is reluctantly passing, and Dale shakes his head, refusing. His stomach has been nervous all this spring. Pastrami and milk don't mix.

My dear friend and neighbour Myron Kriegman takes a lusty swallow, licks his smiling lips, and continues in his rapid rasping voice. 'O.K.; still, you say, you have to begin with *some*thing before you have a Higgs field; how do you get to almost nothing from abso*lute*ly nothing? Well, the answer turns out to be good old simple geometry. You're a mathematician, you'll dig this. What do we know about the simplest structures yet, the quarks? We know – come on, fella, *think*.'

Dale gropes. The party noise has increased, a corner high in his stomach hurts, Esther is laughing on the other side of the living room, beneath the knob-and-spindle header of the archway, exhaling smoke in a plume, her little face tipped back jauntily. 'They come in colors and flavors,' he says, 'and carry positive or negative charges in increments of a third –'

Kriegman pounces: 'You've got it! They invariably occur in threes, and cannot be pried apart. Now what does that suggest to you? Think. Three things, inseparable.'

Father, Son, and Holy Ghost floats across Dale's field of inner vision but does not make it to his lips. Nor does Id, Ego, and Superego. Nor do Kriegman's three daughters.

'The three dimensions of space!' Kriegman proclaims. 'They can't be pried apart either. Now, let's ask ourselves, what's so hot about three dimensions? Why don't we live in two, or four, or twenty-four?'

Odd that the man would mention those almost-magic, almost-revelatory numbers that Dale used to circle painstakingly in red; he now sees them to have been illusions, ripples in nothingness such as Kriegman is rhapsodizing about.

'You're not thinking. Because,' the answer gleefully runs, 'you need no more or less than three dimensions to make a *knot*, a knot that tightens on itself and won't pull apart, and that's what the ultimate particles are – knots in space-time. You can't make a knot in two dimensions because there's no over or under, and – here's the fascinating thing, see if you can picture it – you can make a ravelling in four dimensions but it isn't a knot, it won't hold, it will just pull apart, it won't per*sist*. Hey, you're going to ask me – I can see it in your face – what's this concept, persistence? For persistence you need time, right? And that's the key right there: without time you don't have anything, and if time was two-dimensional instead of one-, you wouldn't have anything either, since you could turn around in it and there wouldn't be any causality. Without causality, there wouldn't be a universe, it would keep reversing itself. I know this stuff must be pretty elementary to you, I can see from the way you keep looking over my shoulder.'

'No, I just –'

'If you've changed your mind about wanting a drink, it's not

Esther's going to get it for you, you should ask one of the girls.'

Dale blushes, and tries to focus on this tireless exposition, though he feels like a knot in four dimensions, unravelling. 'I beg your pardon,' he says, 'how did you say we get from nothing to something?'

Kriegman lightly pats himself on the top of his head to make sure the garland is still in place. 'O.K. Good question. I was just filling in the geometry so you can see the necessity behind space-time as it is and don't go getting all teleological on me. A lesser number of spatial dimensions, it just so happens, couldn't provide enough juxtapositions to get molecules of any complexity, let alone, say, brain cells. More than four, which is what you have with space-time, the complexity increases but not significantly: four is plenty, sufficient. O.K.?'

Dale nods, thinking of Esther and myself, himself and Verna. Juxtapositions.

'So,' says Kriegman. 'Imagine nothing, a total vacuum. But wait! There's something in it! Points, po*ten*tial geometry. A kind of dust of structureless points. Or, if that's too woolly for you, try "a Borel set of points not yet assembled into a manifold of any particular dimensionality." Think of this dust as swirling; since there's no dimension yet, no nearness or farness, it's not exactly swirling as you and I know swirling but, anyway, some of them blow into straight lines and then vanish, because there's nothing to hold the structure. Same thing if they happen, by chance – all this is chance, blind chance it has to be, Jesus' – Kriegman is shrinking, growing stooped; his chins are melting more solidly into his chest; he bobs like a man being given repeated blows on the back of his head – 'if they configurate into two dimensions, into three, even into four where the fourth isn't time; they all vanish, just accidents in this dust of points, nothing could be said to exist, until – even the word "until" is deceptive, implying duration, which doesn't exist yet – until bingo! Space-time. Three spatial dimensions, plus time. It knots. It freezes. The seed of the universe has come into being. Out of nothing. Out of nothing and brute geometry, laws that can't be otherwise, nobody handed them to Moses, nobody had to. Once you've got that seed, that little itty-bitty mustard seed – ka-*boom!* Big Bang is right around the corner.'

'But –' Dale is awed not so much by what this man says as by his fervor, the light of faith in his little tripartite spectacles, the tan monotone of his face and its cascading folds, his receding springy hair, his thick eyebrows thrust outward and up like tiny rhinoceros horns. This man is living, he is on top of his life, life is no burden to him. Dale feels crushed beneath his beady, shuttling, joyful and unembarrassed gaze. 'But,' he weakly argues, ' "dust of points," "freezes," "seed" – this is all metaphor.'

'What isn't?' Kriegman says. 'Like Plato says, shadows at the back of the cave. Still, you can't quit on reason; next thing you'll get somebody like Hitler or Bonzo's pal running things. Look. You know computers. Think binary. When matter meets antimatter, both vanish, into pure energy. But both existed; I mean, there was a condition we'll call "existence." Think of one and minus one. Together they add up to zero, nothing, *nada, niente,* right? Picture them together, then picture them *separating* – peeling apart.' He hands Dale his drink and demonstrates separating with his thick hairy hands palm to palm, then gliding upward and apart. 'Get it?' He makes two fists at the level of his shoulders. 'Now you have something, you have *two* somethings, where once you had nothing.'

'But in the binary system,' Dale points out, handing back the squeezable glass, 'the alternative to one isn't minus one, it's zero. That's the beauty of it, mechanically.'

'O.K. Gotcha. You're asking me, What's this minus one? I'll tell you. It's a *plus one moving backward in time.* This is all in the space-time foam, inside the Planck duration, don't forget. The dust of points gives birth to time, and time gives birth to the dust of points. Elegant, huh? It *has* to be. It's blind chance, plus pure math. They're proving it, every day. Astronomy, particle physics, it's all coming together. Relax into it, young fella. It feels great. Space-time foam.'

Kriegman is joking; Dale prefers him zealous, evangelical on behalf of nonbelief. Esther has vanished from the archway. New guests keep arriving: Noreen Davis, the black receptionist who so smiling gave him those forms seven months ago, with her bald co-worker in the Divinity School front office, and somebody who looks like Amy Eubank but can't be, his recognition apparatus must be out of whack. He masochistically asks Kriegman, 'How about the origin of life? Those odds are pretty impossible, too. I mean, to get a self-replicating organism with its own energy system.'

Kriegman snorts; he twists his face downward as if suddenly very shy; his whole body beneath its garland, in its dirty corduroy jacket with patched elbows and loose buttons, appears to melt and then to straighten again into a bearing almost military. 'Now that just happens to be right up my alley,' he tells Dale. 'That other stuff was just glorified bullshit, way out of my field. I don't know what the hell a Borel set of points are. But I happen to know *exact*ly how life arose; it's brand-new news, at least to the average layman like yourself. Clay. Clay is the answer. Crystal formation in fine clays provided the template, the scaffolding, for the organic compounds and the primitive forms of life. All life did, you see, was take over the phenotype that crystalline clays had evolved on their own, the genetic pass-down factor being entirely controlled by the crystal growth and epitaxy, and the mutation factor deriving from crystal defects, which supply, you don't need me to tell you, the stable alternative configurations you need for information storage. So, you're going to ask, where's the evolution? Picture the pore space of a sandstone, young fella. Every rainstorm, all sorts of mineral solutions are percolating through. Various types of replicating crystals are present, each reproducing its characteristic defects. Some fit together so tightly they form an impervious plug: this is no good. Others are so loose they're washed away when the rains come: this is no good either. But a third type both hangs in there and lets the geochemical solutions, let's even call them nutrients, wash through: this is good. This type of crystal multiplies and grows. It *grows*. Now in that sandstone pore you have a sticky, permeable paste that replicates itself. You have a prototype of life.' Kreigman takes a long swallow of my Almaden and smacks his lips. A half-empty glass sits abandoned on the walnut end table beside the red settee, and my beloved neigbor deftly swaps it with his own, emptied glass.

'But –' Dale says, expecting to be interrupted.

'But, you're going to say, how about us? How were the organic molecules introduced? And why? Well, not to get too technical, some of the amino acids, di- and tricarboxylic acids, make some metal ions, like aluminium, more soluble. This gives us a proto-enzyme. Others, like the polyphosphates, are especially adhesive, which, like I say, has survival value in this prezoic world we're trying to picture. Heterocyclic bases like adenine have a tendency to stick *between* the layers of clay; pretty soon, relatively speaking, you're going to get some RNA-like polymer, with its negatively charged backbone, interacting with the edges of clay particles, which tend to bear a positive charge. *Then* – listen, I know I'm boring the pants off you, I can see from your eyes you're dying to mix it up with somebody over my shoulder, maybe one of my girls. Miriam's the one you might take a shine to, if you don't mind a little Sufi progapanda;

it's the no-alcohol part of it that I couldn't hack. Then, as I was saying, once you've got something like RNA in *not* the primordial soup this time – nobody in the know ever was too comfortable with that crackbrained theory: too – what's the word? – soupy – but a nice crisp paste of clay genes, organic replication is right around the corner, first as a subsystem, a kind of optional extra parallel with the crystal growth, and then taking over with that gene swap I mentioned earlier, and the clay genes falling away, since the organic molecules, mostly carbon, can do the job better, once they're established. Believe me, pal, it fills a lot of theoretical holes. Nothing to matter, dead matter to life, smooth as silk. God? Forget the old bluffer.'

Esther has returned to the living room, far on the other side, and has taken up talking to a young man Dale doesn't know, a graduate student in some professor's entourage, a fair harem boy with messy lank hair that he keeps flicking back with his fingers; Esther's little head, its glowing wide brow and folded gingery-red wings, is tilted amusedly, as it was with Dale at Thanksgiving last year. 'How about life to mind?' he asks Kriegman. His own voice in the bones of his head sounds far away.

Kriegman snorts. 'Don't insult my intelligence,' he says. His smile has dried up. His pants have suddenly been bored off. 'Mind is just a manner of speaking. It's what the brain does. The brain is what's evolved to operate our hands, mostly. If what you've given me is all there is to your theories, young fella, you've got a long way to go.'

'I know,' Dale says, humbly. In his sick-Christian way he relishes the taste of ashes in his mouth, the sensation of having been intellectually flattened.

'You got a girl friend?'

The abrupt gruff question dumbfounds Dale.

'Better get one,' Kriegman advises him. 'It might clean out the cobwebs.'

133

William Blake, too, cared for little for scientists and philosophers and their confined and trivial view (as he saw it) of the world.

> Mock on, Mock on Voltaire, Rousseau:
> Mock on, Mock on: 'tis all in vain!

You throw the sand against the wind,
And the wind blows it back again.

And every sand becomes a Gem
Reflected in the beams divine;
Blown back they blind the mocking Eye,
But still in Israel's paths they shine.

The Atoms of Democritus
And Newton's Particles of light
Are sands upon the Red sea shore,
Where Israel's tents do shine so bright.

134

Leopold Infeld was a mathematician and theoretical physicist, associated for some years with Einstein. His autobiography, Quest, *is a compelling, sometimes painfully moving document about the agonies of trying to follow a career and stay alive in a hostile society.*

A little over a year had passed since my return from England. The question of the professorship at Wilno was still undecided. I regarded my chances as fairly good. There were three steps prescribed by law: *one*, the dean must write to all professors of the same subject in Poland, asking whom they would recommend; *two*, the faculty selects one among these recommended candidates; *three*, the Ministry of Education approves or (very seldom) disapproves the choice. I knew that only the first stage of this involved procedure was completed.

During the Christmas vacation I went to Cracow to see my family. Before going I wrote a postcard to W., the new professor of theoretical physics in Cracow, announcing my visit. We knew each other well. He was a type often met among scientists: very quick to grasp the essential point in discussion but too lazy for the persistent, continuous thought by which alone original results can be achieved. Isolated as he was from other physicists, Professor W. was glad of the opportunity of a scientific discussion and made an appointment with me in a café.

'What are you working on?' was his first question.

'Still on the unitary field theory. Recently I discovered that Maxwell's theory can be generalized in many different ways. This is not astonishing. But I no longer believe that the generalization presented by Born is the simplest. I don't like the arbitrariness of the whole problem.' I took a paper napkin and began to write some formulae to explain my last results. Then I changed the subject:

'By the way, how is Wilno? I hope you have already answered their question about a recommendation.'

'Yes, I did,' answered W. very quietly, absent-mindedly looking at the formulae. 'I recommended Professor C., your superior, from Lwow.'

'What?'

He looked up, holding the paper napkin in his hand, astonished at my violent reaction.

'I thought you knew. Didn't Doctor C. tell you? They decided that they wanted Professor C. from Lwow, and worked upon him until he accepted even though it means less prestige for him. They asked me to make it easy for them by recommending Doctor C. It is an old custom that, together with the official letter asking for a recommendation, private letters are sent suggesting a definite name for recommendation. As it was a purely formal affair and I knew that my letter wouldn't influence the issue, I thought I might as well help them.'

He explained the incident as quietly and simply as though it were the most natural thing in the world. He was astonished that I had ever expected to become a professor in Wilno. He regarded anti-Semitism as something which one must count on and be prepared for, like bad weather. My strong reaction seemed to him out of place. I ought to have learned better in the thirty-eight years of my life.

University professor though he was, I could not help bursting out with a speech which I still do not regret:

'You played a dirty trick on me, which was all the worse because you have pretended to be my friend. You know very well that I am certainly no less fit for this position than Professor C. But Wilno again decided to push C. above me, and it did not occur to you to be loyal to me. You helped them to play the trick and you regard your behaviour as natural. You are even astonished that I should dare to reproach you. In this small matter, in which you could have shown your decency and at least passively opposed anti-Semitism, you accepted the anti-Semitic point of view, a view which sooner or later will turn against you. It begins with one hundred per cent Jews but it does not stop there. The next step is against the fifty per cent Jews like yourself. Acting as you did, you threw away the opportunity to show some decency. You understand that now there is no possibility of friendly relations between us.'

Professor W. became red in the face.

'I guess you are right. But I did not see the problem in this way. It just did not enter my head.'

Without awaiting further explanation I got up and left the table. Two days later I returned to Lwow. I had dinner with Professor C., and between the soup and the meat I remarked:

'I saw Professor W. in Cracow and learned some interesting gossip about Wilno.'

Professor C. blushed, looked down at his napkin and mumbled:

'He told you, most probably, that I am going to Wilno.'

'Yes. I decided to talk to you about it. I have some feeling of tidiness and should like to make my point of view clear, especially as our relations as colleagues here in Lwow have been correct. To stay or to leave Lwow is your personal affair in which I have no right to interfere, although I am a little surprised that you are willing to leave Lwow for an inferior university. But I want you to know that by accepting Wilno you kill my only chance of getting a professorship. I want you to know that I will have to wait twenty years before the chance is repeated, and then I will certainly miss it once more because I will be too old.'

He did not raise his eyes from the table.

'I decided to go to Wilno because I have some relatives there.'

I said: 'I understand. This is an important reason, although I myself prefer to run away from places where I have relatives.'

He felt how silly his answer sounded and added with a sudden outburst:

'They asked me to accept the professorship there. I recommended you, but they told me that they wouldn't take you.'

'Why?'

I wanted to force him to give the simple answer: 'Because you are a Jew.' But instead I heard:

'Because they do not like your writing for the government paper *Gazeta Polska*.'

The intrigue was carefully staged, much better than in the case of my docentship. First Professor C. was convinced that he ought to accept the invitation to Wilno to save the Polish character of the university. The rest was easy. C. would go to Wilno. His chair would then be free in Lwow. Professor Loria would have neither the courage nor the opportunity to push me in Lwow. There was then in Lwow a theoretical physicist, older than I, who was already a professor of the polytechnic school and who could simply be transferred to the university, as his chair of theoretical physics had recently been abolished by the Ministry of Education. The whole strategic plan was worked out in advance. It was like a series of moves in a chess game:

Professor N. from Cracow retires.

Professor W. from Wilno goes to Cracow.

Professor C. from Lwow goes to Wilno.

Professor R. from Lwow Polytechnic School goes to Lwow University.

The chair of theoretical physics in Lwow Polytechnic School is liquidated.

No loopholes anywhere. No Jewish docents wanted. The plan was perfect and it demonstrated clearly that I was superfluous in Polish academic life.

Before making my decision I discussed the situation with Loria. He told me:

'You know how much I should like to help you because I know that you deserve it and that an injustice is being done to you. Anti-Semitism has increased tremendously of late. Our neighbour has set examples which people here would like to imitate and duplicate. The Jewish problem of three and a half million people is a hopeless one. They will be squeezed out economically if the social attitude doesn't change. The only hope is to belong to that thin layer for which exceptions are made. In this respect you are a privileged person. You have a good income from working here and writing, and you can easily afford to wait until the atmosphere changes and we shall be able to do something. But I cannot do anything at this moment. I don't believe in fighting for a hopeless cause.'

'It is not so simple,' I argued. 'First I don't believe that it will be possible to retain my privileged position for any length of time. I could still be quite happy here with the docentship and my newspaper work. But anti-Semitism is like leprosy. Sooner or later it will affect the whole cultural life of our country if it is not checked by social change. The attack on the position of Jews like me will come later, but it must come. And there is another argument against my waiting here hopelessly for a change, an argument more general and basic. I am beginning to be ashamed of my privileged position. The waves of hate are too strong. It does not help me to say that they did not splash me. I know that to go away is cowardice. But so it is to stay here and do nothing about it.'

Loria disagreed with my last remark:

'You cannot do anything about the Jewish problem. In five hundred years historians will say, "After the great war the human race went through fifty years of darkness and barbarism." This will perhaps be their only comment. But these are the fifty years of our lives, the only ones we have. It seems to me that the only solution is to do little decent, seemingly insignificant deeds. I was never considered much of a fighter. But by saving you from teaching in a gymnasium I did something good, something that was valuable. To be decent in the small field in which one is allowed to act means much and – believe me – it is quite unusual among university professors.'

I felt that Loria wanted to keep me in Poland and I fought his arguments: 'I simply cannot bear this atmosphere of hate, of intrigue and discrimination any longer. You could say that I see it in such an exaggerated form now only because it has beaten me personally, that I should have closed my eyes to it had I been offered the silly chair in Wilno. You are right, and I despise myself for it. In some ways I am glad that I received this blow, that it saved me from self-satisfaction and snobbishness. I want to leave Poland. I cannot bear the feeling of being unwanted.'

135

Jeremy Bernstein, in The Life it Brings, *reflects on one of the many small tragedies that afflicted American academics during the period of Senator McCarthy's ascendancy. It scarred others besides Oppenheimer.*

At the same time, Kenneth Bainbridge – a gentle and kind experimental physicist who had played a key role at Los Alamos during the war, and who had become the chairman of the department – suggested that I take a reading course with Wendell Furry in the theory of electromagnetic radiation. He told me that Furry had been one of Oppenheimer's students and that I could learn a great deal from him. So I signed up with Furry, who had taught the freshman physics course I took when I was a junior.

I think it is important to the sequel for me to describe Wendell Furry. He was a solid-looking man with a large stomach that spilled over his pants. He looked and talked like a small-town Midwesterner who might own a garage. He seemed to cultivate his image of being something of an eccentric, a bit of a bumpkin. He was, however, especially in the 1930s, a superb theoretical physicist. Even now I keep learning about contributions Furry made, especially to the quantum theory of radiation. He seemed the most apolitical person imaginable. Both during the course I took with him as a junior and during the reading course, when I was with him alone for many hours, there was never the remotest hint that he had any interest in politics or, indeed, in anything very worldly. It came as a shock that Furry had been a member of the Communist Party for some years after 1938, and that even as I was taking his course, he and the university were under siege by Joseph McCarthy.

One of the strangest documents I have saved from those years is a reprint of the stenographic transcript of the hearings of the Permanent Subcommittee on Investigations of the Committee on Government Operations of the United States Senate, Boston, Massachusetts, January 15, 1954. What is especially odd about the copy I have is that it is signed by the principal participants. Furry wrote, 'Best regards, W. H. Furry'; Leon J. Kamin, the other witness, wrote, simply, in a careful script, 'Leon J. Kamin'; and McCarthy signed with a bold flourish, in very large letters, 'Joseph R. McCarthy.' I have no recollection of asking for the autographs, yet there they are. It seems very peculiar that McCarthy would have signed so flamboyantly, just beneath the signatures of the two men he had pilloried. But he did.

This was Furry's second appearance before the committee. He had appeared on November 4, 1953, and had refused, on constitutional grounds, to answer any questions at all. Meanwhile, the university had taken a position that sounded morally acceptable but that put witnesses like Furry in serious legal jeopardy. The university maintained that if they wanted to retain their jobs, faculty members called to testify would have to reveal their own activities but would not have to 'name names' – reveal the identities of their associates. This placed a man like Furry, for example, in danger of being cited for contempt, since in agreeing to answer some questions about his Communist past, he had in effect given up his immunity.

McCarthy, of course, understood the situation and exploited it for all it was worth. Furry began the hearings by making a statement:

> Our forefathers wrote into the Constitution the privilege of the Fifth Amendment to provide protection which good citizens may sometimes sorely need. Innocent people who feel the threat of false, mistaken, or overzealous prosecution because of unpopular opinions have every right to invoke this protection. It is clear, however, that widespread misrepresentation has produced in many minds a distorted idea of the meaning of the constitutional privilege. Though its real purpose has always been to shield the innocent, many people have been misled into believing that the exercise of the privilege is an admission of guilt. I have now come to the belief that for me to continue to claim my constitutional privilege would bring undue harm to me and to the great institution with which I am connected.
>
> Although I am sure that my past claims to the privilege have been both legally justified and morally right, I now intend to waive my constitutional rights and give this committee all the evidence it may legitimately seek concerning my own activities and associations. I hope that by telling my own political history I can help to dispel suspicion and contribute to public understanding.
>
> Experience has taught me that the inquiry is likely to concern other persons than myself. I feel obliged to state now that I shall respectfully refuse to answer questions that bring in the names of other people. I wish to make it clear, however, that if I knew of any person whose conduct as I saw it was criminal, I should feel bound to reveal the facts. I am not seeking to protect the guilty from prosecution. I wish merely to shield the innocent from persecution. I hope that on this matter the committee will respect my conscience.

In his response McCarthy showed what he thought of Furry's conscience. 'May I say, Mr. Furry,' he said, 'that a man who has been a member of the Communist conspiracy is not exactly the last word on who is guilty and who is innocent.' Furry claimed, and I have every reason to believe him, that he had not been a member of the Communist Party for many years when this confrontation took place. In fact, from his testimony, one got the impression that Furry was thoroughly disillusioned with Communism by this time. McCarthy, however, stated, 'It is not up to you to decide that; it is not up to us to decide that. It is up to this committee to expose the facts. It is up to the law-enforcing agencies to decide who should be prosecuted.'

There then followed a kind of deadly cat-and-mouse game, in which McCarthy tried again and again to trick or coerce Furry into naming names. At one point the dialogue resembled an Ionesco play. Furry admitted that there had been about 'half a dozen' members of the Communist Party working with him on radar during the war. He added, 'By my observation they were among the more security-minded members of this laboratory. They never in any way departed from the rules, to my knowledge, or showed any inclination to regard any outside connections, including that with the Communist Party, as having any bearing on their work. They were devoted to the war effort, they worked loyally, and I shall not reveal their names.' McCarthy decided that Furry might talk if he gave the six men numbers instead of names. 'Let's take them by number instead of name,' he said, overlooking the probability that his numbers and Furry's numbers might be entirely different. The dialogue went as follows:

THE CHAIRMAN: Let's take Number 2. When did you last see Number 2?

MR. FURRY: Of course, I have to decide who is Number 2.

THE CHAIRMAN: You pick out Number 2.

MR. FURRY: And there are a number of these people that I have really very little notion when I last saw them. It has been some time within the last few years.

THE CHAIRMAN: You said there were six. You were a Communist and they were Communists. You say you will not give the name. I am trying to identify them by number. Let's take Number 2. When did you last see him?

MR. FURRY: Well, for Number 2, I will take a man of whom I am not sure just when I last saw him, but it was within the last year or two, and I know from professional connections just where he is working.

THE CHAIRMAN: Where is he working?

MR. FURRY: In an American university and I am pretty sure on nothing connected with government work, certainly no secret work.

THE CHAIRMAN: Is he teaching?

MR. FURRY: Yes. It is my impression that his work is just like mine at the present, teaching and research of what you might call the free-enterprise kind, which still exists in this country.

THE CHAIRMAN: What is his name?

MR. FURRY: I refuse to state.

THE CHAIRMAN: You are ordered to state it.

MR. FURRY: I refuse.

THE CHAIRMAN: Number 3 . . .

And so on, through the whole list.

After some more parrying, McCarthy turned his attention to the university. He asked, 'Do you know of any one connected with Harvard who is or was a member of the Communist Party?' To which Furry replied, 'Sir, I am not sorting people for the committee.'

THE CHAIRMAN: Answer the question.
MR. FURRY: Well, I would like to make this statement, and that is that I have never at any time known anyone who held a permanent position on the Harvard faculty, with the exception of myself, or who has since come to hold or who now holds a permanent position on the Harvard faculty, to be a member of the Communist Party. Apart from that, I will refuse to answer the question.

Finally McCarthy erupted in rage.

This, in the opinion of the chair, is one of the most aggravated cases of contempt that we have had before us, as I see it. Here you have a man teaching at one of our large universities. He knows there were six Communists handling secret government work, radar work, atomic work. He refuses to give either the committee or the FBI or anyone else the information which he has. To me it is inconceivable that a university which has had the reputation of being a great university would keep this type of creature on teaching our children. Because of men like this who have refused to give the government the information which they have in their own minds about Communists who are working on our secret work, many young men have died in the past, and if we lose a war in the future it will be the result of the lack of loyalty, complete [immorality] [unmorality] of these individual who continue to protect the conspirators.

With that, the hearing ended. On her way out of the hearing room, one Boston woman spat on Furry.

I never got to know Furry well enough to know how deeply, and in what ways, all of this affected him, but he certainly struck me as a beaten man. His situation was not made happier, I am sure, by the brilliance of Schwinger, so much younger and so talented. I have a vivid memory of sitting in one of Schwinger's lectures, a year or so later, and seeing Furry in the middle of the classroom, auditing. What was strange was that he was ostentatiously turning over the pages of *Time* magazine during the lecture. If Schwinger noticed, he did not let on. During my reading course with Furry, which preceded these public hearings but coincided with his first appearance before the committee, I was completely unaware of Furry's personal ordeal. It is true that Furry was away from time to time in Washington, but many people in the physics department went to Washington. At the end of the semester I produced a term paper on the electron, which I still have. It came back with a single comment in red pencil by Furry on a minor matter of notation. This was a few weeks before his public hearing, and I wonder whether he had had time to read the whole thing.

I do not think that Furry was ever tried for contempt of Congress. I am not sure why, except that Harvard, with all its well-placed alumni and all of its faculty engaged in vital government service, was a rather formidable opponent. McCarthy did not do well when he went after powerful targets like the army. He died in 1957. Furry remained at Harvard and was for a while the chairman of the department. He later retired, and died a few years ago. I wonder whether the students who came in contact with him in later years had any notion of what he had gone through that fall and winter.

136

Here is Einstein (from R.W. Clark's biography) drawn into a political and personal imbroglio.

Soon after his return from Leiden to Berlin, Einstein discovered that his old friend Friedrich Adler not only thought along anti-war lines similar to his own but had taken up arms to reinforce them. In 1912 Adler had left Switzerland for Austria and here, desperate at the Government's refusal to convene Parliament and thus put its action to the test of public debate, he took what seemed to be the most reasonable logical action. In October 1916, he walked into the fashionable Hotel Meissel and Schadn and at point-blank range shot dead the Prime Minister, Count Stürgkh.

When Adler was put on trial Einstein wrote offering to give evidence as a character witness, an offer which Adler, in keeping with his subsequent actions, appears to have loftily declined as being unnecessary. Awaiting trial, he settled down into a succession of prisons and military fortresses to write a long thesis on relativity, 'Local Time, System Time, Zone Time'.

On July 14th, 1917, Adler wrote to Einstein asking for advice on his work. Einstein replied cordially, and a typewritten draft of the manuscript soon arrived in Berlin. Meanwhile, other copies were being sent to psychiatrists who were asked whether Adler was mentally deranged. 'The experts, especially the physicists, were placed in a very difficult situation,' says Philipp Frank, who himself received a copy. 'Adler's father and family desired that this work should be made the basis for the opinion that Adler was mentally deranged. But this would necessarily be highly insulting to the author, since he believed that he had produced an excellent scientific achievement. Moreover, speaking objectively, there was nothing in any way abnormal about it except that his arguments were wrong.' Einstein held much the same views, noting that it was based on 'very shaky foundations'.

Whether or not Adler's critical study of relativity influenced his fate is unclear. But although he was condemned to death, this sentence was commuted to eighteen months, probably the most lenient punishment in history for assassination of a Prime Minister. The eighteen months do not seem to have been notably rigorous; and a letter to Einstein, written from the military fortress of Stein-an-Donau in July 1918, reveals a prisoner happily immersed in the problems of science who could end his letter with the comment that in these difficult times conditions were much better inside prison walls than outside them.

137

A small group of socially concerned, or, depending on your point of view, publicity-hungry scientists were responsible for causing intense public alarm, on which politicians battened with alacrity, by raising the spectre of cancer-bearing bacteria, which would escape from molecular biology laboratories and into the drinking water. This all but brought recombinant DNA research on the East Coast of the USA to a halt. Here, from Stephen Hall's Invisible Frontiers, *is a description of a clash between scientists and the representatives of the People.*

When the debate finally went public, it began with a local high school choir filling the City Council chambers with adolescent strains of 'This Land Is Your Land' and a sign reading 'No Recombination Without Representation.' Onlookers and kibitzers packed the gallery, jammed the balcony, spilled out into the hallway on the evening of June 23, 1976. Along a row of chairs to one side, shoulder to shoulder, sat Mark Ptashne, Matthew Meselson, David Baltimore, and Wally Gilbert – molecular biology's equivalent of baseball's Murderers' Row. Television lights and cameras, for both television news and archival footage, were trained on the councillors and on the witness table. Having gaveled the meeting to order, Al Vellucci basked in the glow.

It is difficult to imagine a city council meeting, in any place, at any time, creating more hullaballoo and attracting more national attention and generating more ink and comment than the Cambridge DNA debates. Rare are the town meetings where the proceedings are treated as national events, attracting the attention of the *New York Times* and other national publications, beamed around the world by the wire services. 'Creation of life' experiments held sensational fascination for people far beyond Cambridge. There was also appealing confrontational value: the intellectual titans of Harvard and MIT would be forced to explain themselves and their work before a committee of lay citizens. 'At times unnervingly like an inquisition' in the opinion of one chronicler, 'a very good thing' in the words of a participant, it was an extraordinary confrontation of science and society. And it caught the biologists by complete surprise. 'Everybody assumed the whole thing would just sort of go away,' Walter Gilbert recalls. 'Once the city councillors discovered that television cameras would be called in, and the whole meeting televised, then it all went to pot quickly.' 'If the first hearing on the recombinant DNA debate had had a hundred people from Harvard supporting it and a hundred people opposing it, and *didn't* have 60-some television cameras and radio stations and reporters, it probably would have died a natural death that night,' says a former City Council member.

Mayor Vellucci set the tone, for better or worse, at the outset when he issued the following warning to the assembled scientists: 'Refrain from using the alphabet. Most of us in this room, including myself, are lay people. We don't understand your alphabet. So you will spell it out for us so we'll know exactly what you're talking about, because we are here to listen.' Vellucci was not referring to the genetic alphabet of A, T, C, and G; rather, this was an invitation to the scientists to keep it simple and straightforward.

A limit of ten minutes was imposed on each speaker, but the first two witnesses – Mark Ptashne of Harvard and Maxine Singer, a representative of the National Institutes of Health – spent more than two hours fielding questions. Much of the flavor and substance of the dispute emerged in those first few hours.

Ptashne explained the background of the P3 lab at Harvard, citing National Cancer Institute figures that there were already hundreds of P3 laboratories in use in the country, including one at the Harvard Medical School across the river in Boston and another at MIT, about a mile down Massachusetts Avenue from City Hall in Cambridge. 'Most molecular biologists – not all, but most – expect that the information being derived from these experiments that are now ongoing will profoundly advance our understanding of life processes,' he said. Moments later, he added, 'We must realise that unlike other real risks involved in experimentation, the risks in this case are purely hypothetical.' In perhaps the most aggressive assertion on behalf of recombinant DNA made that evening, Ptashne went on to add:

'Nevertheless, we cannot say that there is absolutely no risk involved in these experiments. But then, Mr. Mayor, I ask you to consider that [that] statement – little risk – can be made about few human activities. It certainly cannot be made for many of the experiments performed every day in biological and chemical laboratories. The degree of risk involved in carefully regulated recombinant DNA experiments is almost virtually, in my estimation, less than that in maintaining a household pet.' In an audience stacked with partisans of both stripes, the comment was greeted with hisses of derision and swells of applause. What was stunning, at least to Councillor Ackermann, was the unusual political schism revealed in this partisanship. 'There were more students for it,' she recalled in surprise, 'than against it.' One more reminder that this wasn't like Vietnam.

If the atmosphere occasionally tilted toward the carnivalesque, the impresario of the hour was Vellucci. Flamboyant and grave, shrewd and bumpkinish, good cop and bad all at once, by sheer dint of style and personality he wrestled the agenda onto turf of his liking and scored political points with calculated yelps of Everyman outrage. He chided Daniel Branton, head of Harvard's biosafety committee, for failing to contact public health officials about the Harvard meetings on the lab, and reacted with chagrin and anger when Matthew Meselson, then chairman of Harvard's biochemistry and molecular biology department, conceded that chemicals from experiments were sometimes flushed down the Bio Lab drains. After the attractive Maxine Singer introduced herself, Vellucci inquired for her phone number, a request overlooked for other witnesses; at another point, he observed to no one in particular. 'There are sharp minds in this audience,' and then added in an aside to city councillor Leonard Russell, 'You and me are a couple of sharp minds.'

Singer explained the new NIH guidelines, terming the likelihood 'extraordinarily low' that P3 experiments could cause serious trouble. Vellucci then weighed in with his 'statements of questions.' It was a masterful populist *tour de force*, not so much for its logic or rhetoric (or grammar, for that matter), but for its unabashed, bald-faced exercise of power in setting the terms of the public debate.

'One,' Vellucci began. 'Did anyone of this group bother at any time to write to the mayor and the City Council to inform us you intended to carry out these experiments in the City of Cambridge, and you just said that you had public hearings.

'You plan to use *E. coli* in your experiments. Do I have *E. coli* inside my body right now? That's a question. Don't answer, but you may, as you go along.

'Does everyone in this room have *E. coli* inside their bodies right now?

'Can you make an absolute, one hundred percent guarantee that there is no possible risk which might arise from this experimentation? Is there zero risk of danger? Answer that question later, too, please.

'Would recombinant DNA experiments be safer if they were done in a maximum security lab, a P4 lab, in an isolated, nonpopulated area of the country? Question.

'Would this be safer than using a P3 lab in one of the most densely populated cities in the nation? Question.

'Is it true that in the history of science mistakes have been made, or known to happen? Question.

'Do scientists ever exercise poor judgment? Question.

'Do they ever have accidents? Question.

'Do you possess enough foresight and wisdom to decide which direction the future of mankind should take? Question.

'The great war poet Joyce Kilmer once wrote, "Poems are made by fools like me, but only God can make a tree." I have made references to Frankenstein over the past week, and some people think this all a big joke. That was my way of describing what happens when genes are put together in a new way. . . .'

At this point, a flicker of a smirk spread on Mark Ptashne's face. Vellucci's radar picked it up immediately. 'This is a deadly matter, sir,' he said sharply. 'Ma'am, sir. Harvard University. This is a serious matter. It is not a laughing matter, please believe me. It is not a laughing matter, and this is for the National Institutes [Maxine Singer]: this is not a laughing matter. If worse comes to worse, we could have a major disaster on our hands. I guarantee everyone in this room that if that happens no one will be laughing then.

'Protecting the health and safety of the people of Cambridge is a solemn trust,' the mayor intoned. 'I intend to treat that trust with complete dedication, and, madam, it was only twelve years ago that I sat in that seat with the City Council full to capacity, as full as it is tonight, fighting the coming of the NASA site in Cambridge. And I predicted that that whole thing would collapse, and it collapsed.

'And now,' Vellucci vowed, 'tonight I come here with the same fight in me.'

Vellucci was not the only city official to vent his spleen that evening. One can infer from the comments of other city councillors an uncomfortable mix of bewilderment, suspicion, resentment, paranoia, and alarm over the entire issue. After two hours of testimony, Barbara Ackermann made the following observation to Daniel Branton of Harvard's biosafety committee. 'I would say that a committee that was dealing in matters which are even one millionth of one percent dangerous ought to involve at least half public representation.' If risk is as much a social calculation as a statistical assertion, this was the new math of public hazard.

The prevailing fears were perhaps best articulated by a tall, bearded councillor named David Clem, who found it paradoxical that the NIH, the federal agency charged with funding recombinant DNA research and therefore 'promulgating' it, now assumed the mantle of regulator as well. 'I really don't give a damn about a P3 laboratory at Harvard University,' he said, 'because I can't visit that laboratory and discover whether its P1, P2, or P4 or whatever. I don't have the expertise to analyze or investigate any type of laboratory facility at Harvard University. But,' he added, indignation growing in his voice, 'it strikes me as very to the point that there is an important principle in this country that the people who have a vested self-interest in certain types of activities should not be the ones who are charged not only with promulgating it but regulating it.' To a huge outburst of applause, Clem continued, 'This country missed the boat with nuclear research and the Atomic Energy Commission and we're going to find ourselves in one hell of a bind because we are allowing one agency with a vested interest to initiate, fund, and encourage research and yet we are assuming that they are nonbiased and have the ability to regulate that, and more importantly, to enforce their regulation.'

Soon after, to competing choruses of applause and boos, Alfred

Vellucci – who was just there 'to listen' – introduced a resolution. 'Whereas,' he read, 'there is still considerable doubt concerning the safety of experimentation dealing with recombinant DNA, therefore be it resolved: that the Cambridge City Council insists that no experimentation involving recombinant DNA should be done within the City of Cambridge for at least two years, and be it further resolved: that the Cambridge City Council will do anything and everything in its power to enforce this resolution.'

Ptashne's reaction was instantaneous. 'If you pass that resolution,' he said, visibly agitated, 'virtually every experiment done by members of the biochemistry department at Harvard will have to stop and virtually every experiment done by about half the members of the biology department would have to stop, including experiments that no one, sir, *no one,* has ever claimed had the slightest danger inherent in them – namely recombinant experiments done under P1 conditions. And so on such an important issue, it seems to me you have to clarify much more clearly what the issues are before proposing such a resolution.'

'What I want to alert you to,' Alfred Vellucci replied, 'when I was a little boy I used to fish in the Charles River and I woke up one morning and found millions of fish dead in the Charles River, and you tonight tell me that you've dumped chemicals into the sewer system of Cambridge and the sewer system overflows into the Charles River.' Peremptory and dismissive, he grumbled, 'Carry on.'

Round one to Vellucci.

138

Here is a hazard to the survival of the species, ice-nine, dreamed up by Kurt Vonnegut for his novel, Cat's Cradle.

'Do you mean,' I said to Dr. Breed, 'that nobody in this Laboratory is ever told what to work on? Nobody even *suggests* what they work on?'

'People suggest things all the time, but it isn't in the nature of a pure-research man to pay any attention to suggestions. His head is full of projects of his own, and that's the way we want it.'

'Did anybody ever try to suggest projects to Dr. Hoenikker?'

'Certainly. Admirals and generals in particular. They looked upon him as a sort of magician who could make America invincible with a wave of his wand. They brought all kinds of crackpot schemes up here – still do. The only thing wrong with the schemes is that, given our present state of knowledge, the schemes won't work. Scientists on the order of Dr. Hoenikker are supposed to fill the little gaps. I remember, shortly before Felix died, there was a Marine general who was hounding him to do something about mud.'

'Mud?'

'The Marines, after almost two hundred years of wallowing in mud, were sick of it,' said Dr. Breed. 'The general, as their spokesman, felt that one of the aspects of progress should be that Marines no longer had to fight in mud.'

'What did the general have in mind?'

'The absence of mud. No more mud.'

'I suppose,' I theorized, 'it might be possible with mountains of some sort of chemical, or tons of some sort of machinery . . .'

'What the general had in mind was a little pill or a little machine. Not only were the Marines sick of mud, they were sick of carrying cumbersome objects. They wanted something *little* to carry for a change.'

'What did Dr. Hoenikker say?'

'In his playful way, and *all* his ways were playful, Felix suggested that there might be a single grain of something – even a microscopic grain – that could make infinite expanses of muck, marsh, swamp, creeks, pools, quicksand, and mire as solid as this desk.'

Dr. Breed banged his speckled old fist on the desk. The desk was a kidney-shaped, sea green steel affair. 'One Marine could carry more than enough of the stuff to free an armored division bogged down in the everglades. According to Felix, one Marine could carry enough of the stuff to do that under the nail of his little finger.'

'That's impossible.'

'You would say so, I would say so – practically everybody would say so. To Felix, in his playful way, it was entirely possible. The miracle of Felix – and I sincerely hope you'll put this in your book somewhere – was that he always approached old puzzles as though they were brand new.'

'I feel like Francine Pefko now,' I said, 'and all the girls in the Girl Pool, too. Dr. Hoenikker could never have explained to me how something that could be carried under a fingernail could make a swamp as solid as your desk.'

'I told you what a good explainer Felix was . . .'

'Even so . . .'

'He was able to explain it to me,' said Dr. Breed, 'and I'm sure I can explain it to you. The puzzle is how to get Marines out of the mud – right?'

'Right.'

'All right,' said Dr. Breed, 'listen carefully. Here we go.'

'There are several ways,' Dr. Breed said to me, 'in which certain liquids can crystallize – can freeze – several ways in which their atoms can stack and lock in an orderly, rigid way.'

That old man with spotted hands invited me to think of the several ways in which cannonballs might be stacked on a courthouse lawn, of the several ways in which oranges might be packed into a crate.

'So it is with atoms in crystals, too; and two different crystals of the same substance can have quite different physical properties.'

He told me about a factory that had been growing big crystals of ethylene diamine tartrate. The crystals were useful in certain manufacturing operations, he said. But one day the factory discovered that the crystals it was growing no longer had the properties desired. The atoms had begun to stack and lock – to freeze – in different fashion. The liquid that was crystallizing hadn't changed, but the crystals it was forming were, as far as industrial applications went, pure junk.

How this had come about was a mystery. The theoretical villain, however, was what Dr. Breed called 'a seed.' He meant by that a tiny grain of the undesired crystal pattern. The seed, which had come from God-only-knows-where, taught the atoms the novel way in which to stack and lock, to crystallize, to freeze.

'Now think about cannonballs on a courthouse lawn or about oranges in a crate again,' he suggested. And he helped me to see that the pattern of the bottom layers of cannonballs or of oranges determined how each subsequent layer would stack and lock. 'The bottom layer is the seed of how every cannonball or every orange that comes after is going to behave, even to an infinite number of cannonballs or oranges.'

'Now suppose,' chortled Dr. Breed, enjoying himself, 'that there were many possible ways in which water could crystallize, could freeze. Suppose that the sort of ice we skate upon and put into highballs – what we might call *ice-one* – is only one of several types of ice. Suppose water always froze as *ice-one* on Earth because it had never had a seed to teach it how to form *ice-two, ice-three, ice-four . . . ?* And suppose,' he rapped on his desk with his old hand again, 'that there were one form, which we will call *ice-nine* – a crystal as hard as this desk – with a melting point of, let us say, one-hundred degrees Fahrenheit, or, better still, a melting point of one-hundred-and-thirty degrees.'

'All right, I'm still with you,' I said.

Dr. Breed was interrupted by whispers in his outer office, whispers loud and portentous. They were the sounds of the Girl Pool.

The girls were preparing to sing in the outer office.

And they did sing, as Dr. Breed and I appeared in the doorway. Each of about a hundred girls had made herself into a choirgirl by putting on a collar of white bond paper, secured by a paper clip. They sang beautifully.

I was surprised and mawkishly heartbroken. I am always moved by that seldom-used treasure, the sweetness with which most girls can sing.

The girls sang 'O Little Town of Bethlehem.' I am not likely to forget very soon their interpretation of the line:

'The hopes and fears of all the years are here with us tonight.'

When old Dr. Breed, with the help of Miss Faust, had passed out the Christmas chocolate bars to the girls, we returned to his office.

There, he said to me, 'Where were we? Oh yes!' And that old man asked me to think of United States Marines in a Godforsaken swamp.

'Their trucks and tanks and howitzers are wallowing,' he complained, 'sinking in stinking miasma and ooze.'

He raised a finger and winked me. 'But suppose, young man, that one Marine had with him a tiny capsule containing a seed of *ice-nine*, a new way for the atoms of water to stack and lock, to freeze. If that Marine threw that seed into the nearest puddle . . . ?'

'The puddle would freeze?' I guessed.

'And all the muck around the puddle?'

'It would freeze?'

'And all the puddles in the frozen muck?'

'They would freeze?'

'And the pools and the streams in the frozen muck?'

'They would freeze?'

'You *bet* they would!' he cried. 'And the United States Marines would rise from the swamp and march on!'

'There *is* such stuff?' I asked.

'No, no, no, no,' said Dr. Breed, losing patience with me again. 'I only told you all in this in order to give you some insight into the extraordinary novelty of the ways in which Felix was likely to approach an old problem. What I've just told you is what he told the Marine general who was hounding him about mud.

'Felix ate alone here in the cafeteria every day. It was a rule that no one was to sit with him, to interrupt his chain of thought. But the Marine general barged in, pulled up a chair, and started talking about mud. What I've told you was Felix's offhand reply.'

'There – there really *isn't* such a thing?'

'I just told you there wasn't!' cried Dr. Breed hotly. 'Felix died shortly after that! And, if you'd been listening to what I've been trying to tell you about pure research men, you wouldn't ask such a question! Pure research men work on what fascinates them, not on what fascinates other people.'

'I keep thinking about that swamp. . . .'

'You can *stop* thinking about it! I've made the only point I wanted to make with the swamp.'

'If the streams flowing through the swamp froze as *ice-nine*, what about the rivers and lakes the streams fed?'

'They'd freeze. But there is no such thing as *ice-nine*.'

'And the oceans the frozen rivers fed?'

'They'd freeze, of course,' he snapped. 'I suppose you're going to rush to market with a sensational story about *ice-nine* now. I tell you again, it does not exist!'

'And the springs feeding the frozen lakes and streams, and all the water underground feeding the springs?'

'They'd freeze, damn it!' he cried. 'But if I had known that you were a member of the yellow press,' he said grandly, rising to his feet, 'I wouldn't have wasted a minute with you!'

'And the rain?'

'When it fell, it would freeze into hard little hobnails of *ice-nine* –

and that would be the end of the world! And the end of the interview, too! Good-bye!'

Dr. Breed was mistaken about at least one thing: there was such a thing as *ice-nine*.

And *ice-nine* was on earth.

Ice-nine was the last gift Felix Hoenikker created for mankind before going to his just reward.

He did it without anyone's realizing what he was doing. He did it without leaving records of what he'd done.

True, elaborate apparatus was necessary in the act of creation, but it already existed in the Research Laboratory. Dr. Hoenikker had only to go calling on Laboratory neighbors – borrowing this and that, making a winsome neighborhood nuisance of himself – until, so to speak, he had baked his last batch of brownies.

He had made a chip of *ice-nine*. It was blue-white. It had a melting point of one-hundred-fourteen-point-four-degrees Fahrenheit.

Felix Hoenikker had put the chip in a little bottle; and he put the bottle in his pocket. And he had gone to his cottage on Cape Cod with his three children, there intending to celebrate Christmas.

Angela had been thirty-four. Frank had been twenty-four. Little Newt had been eighteen.

The old man had died on Christmas Eve, having told only his children about *ice-nine*.

His children had divided the *ice-nine* among themselves.

139

Vonnegut's interest in science stems from his closeness to his brother, a meteorologist. In an address to the American Physical Society, he explains how the inspiration for Cat's Cradle *came to him.*

You have summoned me here in my sunset years as a writer. I am forty-six. F. Scott Fitzgerald was dead when he was my age. So was Anton Chekhov. So was D. H. Lawrence. So was George Orwell, a man I admire almost more than any other man. Physicists live longer than writers, by and large. Copernicus died at seventy. Galileo died at seventy-eight, Isaac Newton died at eighty-five. They lived that long even before the discovery of all the miracles of modern medicine. Think of how much longer they might have lived with heart transplants.

You have called me a humanist, and I have looked into humanism some, and I have found that a humanist is a person who is tremendously interested in human beings. My dog is a humanist. His name is Sandy. He is a sheep dog. I know that Sandy is a dud name for a sheep dog, but there it is.

One day when I was a teacher of creative writing at the University of Iowa, in Iowa City, I realized that Sandy had never seen a truly large carnivore. He had never smelled one, either. I assumed that he would be thrilled out of his wits. So I took him to a small zoo they had in Iowa City to see two black bears in a cage.

'Hey, Sandy,' I said to him on the way to the zoo, 'wait till you see. Wait till you smell.'

Those bears didn't interest him at all, even though they were only three inches away. The stink was enough to knock me over. But Sandy didn't seem to notice. He was too busy watching people.

Most people are mainly interested in people, too. Or that has been my experience in the writing game. That's why it was so intelligent of us to send human beings to the moon instead of instruments. Most people aren't very interested in instruments. One of the things that I tell beginning writers is this: 'If you describe a landscape, or a cityscape, or a seascape, always be sure to put a human figure somewhere in the scene. Why? Because readers are human beings, mostly interested in human beings. People are humanists. *Most* of them are humanists, that is.'

Shortly before coming to this meeting from Cape Cod, I received this letter:

Dear Mr. Vonnegut,

I saw with interest the announcement of the talk entitled 'The Virtuous Scientist,' to be delivered by you and Eames and Drexler at the New York A.P.S. meeting. Unfortunately, I will not be present at the New York meeting this year. However, as a humanistic physicist, I would very much appreciate receiving a copy of the talk. Thanking you in advance.

Sincerely,
GEORGE F. NORWOOD, JR.,
assistant professor of physics,
University of Miami,
Coral Gables, Florida.

If Professor Norwood really is a humanistic physicist, then he is exactly my idea of what a virtuous physicist should be. A virtuous physicist is a humanistic physicist. Being a humanistic physicist, incidentally, is a good way to get *two* Nobel Prizes instead of one. What does a humanistic physicist do? Why, he watches people, listens to them, thinks about them, wishes them and their planet well. He wouldn't knowingly hurt people. He wouldn't knowingly help politicians or soldiers hurt people. If he comes across a technique that would obviously hurt people, he keeps it to himself. He knows that a scientist can be an accessory to murder most foul. That's simple enough, surely. That's surely clear.

I was invited here, I think, mostly because of a book of mine called *Cat's Cradle*. It is still in print, so if you rush out to buy it, you will not be disappointed. It is about an old-fashioned scientist who isn't interested in people. In the midst of a terrible family argument, he asks a question about turtles. Nobody has been talking about turtles. But the old man suddenly wants to know: When turtles pull in their heads, do their spines buckle or contract?

This absentminded old man, who doesn't give a damn for people, discovers a form of ice which is stable at room temperature. He dies, and some idiots get possession of the substance, which I call Ice-9. The idiots eventually drop some of the stuff into the sea, and the waters of the earth freeze – and that is the end of life on earth as we know it.

I got this lovely idea while I was working as a public-relations man at General Electric. I used to write publicity releases about the research laboratory there, where my brother worked. While there, I heard a story about a visit H. G. Wells had made to the laboratory in the early Thirties.

General Electric was alarmed by the news of his coming, because they did not know how to entertain him. The company told Irving Langmuir, who was a most important man in Schenectady, the only Nobel Prize winner in private industry, that he was going to have to entertain Wells. Langmuir didn't want to do it, but he dutifully tried to imagine diversions that would delight Mr. Wells. He made up a science-fiction story he hoped Mr. Wells would want to write. It was about a form of ice which was stable at room temperature. Mr. Wells was not stimulated by the story. He later died, and so did Langmuir. After Langmuir died, I thought to myself, well, I think maybe I'll write a story.

While I was writing that story about Ice-9, I happened to go to a cocktail party where I was introduced to a crystallographer. I told him about this ice which was stable at room temperature. He put his cocktail glass on the mantelpiece. He sat down in an easy chair in the corner. He did not speak to anyone or change expression for half an hour. Then he got up, came back over to the mantelpiece, and picked up his cocktail glass, and he said to me, 'Nope.' Ice-9 was impossible.

Be that as it may, other scientific developments have been almost that horrible. The idea of Ice-9 had a certain moral validity at any rate, even though scientifically it had to be pure bunk.

140

Fred Hoyle, sometime Plumian Professor of Astronomy in Cambridge, has written several novels, the best of which is his tale of an extraterrestrial intelligence, The Black Cloud. *The hero, Charles Kingsley, is both intellectually commanding and intrepid. He is moreover the Plumian Professor of Astronomy. Could Hoyle have intended him for a self-portrait?*

Kingsley and the Astronomer Royal arrived in Los Angeles early on the morning of 20 January. Marlowe was waiting to meet them at the airport. After a quick breakfast in a drug store they hit the freeway system to Pasadena.

'Goodness me, what a difference from Cambridge,' grunted Kingsley. 'Sixty miles an hour instead of fifteen, blue skies instead of endless rain and drizzle, temperature in the sixties even as early in the day as this.'

He was very weary after the long flight, first across the Atlantic, then a few hours' waiting in New York – too short to be able to do anything interesting, yet long enough to be tiresome, the epitome of air travel, and lastly the trip across the U.S.A. during the night. Still it was a great deal better than a year at sea getting round the Horn, which is what men had to do a century ago. He would have liked a long sleep, but if the Astronomer Royal was willing to go straight to the Observatory, he supposed he ought to go along too.

After Kingsley and the Astronomer Royal had been introduced to those members of the Observatory that they had not previously met, and after greetings with old friends, the meeting started in the library. With the addition of the British visitors it was the same company that had met to discuss Jensen's discovery the previous week.

Marlowe gave a succinct account of this discovery, of his own observations, and of Weichart's argument and startling conclusion.

'And so you see,' he concluded, 'why we were so interested to receive your cablegram.'

'We do indeed,' answered the Astronomer Royal. 'These photographs are most remarkable. You give the position of the centre of the cloud as Right Ascension 5 hours 49 minutes, Declination minus 30 degrees 16 minutes. That seems to be in excellent agreement with Kingsley's calculations.'

'Now would you two care to give us a short account of your investigations?' said Herrick. 'Perhaps the Astronomer Royal could tell us about the observational side and then Dr Kingsley could say a little about his calculations.'

The Astronomer Royal gave a description of the displacements that had been discovered in the positions of the planets, particularly of the outer planets. He discussed how the observations had been carefully checked to make sure that they contained no errors. He did not fail to give credit to the work of Mr George Green.

'Heavens, he's at it again,' thought Kingsley.

The rest of the company heard the Astronomer Royal out with interest, however.

'And so,' he concluded, 'I'll hand over to Dr Kingsley, and let him outline the basis of his calculations.'

'There is not a great deal to be said,' began Kingsley. 'Granting the accuracy of the observations that the Astronomer Royal has just told us about – and I must admit to having been somewhat reluctant at first to concede this – it was clear that the planets were being disturbed by the gravitational influence of some body, or material, intruding into the solar system. The problem was to use the observed disturbances to calculate the position, mass, and velocity of the intruding material.'

'Did you work on the basis that the material acted as a point mass?' asked Weichart.

'Yes, that seemed to be the best thing to do, at any rate to begin with. The Astronomer Royal did mention the possibility of an extended cloud. But I must confess that psychologically I've been thinking in terms of a condensed body of comparatively small size. I've only just begun to assimilate the cloud idea, now that I've seen these photographs.'

'How far do you think your wrong assumption affected the calculations?' Kingsley was asked.

'Hardly at all. So far as producing planetary disturbances is concerned, the difference between your cloud and a much more condensed body would be quite small. Perhaps the slight differences between my results and your observations arise from this cause.'

'Yes, that's quite clear,' broke in Marlowe amid aniseed smoke. 'How much information did you need to get your results? Did you use the disturbances of all the planets?'

'One planet was enough. I used the observations of Saturn to make the calculations about the Cloud – if I may call it that. Then having determined the position, mass, etc., of the Cloud, I inverted the calculation for the other planets and so worked out what the disturbances of Jupiter, Mars, Uranus, and Neptune ought to be.'

'Then you could compare your results with the observations?'

'Exactly so. The comparison is given in these tables that I've got here. I'll hand them round. You can see that the agreement is pretty good. That's why we felt reasonably confident about our deductions, and why we felt justified in sending our cable.'

'Now I'd like just to know how your estimates compare with mine,' asked Weichart. 'It seemed to me that the Cloud would take about eighteen months to reach the Earth. What answer do you get?'

'I've already checked that, Dave,' remarked Marlowe. 'It agrees very well. Dr Kingsley's values give about seventeen months.'

'Perhaps a little less than that,' observed Kingsley. 'You get seventeen months if you don't allow for the acceleration of the Cloud as it approaches the Sun. It's moving at about seventy kilometres per second at the moment, but by the time it reaches the Earth it'll have speeded to about eighty. The time required for the Cloud to reach the Earth works out at nearly sixteen months.'

Herrick quietly took charge of the discussion.

'Well, now that we understand each other's point of view, what conclusions can we reach? It seems to me that we have both been under some misapprehension. For our part we thought of a much larger cloud lying considerably outside the solar system, while, as Dr Kingsley says, he thought of a condensed body within the solar system. The truth lies somewhere between these views. We have to do with a rather small cloud that is already within the solar system. What can we say about it?'

'Quite a bit,' answered Marlowe. 'Our measurement of the

angular diameter of the Cloud as about two and a half degrees, combined with Dr Kingsley's distance of about 21 astronomical units, shows that the Cloud has a diameter about equal to the distance from the Sun to the Earth.'

'Yes, and with this size we can immediately get an estimate of the density of the material in the Cloud,' went on Kingsley. 'It looks to me as though the volume of the cloud is roughly 10^{40}c.c. Its mass is about $1{\cdot}3 \times 10^{30}$ gm, which gives a density of $1{\cdot}3 \times 10^{-10}$ gm. per cm^3.'

A silence fell on the little company. It was broken by Emerson.

'That's an awful high density. If the gas comes between us and the Sun it'll block out the Sun's light completely. It looks to me as if it's going to get almighty cold here on Earth!'

'That doesn't necessarily follow,' broke in Barnett. 'The gas itself may get hot, and heat may flow through it.'

'That depends on how much energy is required to heat the Cloud,' remarked Weichart.

'And on its opacity, and a hundred and one other factors,' added Kingsley. 'I must say it seems very unlikely to me that much heat will get through the gas. Let's work out the energy required to heat it to an ordinary sort of temperature.'

He went out to the blackboard, and wrote:

Mass of Cloud $1{\cdot}3 \times 10^{30}$ grams.
Composition of Cloud probably hydrogen gas, for the most part in neutral form.
Energy required to lift temperature of gas by T degrees is

$$1{\cdot}5 \times 1{\cdot}3 \times 10^{30}\ RT \text{ ergs}$$

where R is the gas constant. Writing L for the total energy emitted by the Sun, the time required to raise the temperature is

$$1{\cdot}5 \times 1{\cdot}3 \times 10^{30}\ RT/L \text{ seconds}$$

Put $R = 8{\cdot}3 \times 10^7$, $T = 300$, $L - 4 \times 10^{33}$ ergs per second gives a time of about $1{\cdot}2 \times 10^7$ seconds, i.e. about 5 months.

'That looks sound enough,' commented Weichart. 'And I'd say that what you've got is very much a minimum estimate.'

'That's so,' nodded Kingsley. 'And my minimum is already very much longer than it will take the Cloud to pass us by. At a speed of 80 kilometres per second it'll sweep across the Earth's orbit in about a month. So it looks to me pretty certain that if the Cloud does come between us and the Sun it'll cut out the heat from the Sun quite completely.'

'You say *if* the Cloud comes between us and the Sun. Do you think there's a chance it may miss us?' asked Herrick.

'There's certainly a chance, quite a chance I'd say. Look here.'

Kingsley moved again to the blackboard.

'Here's the Earth's orbit round the Sun. We're here at the moment. And the Cloud, to draw it to scale, is over here. If it's moving like this, dead set for the Sun, then it'll certainly block the Sun. But if it's moving this second way, then it could well miss us altogether.'

'It looks to me as if we're rather lucky,' Barnett laughed uneasily.

The Thrill of the Chase: Discovery

141

Edgar Allan Poe, in his essay, Eureka, *quotes Kepler's famous declaration of the fierce pride that he took in his discovery.*

'Yes, Kepler was essentially a *theorist*; but this title, *now* of so much sanctity, was, in those ancient days, a designation of supreme contempt. It is only *now* that men begin to appreciate that divine old man – to sympathize with the prophetical and poetical rhapsody of his ever-memorable words. For *my* part,' continues the unknown correspondent, 'I glow with a sacred fire when I even think of them, and feel that I shall never grow weary of their repetition: – in concluding this letter, let me have the real pleasure of transcribing them once again: - "*I care not whether my work be read now or by posterity. I can afford to wait a century for readers when God himself has waited six thousand years for an observer. I triumph. I have stolen the golden secret of the Egyptians. I will indulge my sacred fury.*" '

142

Michael Frayn, in The Tin Men, *conjures up the following conceit of a breakthrough in computer research.*

If Goldwasser was remembered for nothing else, Macintosh once told Rowe, he would be remembered for his invention of UHL.

UHL was Unit Headline Language, and it consisted of a comprehensive lexicon of all the multi-purpose monosyllables used by headline-writers. Goldwasser's insight had been to see that if the grammar of "ban", "dash", "fear", and the rest was ambiguous they could be used in almost any order to make a sentence, and that if they could be used in almost any order to make a sentence they could be easily randomised. Here then was one easy way in which a computer could find material for an automated newspaper – put together a headline in basic UHL first and then fit the story to it.

UHL, Goldwasser quickly realised, was an ideal answer to the problem of making a story run from day to day in an automated paper. Say, for example, that the randomiser turned up

STRIKE THREAT

By adding one unit at random to the formula each day the story could go:

STRIKE THREAT BID
STRIKE THREAT PROBE
STRIKE THREAT PLEA

And so on. Or the units could be added cumulatively:

STRIKE THREAT PLEA
STRIKE THREAT PLEA PROBE
STRIKE THREAT PLEA PROBE MOVE
STRIKE THREAT PLEA PROBE
MOVE SHOCK
STRIKE THREAT PLEA PROBE
MOVE SHOCK HOPE
STRIKE THREAT PLEA PROBE
MOVE SHOCK HOPE STORM

Or the units could be used entirely at random:

LEAK ROW LOOMS
TEST ROW LEAK
LEAK HOPE DASH BID
TEST DEAL RACE
HATE PLEA MOVE
RACE HATE PLEA MOVE DEAL

Such headlines, moreover, gave a newspaper a valuable air of dealing with serious news, and helped to dilute its obsession with the frilly-knickeredness of the world, without alarming or upsetting the customers. Goldwasser had had a survey conducted, in fact, in which 457 people were shown the headlines.

ROW HOPE MOVE FLOP
LEAK DASH SHOCK
HATE BAN BID PROBE

Asked if they thought they understood the headlines, 86.4 per cent said yes, but of these 97.3 per cent were unable to offer any explanation of what it was they had understood. With UHL, in other words, a computer could turn out a paper whose language was both soothingly familiar and yet calmingly incomprehensible.

Goldwasser sometimes looked back to the time when he had invented UHL as a lost golden age. That was before Nobbs had risen to the heights of Principal Research Assistant, and with it his beardedness and his belief in the universal matehood of man. In those days Goldwasser was newly appointed Head of his department. He had hurried eagerly to work each day in whatever clothes first came to hand in his haste. He had thought nothing of founding a new inter-language before lunch, arguing with Macintosh through the midday break, devising four news categories in the afternoon, then taking Macintosh and his new wife out to dinner, going on to a film, and finishing up playing chess with Macintosh into the small hours. In those days he had been fairly confident that he was cleverer than Macintosh. He had even been fairly confident that Macintosh had thought he was cleverer than Macintosh. In those days Macintosh had been his Principal Research Assistant.

It was difficult not to believe the world was deteriorating when one considered the replacement of Macintosh by Nobbs. Goldwasser sometimes made a great effort to see the world remaining – as he believed it did – much as it always was, and to see Nobbs as a potential Macintosh to a potential incoming Goldwasser. It was not easy. Now that Macintosh had gone on to become Head of the Ethics Department, Goldwasser no longer invented his way through a world of clear, cerebral TEST PLEA DASH SHOCK absolutes. Now his work seemed ever more full of things like the crash survey.

The crash survey showed that people were not interested in reading about road crashes unless there were at least ten dead. A road crash with ten dead, the majority felt, was slightly less interesting than a rail crash with one dead, unless it had piquant details – the ten dead turning out to be five still virginal honeymoon couples, for example, or pedestrians mown down by the local J.P. on his way home from a hunt ball. A rail crash was always entertaining, with or without children's toys still lying pathetically among the wreckage. Even a rail crash on the Continent made the grade provided there were at least five dead. If it was in the United States the minimum number of dead rose to twenty; in South America 100; in Africa 200; in China 500.

But people really preferred an air crash. Here, curiously enough people showed much less racial discrimination. If the crash was outside Britain, 50 dead Pakistanis or 50 dead Filipinos were as entertaining as 50 dead Americans. What people enjoyed most was about 70 dead, with some 20 survivors including children rescued after at least one night in open boats. They liked to be backed up with a story about a middle-aged housewife who had been booked to fly aboard the plane but who had changed her mind at the last moment.

Goldwasser was depressed for a month over the crash survey. But he could not see any way of producing a satisfactory automated newspaper without finding these things out. Now he was depressed all over again as he formulated the questions to be asked in the murder survey.

His draft ran:

1. Do you prefer to read about a murder in which the victim is (*a*) a small girl (*b*) an old lady (*c*) an illegitimately pregnant young woman (*d*) a prostitute (*e*) a Sunday school teacher?

2. Do you prefer the alleged murderer to be (*a*) a Teddy boy (*b*) a respectable middle-aged man (*c*) an obvious psychopath (*d*) the victim's spouse or lover (*e*) a mental defective?

3. Do you prefer a female corpse to be naked, or to be clad in underclothes?

4. Do you prefer any sexual assault involved to have taken place before or after death?

5. Do you prefer the victim to have been (*a*) shot (*b*) strangled (*c*) stabbed (*d*) beaten to death (*e*) kicked to death (*f*) left to die of exposure?

6. Do you prefer the murder to have taken place in a milieu which seems (*a*) exotic (*b*) sordid (*c*) much like your own and the people next door's?

6. If (*c*) do you prefer the case to reveal that beneath the surface life was (*a*) as apparently respectable as on the surface (*b*) a hidden cesspit of vice and degradation?

Nobbs enjoyed the survey. "'Do you prefer a female corpse to be naked, or to be clad in underclothes?'" he repeated to Goldwasser. "That's what I call a good question, mate. That's what I call a good question."

Goldwasser, his heart heavy, sought refuge from Nobbs and his own questions in several new volumes of IQ tests which a friend in America had sent him. They were beautiful questions, with beautiful answers, and for a few hours they made the world's more horrible contents seem infinitely remote and insignificant. And the results gave his graph for the quarter a slight and credible upward gradient.

Now Ben Jonson's Alchemist, *in the throes of creation.*

SURLY. . . .
That you should hatch gold in a furnace, sir,
As they do eggs in Egypt!
SUBTLE. Sir, do you
Believe that eggs are hatch'd so?
SURLY. If I should?
SUBTLE. Why, I think that the greater miracle.
No egg but differs from a chicken more
Than metals in themselves.
SURLY. That cannot be.
The egg's ordain'd by nature to that end,
And is a chicken *in potentia.*
SUBTLE. The same we say of lead and other metals,
Which would be gold, if they had time.
MAMMON. And that
Our art doth further.
SUBTLE. Ay, for 'twere absurd
To think that nature in the earth bred gold
Perfect in the instant: something went before.
There must be remote matter.
SURLY. Ay, what is that?
SUBTLE. Marry, we say –
MAMMON. Ay, now it heats: stand, father,
Pound him to dust.
SUBTLE. It is, of the one part,
A humid exhalation, which we call
Materia liquida, or the unctuous water;
On the other part, a certain crass and vicious
Portion of earth; both which, concorporate,
Do make the elementary matter of gold;
Which is not yet *propria materia*,
But common to all metals and all stones;
For, where it is forsaken of that moisture,
And hath more driness, it becomes a stone:
Where it retains more of the humid fatness,
It turns to sulphur, or to quicksilver,
Who are the parents of all other metals.

Here, from E. C. Large's Sugar in the Air, *is a group of industrial chemists, seeking their fortune.*

'What *is* the catalytic action?' asked Pry, a little too breezily.

'It is not a matter with which you, as an engineer, need concern yourself,' said Zaareb, tartly, 'it is something that you may very properly leave to me.'

Pry winced and said nothing.

He watched Zaareb and Ackworth making a further supply of the stuff. It was a mixture of fine chalk and a little vegetable jelly which had to be soaked in one solution after another. Some of the solutions were ordinary enough, containing simple salts of iron and magnesium, whilst others were complex organic dye-stuffs with names as long as Welsh railway stations. The chalk and the jelly sopped up the solutions, sometimes swelling up to fill the jar, and then being brought down again in curds with the addition of the next constituent.

'. . . tervalent electrolyte . . . thixotropic . . . reversible coagulum . . . chromophores . . . enolic changes . . .' went on Dr. Zaareb explaining to Ackworth the stages in the preparation. Pry understood a part of it, but not all. It seemed to him that the chemically treated chalk, suspended in jelly, would be rather like the grains of chlorophyll suspended in protoplasmic jelly within the tissues of a plant: the intramolecular spaces in the jelly forming a sort of honeycomb or soft matrix, in which the carbonic acid would be held, and in which the sugar would be gently retained during the embryonic stages of its formation. As nature had a very long start in this business of photosynthesis, it did not seem at all a bad idea to imitate its technique, so far at any rate. But Pry was wise enough, for once, not to offer these notions to Dr. Zaareb and inquire whether they were valid. As an engineer he was not expected ever to have done any elementary botany, in fact he knew so little that he had probably overlooked some vitally significant point, and Zaareb had a withering way of refusing to suffer fools gladly.

What Pry did understand quite clearly was that in the final stages of its preparation the blue catalyst had to stand for ten days, before the addition of a few drops of a solution of phosphates would render it fit and ready for use. In the meantime they would not have enough to go on with. But Zaareb had foreseen that. He took a pint bottle from his bag and left it on the bench.

'Not much danger of Cocaine making anything of this – even if he sees us using it, or even gets hold of some of it – is there?' asked the persevering Mr. Pry. 'So far there's nothing in writing.'

'No, but be careful, get your report off to Mr. MacDuff at once – *before* Cocaine's people get to hear what's in the wind – and when Cocaine sees the blue stuff in the tubes, tell him you are just using a coloured pigment, because green leaves contain pigment, that's been done before, and it will mislead him beautifully.'

Pry wrote the report to MacDuff. The increase in the yield of sugar from a mere trace to something that was still little more than a trace, was not a very great step forward, but he made the most of it, and by the time he had finished, he had at least worked himself up into the liveliest enthusiasm for Zaareb's blue catalyst. An enthusiasm quickened with a streak of jealousy, for he desired that it should be he, himself – that day put in his place as a mere engineer – who would make the next step forward.

During the days that followed Pry took a tall stool into the works, had some of the flood lamps turned on, and sat, in protective goggles, watching the gas bubbling up through the main quartz tubes, with and without Zaareb's blue catalyst. He watched the spectacle in a kind of trance for hours on end, permitting no one to disturb him. The spectacle itself could tell him nothing, at best it was a point of focus for his thoughts. But he stared so long that at last it seemed to him that those bubbles, wobbling unsteadily up through the irradiated liquid, did not *want* to remain carbon dioxide, they wanted to change. And there came upon Pry a wholly unscientific conviction that the bubbles were trembling on the verge of combination, that they were waiting for some variation of the conditions to topple over, and turn steadily into sugar, not traces of sugar, but a strong and useful syrup.

145

Primo Levi may come to be judged by future generations as the greatest writer of our time to draw much of his inspiration from science. Levi graduated in chemistry from Turin University shortly before war *broke out in 1939. As a Jew he lived a precarious life, working as an industrial chemist until he joined a partisan group and was captured and sent to Auschwitz. There chemistry saved his life, for he managed to get into the Buna rubber factory (**see no. 190**) and thus survived. After the war he returned to his profession as a chemist – and he was clearly a good one – in the bleak squalor of Italy in the 1940s, and some of the adventures that befell him are magically related in* The Periodic Table, *in which every chapter bears the name of an element. Here is 'Nitrogen'.*

. . . and finally there came the customer we'd always dreamed of, who wanted us as consultants. To be a consultant is the ideal work, the sort from which you derive prestige and money without dirtying your hands, or breaking your backbone, or running the risk of ending up roasted or poisoned: all you have to do is take off your smock, put on your tie, listen in attentive silence to the problem, and then you'll feel like the Delphic oracle. You must then weigh your reply very carefully and formulate it in convoluted, vague language so that the customer also considers you an oracle, worthy of his faith and the rates set by the Chemists' Society.

The dream client was about forty, small, compact, and obese; he wore a thin mustache like Clark Gable and had tufts of black hairs everywhere – in his ears, inside his nostrils, on the backs of his hands, and on the ridge of his fingers almost down to his fingernails. He was perfumed and pomaded and had a vulgar aspect: he looked like a pimp or, better,a third-rate actor playing the part of a pimp; or a tough from the slummy outskirts. He explained to me that he was the owner of a cosmetics factory and had trouble with a certain kind of lipstick. Good, let him bring us a sample; but no, he said, it was a particular problem, which had to be examined on the spot; it was better for one of us to visit and see what the problem was. Tomorrow at ten? Tomorrow.

It would have been great to show up in a car, but of course if you were a chemist with a car, instead of a miserable returnee, a spare-time writer, and besides just married, you wouldn't spend time here sweating pyruvic acid and chasing after dubious lipstick manufacturers. I put on the best of my (two) suits and thought that it was a good idea to leave my bike in some courtyard nearby and pretend I had arrived in a cab, but when I entered the factory I realized that my scruples about prestige were entirely inappropriate. The factory was a dirty, disorderly shed, full of drafts, in which a dozen impudent, indolent, filthy, and showily made up girls crept about. The owner gave me some explanations, exhibiting pride and trying to look important: he called the lipstick "rouge," the aniline "anelline," and the benzoic aldehyde "adelaide." The work process was simple: a girl melted certain waxes and fats in an ordinary enameled pot, adding a little perfume and a little coloring, then poured the lot into a minuscule ingot mold. Another girl cooled off the molds under running water and extracted from each of the molds twenty small scarlet cylinders of lipstick; other girls took care of the assembly and packing. The owner rudely grabbed one of the girls, put his hand behind her neck to bring her mouth close to my eyes, and invited me to observe carefully the outline of her lips – there, you see, a few hours after application, especially when it's hot, the lipstick runs, it filters up along the very thin lines that even young women have around their lips, and so it forms an ugly web of red threads that blurs the outline and ruins the whole effect.

I peered, not without embarrassment: the red threads were indeed there, but only on the right half of the girl's mouth, as she stood there impassively undergoing the inspection and chewing American gum. Of course, the owner explained: her left side, and the left side of all the other girls, was made up with an excellent French product, in fact the product that he was vainly trying to imitate. A lipstick can be evaluated only in this way, through a practical comparison: every morning all the girls had to make up with the lipstick, on the right with his, and on the left with the other, and he kissed all the girls eight times a day to check whether the product was kiss-proof.

I asked the tough for his lipstick's recipe, and a sample of both products. Reading the recipe, I immediately got the suspicion as to where the defect came from, but it seemed to me more advisable to make certain and let my reply fall a bit from on high, and I requested two days' time "for the analyses." I recovered my bike, and as I pedaled along I thought that, if this business went well, I could perhaps exchange it for a motorbike and quit pedaling.

Back at the lab, I took a sheet of filter paper, made two small red dots with the two samples, and put it in the stove at 80 degrees centigrade. After a quarter of an hour the small dot of the left lipstick was still a dot, although surrounded by a greasy aura, while the small dot of the right lipstick was faded and spread, had become a pinkish halo as large as a coin. In my man's recipe there was a soluble dye; it was clear that, when the heat of the woman's skin (or my stove) caused the fat to melt, the dye followed it as it spread. Instead, the other lipstick must contain a red pigment, well dispersed but insoluble and therefore not migrant: I ascertained this easily by diluting it with benzene and subjecting it to centrifugation, and there it was, deposited on the bottom of the test tube. Thanks to the experience I had accumulated at the lakeshore plant I was able to identify it: it was an expensive pigment and not easy to disperse, and besides, my tough did not have any equipment suited to dispersing pigments. Fine, it was his headache, let him figure it out, him with his harem of girl guinea pigs and his revolting metered kisses. For my part, I had performed my professional services; I made a report, attached an invoice with the necessary tax stamps and the picturesque specimen of filter paper, went back to the factory, handed it over, took my fee, and prepared to say goodbye.

But the tough detained me: he was satisfied with my work and wanted to offer me a business deal. Could I get him a few kilos of alloxan? He would pay a good price for it, provided I committed myself by contract to supply it only to him. He had read, I no longer remember in what magazine, that alloxan in contact with mucous membrane confers on it an extremely permanent red color, because it is not a superimposition, in short a layer of varnish like lipstick, but a true and proper dye, as used on wool and cotton.

I gulped, and to stay on the safe side replied that we would have to see: alloxan is not a common compound nor very well known, I don't think my old chemistry textbook devoted more than five lines to it, and at that moment I remembered only vaguely that it was a derivative of urea and had some connection with uric acid.

I dashed to the library at the first opportunity; I refer to the venerable library of the University of Turin's Chemical Institute, at that time, like Mecca, impenetrable to infidels and even hard to penetrate for such faithful as I. One had to think that the administration followed the wise principle according to which it is good to discourage the arts and sciences: only someone impelled by absolute necessity, or by an overwhelming passion, would willingly subject himself to the trials of abnegation that were demanded of him in order to consult the volumes. The library's schedule was brief and irrational, the lighting dim, the file cards in disorder; in the winter, no heat; no chairs but uncomfortable and noisy metal stools; and finally, the librarian was an incompetent, insolent boor of exceeding ugliness, stationed at the threshold to terrify with his appearance and his howl those aspiring to enter. Having been let in, I passed the tests, and right away I hastened to refresh my memory as to the composition and structure of alloxan. Here is its portrait:

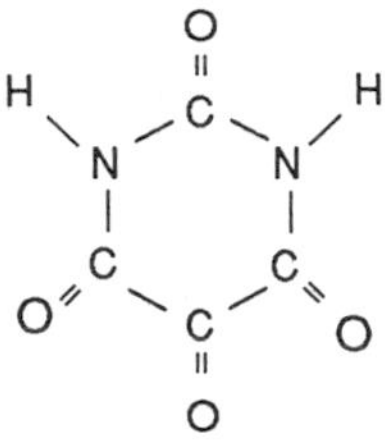

in which O is oxygen, C is carbon, H hydrogen, and N nitrogen. It is a pretty structure, isn't it? It makes you think of something solid, stable, well linked, In fact it happens also in chemistry as in architecture that "beautiful" edifices, that is, symmetrical and simple, are also the most sturdy: in short, the same thing happens with molecules as with the cupolas of cathedrals or the arches of bridges. And it is also possible that the explanation is neither remote nor metaphysical: to say "beautiful" is to say "desirable," and ever since man has built he has wanted to build at the smallest expense and in the most durable fashion, and the aesthetic enjoyment he experiences when contemplating his work comes afterward. Certainly it has not always been this way: there have been centuries in which "beauty" was identified with adornment, the superimposed, the frills; but it is probable that they were deviant epochs and that the true beauty,in which every century recognizes itself, is found in upright stones, ships' hulls, the blade of an ax, the wing of a plane.

Having recognized and appreciated the structural virtue of alloxan, it is urgent that my chemical alter ego, so in love with digressions, get back on the rails, which is that of fornicating with matter in order to support myself – and today, not just myself. I turned with respect to the shelves of the *Zentralblatt* and began to consult it year by year. Hats off to the *Chemisches Zentralblatt*: it is the magazine of magazines, the magazine which, ever since chemistry existed, has reported in the form of furiously concise abstracts all the articles dealing with chemistry that appear in all the magazines in the world. The first years are slender volumes of 300 or 400 pages: today, every year, they dish out fourteen volumes of 1,300 pages each. It is endowed with a majestic authors' index, one for subjects, one for formulas, and you can find in it venerable fossils, such as the legendary memoir in which our father Wöhler tells the story of the first organic synthesis or Sainte-Claire Deville describes the first isolation of metallic aluminium.

From the *Zentralblatt* I ricocheted to *Beilstein*, an equally monumental encyclopedia continually brought up to date in which, as in an Office of Records, each new chemical compound is described as it appears, together with its methods of preparation. Alloxan was known for almost seventy years, but as a laboratory curiosity: the preparation method described had a pure academic value, and proceeded from expensive raw materials which (in those years right after the war) it was vain to hope to find on the market. The sole accessible preparation was the oldest: it did not seem too difficult to execute, and consisted in an oxidizing demolition of uric acid. Just that: uric acid, the stuff connected with gout, intemperant eaters, and stones in the bladder. It was a decidedly unusual raw material, but perhaps not as prohibitively expensive as the others.

In fact subsequent research in the spick and span shelves, smelling of camphor, wax, and century-old chemical labors, taught me that uric acid, very scarce in the excreta of man and mammals, constitutes, however, 50 percent of the excrement of birds and 90 percent of the excrement of reptiles. Fine. I phoned the tough and told him that it could be done, he just had to give me a few days' time: before the month was out I would bring him the first sample of alloxan, and give him an idea of the cost and how much of it I could produce each month. The fact that alloxan, destined to embellish ladies' lips, would come from the excrement of chickens or pythons was a thought which didn't trouble me for a moment. The trade of chemist (fortified, in my case, by the experience of Auschwitz) teaches you to overcome, indeed to ignore, certain revulsions that are neither necessary or congenital: matter is matter, neither noble nor vile, infinitely transformable, and its proximate origin is of no importance whatsoever. Nitgrogen is nitrogen, it passes miraculously from the air into plants, from these into animals, and from animals to us; when its function in our body is exhausted, we eliminate it, but it still remains nitrogen, aseptic, innocent. We – I mean to say we mammals – who in general do not have problems about obtaining water, have learned to wedge it into the urea molecule, which is soluble in water, and as urea we free ourselves of it; other animals, for whom water is precious (or it was for their distant progenitors), have made the ingenious invention of packaging their nitrogen in the form of uric acid, which is insoluble in water, and of eliminating it as a solid, with no necessity of having recourse to water as a vehicle. In an analogous fashion one thinks today of eliminating urban garbage by pressing it into blocks, which can be carried to the dumps or buried inexpensively.

I will go further: far from scandalizing me, the idea of obtaining a cosmetic from excrement, that is, *aurum de stercore* ("gold from dung"), amused me and warmed my heart like a return to the origins, when alchemists extracted phosphorus from urine. It was an adventure both unprecedented and gay and noble besides, because it ennobled, restored, and re-established. That is what nature does: it draws the fern's grace from the putrefaction of the forest floor, and pasturage from manure, in Latin *laetamen* – and does not *laetari* mean "to rejoice"? That's what they taught me in *liceo*, that's how it had been for Virgil, and that's what it became for me. I returned home that evening, told my very recent wife the story of the alloxan and uric acid, and informed her that the next day I would leave on a business trip: that is, I would get on my bike and make a tour of the farms on the outskirts of town (at that time they were still there) in search of chicken shit. She did not hesitate; she likes the countryside, and a wife should follow her husband; she would come along with me. It was a kind of supplement to our honeymoon trip, which for reasons of economy had been frugal and hurried. But she warned me not to have too many illusions: finding chicken shit in its pure state would not be so easy.

In fact it proved quite difficult. First of all, the *pollina* – that's what the country people call it, which we didn't know, nor did we know that, because of its nitrogen content, it is highly valued as a fertilizer for truck gardens – the chicken shit is not given away free, indeed it is sold at a high price. Secondly, whoever buys it has to go and gather it, crawling on all fours into the chicken coops and gleaning all around the threshing floor. And thirdly, what you actually collect can be used directly as a fertilizer, but lends itself badly to other uses: it is a mixture of dung, earth, stones, chicken feed, feathers, and chicken lice, which nest under the chickens' wings. In any event, paying not a little, laboring and dirtying ourselves a lot, my undaunted wife and I returned that evening down Corso Francia with a kilo of sweated-over chicken shit on the bike's carrier rack.

The next day I examined the material: there was a lot of gangue, yet something perhaps could be gotten from it. But simultaneously I got an idea; just at that time, in the Turin subway gallery an exhibition of snakes had opened: Why not go and see it? Snakes are a clean species, they have neither feathers nor lice, and they don't scrabble in the dirt; and besides, a python is quite a bit larger than a chicken. Perhaps their excrement, at 90 percent uric acid, could be obtained in abundance, in sizes not too minute and in conditions of reasonable purity. This time I went alone: my wife is a daughter of Eve and doesn't like snakes.

The director and the various workers attached to the exhibition received me with stupefied scorn. Where were my credentials? Where did I come from? Who did I think I was showing up just like that, as if it were the most natural thing, asking for python shit? Out of the question, not even a gram; pythons are frugal, they eat twice a month and vice versa: especially when they don't get much exercise. Their very scanty shit is worth its weight in gold; besides, they – and all exhibitors and owners of snakes – have permanent and exclusive contracts with big pharmaceutical companies. So get out and stop wasting our time.

I devoted a day to a coarse sifting of the chicken shit, and another two trying to oxidize the acid contained in it into alloxan. The virtue and patience of ancient chemists must have been superhuman, or perhaps my inexperience with organic preparations was boundless. All I got were foul vapors, boredom, humiliation, and a black and murky liquid which irremediably plugged up the filters and displayed no tendency to crystalize, as the text declared it should. The shit remained shit, and the alloxan and its resonant name remained a resonant name. That was not the way to get out of the swamps: by what path would I therefore get out, I the discouraged author of a book which seemed good to me but which nobody read? Best to return among the colorless but safe schemes of inorganic chemistry.

146

Robert Graves's verse (a stanza from The Marmosite's Miscellany*) commemorates August Kekulé, to whom a Freudian vision of snakes eating their tails was vouchsafed during a reverie on the top-deck of a London bus, revealing to him in a flash the structure of benzene.*

The maunderings of a maniac signifying nothing
I hold in respect; I hear his tale out.
Thought comes often clad in the strangest clothing:
So Kekulé the chemist watched the weird rout
Of eager atom-serpents writhing in and out
And waltzing tail to mouth. In that absurd guise
Appeared benzine and anilin, their drugs and their dyes.

147

From chemistry to physics and Mitchell Wilson's novel, Live with Lightning. *Erik is I. I. Rabi (see above, nos. 14, 76, 77, 93).*

Erik turned on the delicate switches which brought the detector circuit to life. On the dial before him, the fragile needle stirred slightly on the white face of the meter. A surge of excitement washed over him and Haviland sensed it at once.

"What happened?"

"You can't be sure. The needle moved."

"Can you get a reading?"

"No. It was too slight. All it did was move."

"That's all right," said Haviland. The excitement caught at him too. "That's fine. Bring the detector up very slowly towards the target."

Erik cranked the mechanism that slid the shielded box smoothly along the bar, but he couldn't remove his eyes from the dial. With great lethargy but unmistakable sureness, the needle continued its slow upward swing as the copper box approached the target. Haviland deserted his desk to stand behind Erik's chair, and then Fabermacher joined them, all intent on the little dial. The delicate length of the pointing needle quivered but moved steadily across the face, five, ten, fifteen, twenty-three, thirty-five, fifty, eighty, and then off scale with a vicious surge when the detector circuit had reached its closest approach to the target. To protect the meter, Erik moved the box back along the bar a little way, but the three men were silent. With the small target as the source an invisible emanation was pouring into the room, flowing into the detector chamber which registered its reaction before them.

Without any consultation, but knowing that the two others

demanded the same proof he himself desired, Erik reversed the entire procedure, moving the detector farther and farther away from the target so that they could see the needle swing slowly back down the scale exactly as it had come up. At the far end, he reversed once again and brought the chamber back towards the center of the dense and invisible cloud of neutrons that must be coming from the target.

"Well, that's the story," said Haviland. He straightened up and then went back to his table. "Start the detector at the zero position once more. Move it up in steps of six inches and let's take down the data."

That was all. No congratulations, no elation that their apparatus was finally completed. At that moment, and for the first time that day, his mind was completely opaque to Erik. Only Fabermacher remained at Erik's shoulder and he was smiling.

"They're real!" he murmured. "My God, they are absolutely real!"

Erik turned. "Didn't you believe that the neutron existed?"

"Oh, I *believed.*" Fabermacher shrugged away the phase. "To me neutrons were symbols, n with a mass of $m_n = 1.008$. But until now I never *saw* them." He glanced at the empty gap between the target and the detector where the hot August breeze churned dust motes and cigarette smoke in the sunbeams. There was no reason to suspect that this particular part of the room contained anything at all unusual, perhaps more lethal than the most intense X rays, but the eye of the detector had seen for them, and so they themselves had seen.

"Please," said Fabermacher to Erik with a slight smile, "let me help take the measurement!"

Erik glanced at Haviland for his permission, and Haviland nodded without responding to Fabermacher's pleading earnestness. He seemed impatient only to get this last piece of data so that they could close down for the day. As Erik read off the numbers, there was no longer any mutuality. Haviland had escaped him once again. He realized how tired he was now, how hungry and yet without any appetite, how tight were his nerves from the incessant noise of the pumps and the constant inward straining away from the high voltage everywhere about him.

When the voltage was down, Haviland resumed the silence with which they had started that morning. But Erik was on the alert for an opening. He shut off the pumps and waited for the apparatus to cool, while Haviland washed his face and chest at the soapstone sink.

"I'm not going to break the vacuum," said Erik. "We can let it stand overnight so that we can start again tomorrow."

Haviland continued his washing. Then he took a fresh towel to dry his face, allowing his wet body to cool.

"Why?" he asked.

"We can actually begin the experiment," said Erik. "We started out to build a neutron generator and here it is – complete. There's nothing else to wait for. Or is there?" he added.

Haviland's face flushed slightly.

"I'd like to think it over for a day or so," he said.

148

And here the slight let-down, often described, in the moment of achievement.

They were both silent for a moment, a little stunned at the sudden release of the pressure on them.

"Is that all there is to it?" Erik asked.

"What did you expect?"

"I don't know. I guess I had the idea that you go along doing the work,and you reach a high point where electricity is discovered or the new invention called the telephone says 'What God hath wrought.' Then you're supposed to stand up and say *Eureka*. The President of the United States marches in, puts his hand on your shoulder and says 'You have done a fine thing, young man.' But this – " He spread his hands. "Where's the big moment?"

Haviland smiled. "I always used to think it would be like that, and even when I got used to the idea that it wasn't, I still hung on to the dream that when I did something really big, the music would play." He thought a moment. "As a matter of fact, there's always a big moment in every experiment, only you never know it until it's all over. And, when you come to think of it, how can you expect it to come at the *end* of the experiment? If anything worth while shows up, it's got to be repeated so many times that all the excitement is washed away. I think the big moment here came when we first got neutrons . . . that afternoon before we actually started the experiment."

"I suppose so," said Erik slowly. "But we were talking about other things that day as well, and perhaps at the time they seemed more important than neutron production."

149

A little pothole on the road of progress is described in Nuel Pharr Davis's Lawrence and Oppenheimer. *The great chemist, G. N. Lewis, was preparing purified heavy water – at that time a most arduous process – and E. O. Lawrence was eagerly waiting for a sample, so that he could accelerate deuterons in his cyclotron.*

Lawrence kept asking Lewis how much heavy water he had until about the first of March Lewis was able to show him a whole cubic centimeter. It was enough to accelerate, but at this point Lewis proved to be no physicist. Worried about whether he had manufactured a poison, he fed the whole sample to a mouse. It brought no good or harm to the mouse, but to Lawrence it almost brought apoplexy. "This was the most expensive cocktail that I think mouse or man ever had!" he complained.

150

Michael Roberts's 'Note on θ, φ and ψ' is a graceful little elegy on the unity of learning.

NOTE ON θ, φ AND ψ

Whereas my lady loves to look
On learned manuscript and book,
Still must she scorn, and scorning sigh,
To think of those I profit by.

Plotinus now, or Plutarch is
A prey to her exegesis,
And while she labours to collate
A page, I grasp a postulate,

And find for one small world of fact
Invariant matrices, compact
Within the dark and igneous rock
Of *Comptes Rendus* or *Proc. Roy. Soc.*

She'll pause a learned hour, and then
Pounce with a bird-like acumen
Neatly to annotate the dark
Of halting sense with one remark;

While I, maybe, precisely seize
The elusive photon's properties
In α's and δ's, set in bronze-
bright vectors, grim quaternions.

Silent we'll sit. We'll not equate
Symbols too plainly disparate,
But hand goes out to friendly hand
That mind and mind may understand

How one same passion burned within
Each learned peer and paladin,
Her Bentley and her Scaliger,
My Heisenberg and Schrödinger.

151

Gregory Benford's theoretical physicist of a few years into the future wrestles with the concept of tachyons – from Timescape.

There was no choice between beauty and truth, really. You had to wind up with both. In art, elegance was a whore of a word, bent a different way by each generation of critics. In physics, though, there was some fragile lesson to be learned from past millennia. Theories were more elegant if they could be transformed mathematically to other frames, other observers. A theory that remained invariant under the most general transformation was the most deft, the nearest to a universal form. Gell-Mann's SU(3) symmetry had arrayed particles into universal ranks. The Lorentz group; isospin; the catalogue of properties labeled Strangeness and Color and Charm – they all cooked gauzy Number into concrete Thing. So to proceed beyond Einstein, one should follow the symmetries.

Markham scratched equations across a yellow pad, searching. He had intended to spend this time plotting his tactics with the NSF, but politics was dross compared with the actual doing of science. He tried different approaches, twisting the compacted tensor notation, peering into the maze of mathematics. He had a guiding principle: nature seemed to like equations stated in covariant differential forms. To find the right expressions –

He worked out the equations governing tachyons in a flat space-time, doing the exercise as a limited case. He nodded. Here were the familiar quantum-mechanical wave equations, yes. He knew where they led. The tachyons could cause a probability wave to reflect back and forth in time. The equations told how this wave function would shuttle, past to future, future to past, a befuddled commuter. Making a paradox meant the wave had no ending, but instead formed some sort of standing wave pattern, like the rippling patterns around an ocean jetty, shifting their troughs and peaks but always returning, an ordering imposed on the blank face of the churning sea. The only way to resolve the paradox was to step in, break up the pattern, like a ship cutting across the troughs, leaving a swirl of sea behind. The ship was the classical observer. But now Markham added the Wickham terms, making the equations symmetric under interchange of tachyons. He rummaged through his briefcase for the paper by Gott that Cathy had given him. Here: *A Time-Symmetric, Matter and Anti-Matter Tachyon Cosmology*. Quite a piece of territory to bite off, indeed. But Gott's solutions were there, luminous on the page. The Wheeler-Feynman forces were there, mixing advanced and retarded tachyon solutions together with non-Euclidean sums. Markham blinked. In his cottony silence he sat very still, eyes racing, imagination leaping ahead to see where the equations would fold and part to yield up fresh effects.

The waves still stood, mutely confused. But there was no role left for the ship, for the classical observer. The old idea in conventional quantum mechanics had been to let the rest of the universe be the observer, let *it* force the waves to collapse. In these new tensor terms, though, there was no way to regress, no way to let the universe as a whole be a stable spot from which all things were measured. No, the universe was coupled in firmly. The tachyon field wired each fragment of matter to every other. Hooking more particles into the network only worsened things. The old quantum theorists, from Heisenberg and Bohr on, had let in some metaphysics at this point, Markham remembered. The wave function collapsed and that was the irreducible fact. The probability of getting a certain solution was proportional to the amplitude of that solution inside the total wave, so in the end you got only a statistical weighting of what would come out of an experiment. But with tachyons that dab of metaphysics had to go.

152

From Paul Preuss's novel of particle physics, Broken Symmetries, *a moment of chilling recognition.*

Frank McDonald was waiting on the control platform; he wore jeans and a plaid flannel shirt, having put the customary lab coat aside. "Everything's set, Ume. That switch will stop the condensor transport. You've got to do it on your own, though."

Narita looked at the scorched control panel. "That's very Japanese of you, Frank. As long as I do it, you're not being disloyal, eh?"

"Something like that. I've known Martin a long time, Ume. He's been wrong a few times, but never about anything important."

"This is important." Narita flipped the warped plastic cover off the T-bar switch, hooked a finger under the crosspiece, and flipped it up. He heard a faint clicking of relays, noted a few small red glass indicator bulbs light up.

"Damn, look at that!" McDonald exclaimed.

"Did it work?" Narita demanded

McDonald glanced at him. "Oh, yeah – the transporter's stopped. I was looking at these sensor circuits I just replaced. I just got another big gamma burst."

"Inside the machine?"

"Well, it's hard to tell. We didn't think so before. Penny Harper was working on it, but frankly . . ." He let the thought die.

Narita looked at the warped and burned instruments. "I was thinking, Frank – what if this collector isn't one hundred percent efficient? What if little drops of holy water have been clinging to the channels in there, accumulating for weeks?"

"What of it? Might be worthwhile cleaning it out, when the ring goes down."

Narita's gaze wandered to the bulging steel door on the other side of the cold, echoing room. He shivered. "I think we got this machine turned off just in time, Frank."

"In time for what?" McDonald sounded skeptical.

"To save the ring, maybe. If Slater is right, that gamma burst you saw just now was the disintegration of an I-particle."

"*One* I-particle? Slater thinks they do that?"

Narita nodded. "Yes, he does. Come on, Frank, I'll buy you a drink. It's cold down here."

153

And now, the real thing, from Gary Taubes's Nobel Dreams – *trouble in the caverns of the CERN laboratory.*

Rubbia wrote André Lagarrique of Gargamelle, saying that the Fermilab experiment had one hundred unambiguous neutral-current events, and that "we are in the phase of final write-ups of the results." He requested that when the European collaboration announced the discovery, they might be kind enough to mention that HPW had equivalent results. The letter could be read either as a bluff or as a delaying tactic.

Lagarrique showed Rubbia's letter to Don Perkins, head of the English contingent at Gargamelle. "I burst out laughing," Perkins recalled. "He said, 'Well, what are you laughing at?' And I said, 'Well, that's Carlo. He's got hundreds of events definitely showing the existence of neutral currents. Then what the hell does he need Gargamelle for? Why doesn't he just go publish them? What's he writing to you for? He needs something. It's ridiculous.'"

Instead of debating for another few months the merits of publishing a paper, as they had appeared ready to do, the Gargamelle collaboration decided to announce their results immediately and take their chances with posterity. Rubbia's bluff had been called.

Rubbia, meanwhile, was having a bad summer all around:

Back in Boston, he was asked by U.S. Immigration to leave the country after his visa had expired and he had failed to file for an extension. The immigration authorities gave him a handful of days to see to his affairs and leave the country. On his last trip to Fermilab, he tried to salvage the situation, demanding that his American colleagues publish their neutral-currents results immediately. Cline and Mann were so peeved by Rubbia's attitude – they objected to being ordered around as though they were graduate students – that, instead, they got nervous and decided to redo their experiment. With more data,and under brutal pressure from the Fermilab management and the physics community to come up with a definitive answer, they concluded prematurely that they did not have evidence for neutral currents. When they told Rubbia, he believed them. He set about spreading the word at CERN that neutral currents were nonexistent and gleefully (as Sulak put it) tried to convince the management to start an investigation into how the Gargamelle people could have propagated such a grevious error. ("If you can't be a giant," Martin Perl, a California physicist once told me, "you want to be the giant killer. That's the name of the game.") With Rubbia's machinations rolling merrily along, the HPW physicists continued refining the second analysis, and realized that they did, indeed, have neutral currents.

By the time it was over, Rubbia and his colleagues were the butt of one of the more biting jokes in the physics community: They were said to have discovered alternating neutral currents. (Dave Cline, whose initials are D.C., picked up the nickname "A.C.D.C.") If Rubbia's backroom politicking accomplished anything, it was to help muddle the situation so thoroughly that none of the Gargamelle people ever got the Nobel Prize their work seemed to deserve.

Within a year, the neutral-currents episode had faded in everyone's memory.

154

John Banville's Kepler *is the latest of several novels in which the enigmatic astronomer appears. (An earlier one, highly regarded in its day but too ponderously paced, I would judge, for most modern readers, was by the German writer, Max Brod.)*

Kepler turned again now to his work on Mars. Conditions around him had improved. Christian Longberg, tired of squabbling, had gone back to Denmark, and there was no more talk of their wager. Tycho Brahe too was seldom seen. There were rumours of plague and Turkish advances, and the stars needed a frequent looking to. The Emperor Rudolph, growing ever more nervous, had moved his imperial mathematician in from Benatek, but even the Curtius house was not close enough, and the Dane was at the palace constantly. The weather was fine, days the colour of Mosel wine, enormous glassy nights. Kepler sometimes sat with Barbara in the garden, or with Regina idly roamed the Hradcany, admiring the houses of the rich and watching the imperial cavalry on parade. But by August the talk of plague had closed the great houses for the season, and even the cavalry found an excuse to be elsewhere. The Emperor decamped to his country seat at Belvedere, taking Tycho Brahe with him. The sweet sadness of summer settled on the deserted hill, and Kepler thought of how as a child, at the end of one of his frequent bouts of illness, he would venture forth on tender limbs into a town made magical by the simple absence of his schoolfellows from its streets.

Mars suddenly yielded up a gift, when with startling ease he refuted Copernicus on oscillation, showing by means of Tycho's data that the planet's orbit intersects the sun at a fixed angle to the orbit of the earth. There were other, smaller victories. At every advance, however, he found himself confronted again by the puzzle of the apparent variation in orbital velocity. He turned to the past for guidance. Ptolemy had saved the principle of uniform speed by means of the *punctum equans*, a point on the diameter of the orbit from which the velocity will appear invariable to an imaginary observer (whom it amused Kepler to imagine, a crusty old fellow, with his brass triquetrum and watering eye and smug, deluded certainty). Copernicus,shocked by Ptolemy's sleight of hand, had rejected the equant point as blasphemously inelegant, but yet had found nothing to put in its place except a clumsy combination of five uniform epicyclic motions superimposed one upon another. These were, all the same, clever and sophisticated manoeuvres, and saved the phenomena admirably. But had his great predecessors taken them, Kepler wondered, to represent the real state of things? The question troubled him. Was there an innate nobility, lacking in him, which set one above the merely empirical? Was his pursuit of the forms of physical reality irredeemably vulgar?

In a tavern on Kleinseit one Saturday night he met Jeppe and the Italian. They had fallen in with a couple of kitchen-hands from the palace, a giant Serb with one eye and a low ferrety fellow from Württemberg, who claimed to have soldiered with Kepler's brother in the Hungarian campaigns. His name was Krump. The Serb rooted in his codpiece and brought out a florin to buy a round of schnapps. Someone struck up on a fiddle, and a trio of whores sang a bawdy song and danced. Krump squinted at them and spat. 'Riddled with it, them are,' he said, 'I know them.' But the Serb was charmed, ogling the capering drabs out of his one oystrous eye and banging his fist on the table in time to the jig. Kepler ordered up another round. 'Ah,' said Jeppe. 'Sir Mathematicus is flush tonight; has my master forgot himself and paid your wages?' 'Something of that,' Kepler answered, and thought himself a gay dog. They played a hand of cards, and there was more drink. The Italian was dressed in a suit of velvet, with a slough hat. Kepler spotted him palming a knave. He won the hand and grinned at Kepler, and then, calling for another jig, got up and with a low bow invited the whores to dance. The candles on the tavern counter shook to the thumping of their feet. 'A merry fellow,' said Jeppe, and Kepler nodded, grinning blearily. The dance became a general rout, and somehow they were suddenly outside in the lane. One of the whores fell down and lay there laughing, kicking her stout legs in the air. Kepler propped himself against the wall and watched the goatish dancers circling in a puddle of light from the tavern window, and all at once out of nowhere, out of everywhere, out of the fiddle music and the flickering light and the pounding of heels, the circling dance and the Italian's drunken eye, there came to him the ragged fragment of a thought. False. What false? That principle. One of the whores was pawing him. Yes, he had it. *The principle of uniform velocity is false.*

155

Lewis Fry Richardson, asked to explain in simple terms a paper that was to become a landmark in atmospheric physics, summed it up in four crisp lines:

Big whirls have little whirls,
That feed on their velocity,

And little whirls have lesser whirls,
And so on to viscosity.

Benjamin Franklin was a highly original amateur scientist, whose discoveries ranged from atmospheric electricity to surface monolayers. John Adams wrote of him: 'Franklin's reputation is more universal than that of Leibniz or Newton, Frederick or Voltaire, and his character more beloved and esteemed than any or all of them'. William Cobbett, on the other hand, called him 'a crafty and lecherous old hypocrite'. Franklin, earlier in his life, was a master printer. He wrote an epitaph for himself, which ran:

The body of
Benjamin Franklin, printer,
(Like the covers of an old book,
Its contents worn out,
And strippt of its leathering and gilding)
Lies here, food for worms!
Yet the work itself shall not be lost,
For it will, as he believed, appear once more,
In a new
And more beautiful edition,
Corrected and amended
By its Author!

(The implication is that he would be reincarnated; this was part of the belief of the religious sect to which he belonged.) The epitaph that Franklin eventually got was grander: 'Eripuit caelo fulmen sceptrumque tyrannis': 'He tore the lightning from the heavens, and the sceptre from the hands of tyrants.'

Franklin's experiments on lightning, divertingly described in his Autobiography, *were insanely dangerous and drew the following comment from Mark Twain: 'What an adroit old adventurer the subject of this memoir was! In order to get a chance to fly his kite on Sunday, he used to hang a key on a string and let on to be fishing for lightning'. Here then is Franklin, ruminating on the reception that the reports of his experiments had received in Europe.*

It was, however, some time before those papers were much taken notice of in England. A copy of them happening to fall into the hands of the Count de Buffon, a philosopher deservedly of great reputation in France, and, indeed, all over Europe, he prevailed with M. Dalibard to translate them into French, and they were printed at Paris. The publication offended the Abbé Nollet, preceptor in Natural Philosophy to the royal family, and an able experimenter, who had form'd and publish'd a theory of electricity, which then had the general vogue. He could not at first believe that such a work came from America, and said it must have been fabricated by his enemies at Paris, to decry his system. Afterwards, having been assur'd that there really existed such a person as Franklin at Philadelphia, which he had doubted, he wrote and published a volume of Letters, chiefly address'd to me, defending his theory, and denying the verity of my experiments, and of the positions deduc'd from them.

I once purpos'd answering the abbé, and actually began the answer; but, on consideration that my writings contain'd a description of experiments which any one might repeat and verify, and if not to be verifi'd, could not be defended; or of observations offer'd as conjectures, and not delivered dogmatically, therefore not laying me under any obligation to defend them; and reflecting that a dispute between two persons, writing in different languages, might be lengthened greatly by mistranslations, and thence misconceptions of one another's meaning, much of one of the abbé's letters being founded on an error of the translation, I concluded to let my papers shift for themselves, believing it was better to spend what time I could spare from public business in making new experiments, than in disputing about those already made. I therefore never answered M. Nollet, and the event gave me no cause to repent my silence; for my friend M. le Roy, of the Royal Academy of Sciences, took up my cause and refuted him; my book was translated into the Italian, German, and Latin languages; and the doctrine it contain'd was by degrees universally adopted by the philosophers of Europe, in preference to that of the abbé; so that he lived to see himself the last of his sect, except Monsieur B—, of Paris, his *élève* and immediate disciple.

What gave my book the more sudden and general celebrity, was the success of one of its proposed experiments, made by Messrs Dalibard and De Lor at Marly, for drawing lightning from the clouds. This engag'd the public attention every where. M. de Lor, who had an apparatus for experimental philosophy, and lectur'd in that branch of science, undertook to repeat what he called the *Philadelphia Experiments*; and, after they were performed before the king and court, all the curious of Paris flocked to see them. I will not swell this narrative with an account of that capital experiment, nor of the infinite pleasure I receiv'd in the success of a similar one I made soon after with a kite at Philadelphia, as both are to be found in the histories of electricity.

156

Emulation and jealousy among scientists is nothing new. But only under the structure of science that now prevails have they become sanctified as the motives that drive scientists ever onwards. Stephen Hall describes in Invisible Frontiers *how two unknown young men decided that they would stake all on a race against one of the great barons of molecular biology and his army of retainers, because here was an opponent worthy of their mettle.*

Biologists traditionally viewed the path from academia to industry as a one-way street, and a downhill one at that; they tended to look upon industry scientists as researchers whose scientific curiosity was conditioned by managers with one eye on the marketplace, whose freedom to publish was curtailed; and there even existed,just beneath the surface, a kind of snobbish machismo in basic research, an unspoken consensus that to tackle anything less than the stickiest, most confounding, most fundamental biological mysteries suggested a certain lack of intellectual ballsiness. In a field where intellectual rewards historically outstripped financial compensations, it meant giving up a lot.

Kleid had more at stake than Goeddel – he would be abandoning what every scientist works toward, his own lab. In exchange, he would earn the dubious honor of commuting to southern California frequently to do the work. For Kleid, the decision was "very chancy," and in the end, he cheerfully abdicated responsibility to Goeddel. "Basically I said to Dave, 'It's your decision. If you want to do it, let's do it. If you don't want to do it, we'll stay here,'" Kleid recalls. He knew that if he turned the project over to Goeddel, it would fly.

Indecisiveness is not one of Goeddel's weak points. He had no academic aspirations, so he didn't mind leaving that bridge aflame behind him. His salary at SRI was paltry compared to what Genentech was offering. Most of all, Goeddel liked the challenge. Here, he realized, was the opportunity to work on a hot project. Here was the opportunity to get in on the ground floor of a business that could have a huge future (although even Goeddel himself admits that personal goals exceeded any loyalty to the company when he decided to sign up). And, perhaps more important, here was the opportunity to go head to head with a scientist of enormous reputation, Wally Gilbert, on a highly visible project. "I was conscious, very conscious, of the competitive aspect," Goeddel says. "Actually, I found that very attractive. We knew we were in a race."

"Why don't we?" Goeddel finally urged Kleid. "You know, they have Itakura. They can make the DNA better than anybody. It's a good chance." Kleid agreed. Just like that, Genentech had its first full-time scientists.

Bits of synthetic DNA, ready to be assembled into a human insulin gene, were waiting for them when they began. Goeddel reported to work in mid-March; thus began a five-month period of work in which he claims to have taken only one day off – to attend the ten-year reunion of his high school class in Poway, California, outside San Diego. Kleid showed up, as he likes to point out, on April Fool's Day; that first day, according to his lab notebook, he signed out at four in the morning.

They worked against molecular biology's traditional deadline: yesterday.

157

Miroslav Holub's poem catches the silent intensity of a late night in the laboratory.

EVENING IN A LAB

The white horse will not emerge from the lake
(of methyl green),
the flaming sheet will not appear
in the dark field condenser.
Pinned down by nine pounds of failure,
pinned down by half an inch of hope
sit and read,

sit as the quietest weaver
and weave and read,

where even verses break their necks,

when all the others have left.

Pinned down by eight barrels of failure,
pinned down by a quarter grain of hope,
sit as the quietest savage beast
and scratch and read.

The white horse will not emerge from the lake
(of methyl green),
the flaming sheet will not appear
in the dark field condenser.

Among cells and needles,
butts and dogs,
among stars,
there, where you wake,
there, where you go to sleep,
where it never was, never is, never mind –
search
and find.

158

And here is a biologist of an earlier epoch, Elie Metchnikoff (portrayed in Olga Metchnikoff's biography of her husband), suddenly revived at a low moment in his life by the advent of a new idea. He divined that there are cells in the circulation which would speed to the site of an infection and engulf intruders, such as bacteria. These were originally termed Fresszellen, *later more genteely rendered into phagocytes.*

I was resting from the shock of the events which provoked my resignation from the University and indulging enthusiastically in researches in the splendid setting of the Straits of Messina.

One day when the whole family had gone to a circus to see some extraordinary performing apes, I remained alone with my microscope, observing the life in the mobile cells of a transparent star-fish larva, when a new thought suddenly flashed across my brain. It struck me that similar cells might serve in the defence of the organism against intruders. Feeling that there was in this something of surpassing interest, I felt so excited that I began striding up and down the room and even went to the seashore in order to collect my thoughts.

I said to myself that, if my supposition was true, a splinter introduced into the body of a star-fish larva, devoid of blood-vessels or of a nervous system, should soon be surrounded by mobile cells as is to be observed in a man who runs a splinter into his finger. This was no sooner said than done.

There was a small garden to our dwelling, in which we had a few days previously organised a 'Christmas tree' for the children on a little tangerine tree; I fetched from it a few rose thorns and introduced them at once under the skin of some beautiful star-fish larvae as transparent as water.

I was too excited to sleep that night in the expectation of the result of my experiment, and very early the next morning I ascertained that it had fully succeeded.

That experiment formed the basis of the phagocyte theory, to the development of which I devoted the next twenty-five years of my life.

159

Renaud Censier, in Georges Duhamel's Pasquier Chronicles, *is a successful bacteriologist. Blomberg, a devotee of Metchnikoff's, is a failure.*

'Luck is all-important, even in matters of science, yes, particularly in science. Why, there are people who devote themselves to research for twenty years and never find anything really significant.'

He threw up his head, shook his hair which was so long that it greased the collar of his coat, and went on:

'I am not thinking of myself, for I really have nothing to complain of.'

Renaud Censier shrugged his shoulders, benevolently bored, slightly embarrassed. But Blomberg went on, dreamily:

'Ten years since Pasteur died! The great "pastorians" are finished. They'll do no more. Believe me, I knew them and respected them. But they're dried up. All except my beloved Élie. He still has fire – the real flame of genius.'

He was the same age as Metchnikoff, and lived so closely in the shadow of this remarkable man that, in his admiration, he imitated his way of speaking, reproducing unconsciously his physical outline and the hundred and one muscular movements which combined to make up the man's attitude and manner.

Renaud Censier put his hand on the old man's arm, a slender, well-shaped, somewhat dried-up hand:

'I am greatly touched by your appreciation, Blomberg. . . '

He expressed himself slowly, separating his phrases, halting sometimes between his syllables with inexplicable pauses which deceived one into thinking he had actually finished speaking. Those who remember him at that period, that is to say about 1905, see in their mind's eye that delicate face, still young looking, his grey hair and light beard contrasting sharply with the tan of his slender face. The neck was not very supple, for he seemed to turn his head slowly, almost sluggishly; but his eyes were swift and turned so sharply that the whites showed up against the dark complexion.

'You'll see,' said Blomberg, with an obstinacy tinged with spitefulness, 'the Pasteur Institute will fade out. Believe me, I shall regret it deeply, and you are the man who will carry on the work; but away from the beaten track. Yes, you are the man of the future. After all, you are only fifty.'

'Fifty-two.'

'Yes, you will be in your prime. I only hope I shall live long enough to be a witness of your conquests.'

Renaud smiled deprecatingly:

'Hardly conquests! The further I go the more I feel that what I reach is only provisional truth, not for all time. You are kind enough to speak of my success . . . but I must confess, Blomberg, that it worries me to read articles in the paper devitalizing my ideas, twisting them to serve all sorts of causes which are very far from those I have at heart. Conquests! Far from it! I am fortunate enough, sometimes, to hit on an idea which may be of service to others – how shall I say? – a sort of stepping-stone or jumping-off place! After all, that is what science needs. Blomberg, I am a skirmisher, a scout; I am not, and do not wish to be, the leader of a school. No. I live and work almost alone, with just one or two students, I have never wished for anything more. You talk of luck, Blomberg – and no doubt you are right. But, for myself, I am only at my ease in semi-retreat. I am obscure by vocation. In spite of what the publicists may say,I shall never be anything but a very little-known man of fame. Do you believe in fate?'

'Well, that depends . . .'

Renaud Censier laughed good-naturedly.

'My humble peasant name will never make a good adjective. You see, some people's fate is settled in advance.'

'Well, neither will Metchnikoff's,' said Blomberg, shaking his head, 'though of course what we are saying is all nonsense.'

His face had clouded over suddenly, and the furrows marked a wry smile. For thirty years he had laboured in the hope of establishing his name securely. He had discovered, and described at interminable length, protoplasmic corpuscules in the animal cells, to which he attributed a parasitical character, and which he called, with touching candour, in his notes and reports 'blombergias' or corpuscules of Blomberg. (He had had to fight indefatigably for the word 'blombergia' to be given the honour of a small letter instead of a capital.) From that starting-point it was but a step to formulate a theory, only a step, nevertheless Blomberg had seen fit to publish a ponderous thesis on what he called 'universal parasitism' or Blomberg theory, after having trifled with 'blombergian' . . . 'blombergism.' Unfortunately the old man was well-nigh the only one to pay tribute to this cult, so that at the slightest mention of the subject, however remote, his spirits would sink, owing to his uncertainty as to whether he should expect homage or raillery.

Censier's expression was free of mockery or criticism, and Blomberg's face cleared.

160

H. G. Wells, in one of his best scientific short stories, 'The Moth', portrays a running scholarly feud, which becomes an unquenchable odium theologicum.

Probably you have heard of Hapley – not W. T. Hapley, the son, but the celebrated Hapley, the Hapley of *Periplaneta Hapliia*, Hapley the entomologist.

If so you know at least of the great feud between Hapley and Professor Pawkins, though certain of its consequences may be new to you. For those who have not, a word or two of explanation is necessary, which the idle reader may go over with a glancing eye, if his indolence so incline him.

It is amazing how very widely diffused is the ignorance of such really important matters as this Hapley-Pawkins feud. Those epoch-making controversies, again, that have convulsed the Geological Society are, I verily believe, almost entirely unknown outside the fellowship of that body. I have heard men of fair general education, even refer to the great scenes at these meetings as vestry-meeting squabbles. Yet the great hate of the English and Scotch geologists has lasted now half a century, and has 'left deep and abundant marks upon the body of the science.' And this Hapley-Pawkins business, though perhaps a more personal affair, stirred passions as profound, if not profounder. Your common man has no conception of the zeal that animates a scientific investigator, the fury of contradiction you can arouse in him. It is the *odium theologicum* in a new form. There are men, for instance, who would gladly burn Professor Ray Lankester at Smithfield for his treatment of the Mollusca in the Encyclopaedia. That fantastic extension of the Cephalopods to cover the Pteropods . . . But I wander from Hapley and Pawkins.

It began years and years ago, with a revision of the Microlepidoptera (whatever these may be) by Pawkins, in which he extinguished a new species created by Hapley. Hapley, who was always quarrelsome, replied by a stinging impeachment of the entire classification of Pawkins.[1] Pawkins in his 'Rejoinder'[2] suggested that Hapley's microscope was as defective as his power of observation, and called him an 'irresponsible meddler' – Hapley was not a professor at that time. Hapley in his retort,[3] spoke of 'blundering collectors,' and described, as if inadvertently, Pawkins' revision as a 'miracle of ineptitude.' It was war to the knife. However, it would scarcely interest the reader to detail how these two great men quarrelled,and how the split between them widened until from the Microlepidoptera they were at war upon

1 "Remarks on a Recent Revision of Microlepidoptera. *Quart. Journ. Entomological Soc.* 1863.

2 "Rejoinder to certain Remarks," etc. *Ibid.* 1864.

3 "Further Remarks," etc. *Ibid.*

every open question in entomology. There were memorable occasions. At times the Royal Entomological Society meetings resembled nothing so much as the Chamber of Deputies. On the whole, I fancy Pawkins was nearer the truth than Hapley. But Hapley was skilful with his rhetoric, had a turn for ridicule rare in a scientific man, was endowed with vast energy, and had a fine sense of injury in the matter of the extinguished species; while Pawkins was a man of dull presence, prosy of speech, in shape not unlike a water-barrel, over conscientious with testimonials, and suspected of jobbing museum appointments. So the young men gathered round Hapley and applauded him. It was a long struggle, vicious from the beginning and growing at last to pitiless antagonism. The successive turns of fortune, now an advantage to the one side and now to another – now Hapley tormented by some success of Pawkins, and now Pawkins outshone by Hapley, belong rather to the history of entomology than to this story.

But in 1891 Pawkins, whose health had been bad for some time, published some work upon the 'mesoblast' of the Death's Head Moth. What the mesoblast of the Death's Head Moth may be does not matter a rap in this story. But the work was far below his usual standard, and gave Hapley an opening he had coveted for years. He must have worked night and day to make the most of his advantage.

In an elaborate critique he rent Pawkins to tatters – one can fancy the man's disordered black hair, and his queer dark eyes flashing as he went for his antagonist – and Pawkins made a reply, halting, ineffectual, with painful gaps of silence, and yet malignant. There was no mistaking his will to wound Hapley, nor his incapacity to do it. But few of those who heard him – I was absent from that meeting – realised how ill the man was.

Hapley got his opponent down, and meant to finish him. He followed with a simply brutal attack upon Pawkins, in the form of a paper upon the development of moths in general, a paper showing evidence of a most extraordinary amount of mental labour, and yet couched in a violently controversial tone. Violent as it was, an editorial note witnesses that it was modified. It must have covered Pawkins with shame and confusion of face. It left no loophole; it was murderous in argument, and utterly contemptuous in tone; an awful thing for the declining years of a man's career.

The world of entomologists waited breathlessly for the rejoinder from Pawkins. He would try one, for Pawkins had always been game. But when it came it surprised them. For the rejoinder of Pawkins was to catch influenza, proceed to pneumonia, and die.

It was perhaps as effectual a reply as he could make under the circumstances, and largely turned the current of feeling against Hapley. The very people who had most gleefully cheered on those gladiators became serious at the consequence. There could be no reasonable doubt the fret of the defeat had contributed to the death of Pawkins. There was a limit even to scientific controversy, said serious people. Another crushing attack was already in the Press and appeared on the day before the funeral. I don't think Hapley exerted himself to stop it. People remembered how Hapley had hounded down his rival, and forgot that rival's defects. Scathing satire reads ill over fresh mould. The thing provoked comment in the daily papers. This it was that made me think that you had probably heard of Hapley and this controversy. But, as I have already remarked, scientific workers live very much in a world of their own; half the people, I dare say, who go along Piccadilly to the Academy each year, could not tell you where the learned societies abide. Many even think that research is a kind of happy-family cage in which all kinds of men lie down together in peace.

In his private thoughts Hapley could not forgive Pawkins for dying. In the first place, it was a mean dodge to escape the absolute pulverisation Hapley had in hand for him, and in the second, it left Hapley's mind with a queer gap in it. For twenty years he had worked hard, sometimes far into the night, and seven days a week, with microscope, scalpel, collecting-net, and pen, and almost entirely with reference to Pawkins. The European reputation he had won had come as an incident in that great antipathy. He had gradually worked up to a climax in this last controversy. It had killed Pawkins, but it had also thrown Hapley out of gear, so to speak, and his doctor advised him to give up work for a time, and rest. So Hapley went down into a quiet village in Kent, and thought day and night of Pawkins, and good things it was now impossible to say about him.

At last Hapley began to realise in what direction the preoccupation tended. He determined to make a fight for it, and started by trying to read novels. But he could not get his mind off Pawkins, white in the face and making his last speech – every sentence a beautiful opening for Hapley. He turned to fiction – and found it had no grip on him. He read the *Island Nights' Entertainments* until his sense of causation was shocked beyond endurance by the Bottle Imp. Then he went to Kipling, and found he 'proved nothing,' besides being irreverent and vulgar. These scientific people have their limitations. Then, unhappily, he tried Besant's *Inner House*, and the opening chapter set his mind upon learned societies and Pawkins at once.

So Hapley turned to chess, and found it a little more soothing. He soon mastered the moves and the chief gambits and commoner closing positions, and began to beat the Vicar. But then the cylindrical contours of the opposite king began to resemble Pawkins standing up and gasping ineffectually against check-mate, and Hapley decided to give up chess.

Perhaps the study of some new branch of science would after all be better diversion. The best rest is change of occupation. Hapley determined to plunge at diatoms, and had one of his smaller microscopes and Halibut's monograph sent down from London. He thought that perhaps if he could get up a vigorous quarrel with Halibut, he might be able to begin life afresh and forget Pawkins. And very soon he was hard at work in his habitual strenuous fashion, at these microscopic denizens of the wayside pool.

It was on the third day of the diatoms that Hapley became aware of a novel addition to the local fauna. He was working late at the microscope, and the only light in the room was the brilliant little lamp with the special form of green shade. Like all experienced microscopists, he kept both eyes open. It is the only way to avoid excessive fatigue. One eye was over the instrument,and bright and distinct before that was the circular field of the microscope, across which a brown diatom was slowly moving. With the other eye Hapley saw, as it were, without seeing. He was only dimly conscious of the brass side of the instrument, the illuminated part of the tablecloth, a sheet of notepaper, the foot of the lamp, and the darkened room beyond.

Suddenly his attention drifted from one eye to the other. The tablecloth was of the material called tapestry by shopmen, and rather brightly coloured. The pattern was in gold, with a small amount of crimson and pale blue upon a grayish ground. At one point the pattern seemed displaced, and there was a vibrating movement of the colours at this point.

Hapley suddenly moved his head back and looked with both eyes. His mouth fell open with astonishment.

It was a large moth or butterfly; its wings spread in butterfly fashion!

It was strange it should be in the room at all, for the windows were closed. Strange that it should not have attracted his attention when fluttering to its present position. Strange that it should match the tablecloth. Stranger far that to him, Hapley, the great entomologist, it was altogether unknown. There was no delusion. It was crawling slowly towards the foot of the lamp.

'New Genus, by heavens! And in England!' said Hapley, staring.

Then he suddenly thought of Pawkins. Nothing would have maddened Pawkins more. . . . And Pawkins was dead!

Something about the head and body of the insect became singularly suggestive of Pawkins, just as the chess king had been.

'Confound Pawkins!' said Hapley. 'But I must catch this.' And looking round him for some means of capturing the moth, he rose slowly out of his chair. Suddenly the insect rose, struck the edge of the lampshade – Hapley heard the 'ping' – and vanished into the shadow.

In a moment Hapley had whipped off the shade, so that the whole room was illuminated. The thing had disappeared, but soon his practised eye detected it upon the wallpaper near the door. He went towards it poising the lampshade for capture. Before he was within striking distance, however, it had risen and was fluttering round the room. After the fashion of its kind, it flew with sudden starts and turns, seeming to vanish here and reappear there. Once Hapley struck, and missed; then again.

The third time he hit his microscope. The instrument swayed, struck and overturned the lamp, and fell noisily upon the floor. The lamp turned over on the table and, very luckily, went out. Hapley was left in the dark. With a start he felt the strange moth blunder into his face.

It was maddening. He had no lights. If he opened the door of the room the thing would get away. In the darkness he saw Pawkins quite distinctly laughing at him. Pawkins had ever an oily laugh. He swore furiously and stamped his foot on the floor.

There was a timid rapping at the door.

Then it opened, perhaps a foot, and very slowly. The alarmed face of the landlady appeared behind a pink candle flame; she wore a night-cap over her gray hair and had some purple garment over her shoulders. 'What *was* that fearful smash?' she said. 'Has anything – '

The strange moth appeared fluttering about the chink of the door. 'Shut that door!' said Hapley, and suddenly rushed at her.

The door slammed hastily. Hapley was left alone in the dark. Then, in the pause, he heard his landlady scuttle upstairs, lock her door, and drag something heavy across the room and put against it.

It became evident to Hapley that his conduct and appearance had been strange and alarming. Confound the moth! and Pawkins! However, it was a pity to lose the moth now. He felt his way into the hall and found the matches, after sending his hat down upon the floor with a noise like a drum. With the lighted candle he returned to the sitting-room. No moth was to be seen. Yet once for a moment it seemed that the thing was fluttering round his head. Hapley very suddenly decided to give up the moth and go to bed. But he was excited. All night long his sleep was broken by dreams of the moth, Pawkins, and his landlady. Twice in the night he turned out and soused his head in cold water.

One thing was very clear to him. His landlady could not possibly understand about the strange moth, especially as he had failed to catch it. No one but an entomologist would understand quite how he felt. She was probably frightened at his behaviour, and yet he failed to see how he could explain it. He decided to say nothing further about the events of last night. After breakfast he saw her in her garden, and decided to go out and talk to reassure her. He talked to her about beans and potatoes, bees, caterpillars, and the price of fruit. She replied in her usual manner, but she looked at him a little suspiciously, and kept walking as he walked, so that there was always a bed of flowers, or a row of beans, or something of the sort, between them. After a while he began to feel singularly irritated at this, and to conceal his vexation went indoors and presently went out for a walk.

The moth, or butterfly, trailing an odd flavour of Pawkins with it, kept coming into that walk, though he did his best to keep his mind off it. Once he saw it quite distinctly, with its wings flattened out, upon the old stone wall that runs along the west edge of the park, but going up to it he found it was only two lumps of gray and yellow lichen. 'This,' said Hapley, 'is the reverse of mimicry. Instead of a butterfly looking like a stone, here is a stone looking like a butterfly!' Once something hovered and fluttered round his head, but by an effort of will he drove that impression out of his mind again.

In the afternoon Hapley called upon the Vicar, and argued with him upon theological questions. They sat in the little arbour covered with brier, and smoked as they wrangled. 'Look at that moth!' said Hapley, suddenly, pointing to the edge of the wooden table.

'Where?' said the Vicar.

'You don't see a moth on the edge of the table there?' said Hapley.

'Certainly not,' said the Vicar.

Hapley was thunderstruck. He gasped. The Vicar was staring at him. Clearly the man saw nothing. 'The eye of faith is no better than the eye of science,' said Hapley awkwardly.

'I don't see your point,' said the Vicar, thinking it was part of the argument.

That night Hapley found the moth crawling over his counterpane. He sat on the edge of the bed in his shirt sleeves and reasoned with himself. Was it pure hallucination? He knew he was slipping, and he battled for his sanity with the same silent energy he had formerly displayed against Pawkins. So persistent is mental habit, that he felt as if it were still a struggle with Pawkins. He was well versed in psychology. He knew that such visual illusions do come as a result of mental strain. But the point was, he did not only *see* the moth, he had heard it when it touched the edge of the lampshade, and afterwards when it hit against the wall, and he had felt it strike his face in the dark.

He looked at it. It was not at all dreamlike, but perfectly clear and solid-looking in the candle-light. He saw the hairy body,and the short feathery antennae, the jointed legs, even a place where the down was rubbed from the wing. He suddenly felt angry with himself for being afraid of a little insect.

His landlady had got the servant to sleep with her that night, because she was afraid to be alone. In addition she had locked the door, and put the chest of drawers against it. They listened and talked in whispers after they had gone to bed, but nothing occurred to alarm them. About eleven they had ventured to put the candle out, and had both dozed off to sleep. They woke up with a start, and sat up in bed, listening in the darkness.

Then they heard slippered feet going to and fro in Hapley's room. A chair was overturned, and there was a violent dab at the wall. Then a china mantel ornament smashed upon the fender. Suddenly the door of the room opened, and they heard him upon the landing. They clung to one another, listening. He seemed to be dancing upon the staircase. Now he would go down three or four steps quickly, then up again, then hurry down into the hall. They heard the umbrella stand go over, and the fanlight break. Then the bolt shot and the chain rattled. He was opening the door.

They hurried to the window. It was a dim gray night; an almost unbroken sheet of watery cloud was sweeping across the moon, and the hedge and trees in front of the house were black against the pale roadway. They saw Hapley, looking like a ghost in his shirt and white trousers, running to and fro in the road, and beating the air. Now he would stop, now he would dart very rapidly at something invisible, now he would move upon it with stealthy strides. At last he went out of sight up the road towards the down. Then, while they argued who should go down and lock the door, he returned. He was walking very fast, and he came straight into the house, closed the door carefully, and went quietly up to his bedroom. Then everything was silent.

'Mrs Colville,' said Hapley, calling down the staircase next morning, 'I hope I did not alarm you last night.'

'You may well ask that!' said Mrs Colville.

'The fact is, I am a sleep-walker, and the last two nights I have been without my sleeping mixture. There is nothing to be alarmed about, really, I am sorry I made such an ass of myself. I will go over the down to Shoreham, and get some stuff to make me sleep soundly. I ought to have done that yesterday.'

But half-way over the down, by the chalk pits, the moth came upon Hapley again. He went on, trying to keep his mind upon chess problems, but it was no good. The thing fluttered into his face, and he struck at it with his hat in self-defence. Then rage, the old rage – the rage he had so often felt against Pawkins – came upon him again. He went on, leaping and striking at the eddying insect. Suddenly he trod on nothing, and fell headlong.

There was a gap in his sensations, and Hapley found himself sitting on the heap of flints in front of the opening of the chalk-pits, with a leg twisted back under him. The strange moth was still fluttering round his head. He struck at it with his hand, and turning his head saw two men approaching him. One was the village doctor. It occurred to Hapley that this was lucky. Then it came into his mind with extraordinary vividness, that no one would ever be able to see the strange moth except himself, and that it behoved him to keep silent about it.

Late that night, however, after his broken leg was set, he was feverish and forgot his self-restraint. He was lying flat on his bed, and he began to run his eyes round the room to see if the moth was still about. He tried not to do this, but it was no good. He soon caught sight of the thing resting close to his hand, by the night-light, on the green tablecloth. The wings quivered. With a sudden wave of anger he smote at it with his fist, and the nurse woke up with a shriek. He had missed it.

'That moth!' he said; and then, 'It was fancy. Nothing!'

All the time he could see quite clearly the insect going round the cornice and darting across the room, and he could also see that the nurse saw nothing of it and looked at him strangely. He must keep himself in hand. He knew he was a lost man if he did not keep himself in hand. But as the night waned the fever grew upon him, and the very dread he had of seeing the moth made him see it. About five, just as the dawn was gray, he tried to get out of bed and catch it, though his leg was afire with pain. The nurse had to struggle with him.

On account of this, they tied him down to the bed. At this the moth grew bolder, and once he felt it settle in his hair. Then, because he struck out violently with his arms, they tied these also. At this the moth came and crawled over this face, and Hapley wept, swore, screamed, prayed for them to take it off him, unavailingly.

The doctor was a blockhead, a just-qualified general practitioner, and quite ignorant of mental science. He simply said there was no moth. Had he possessed the wit, he might still, perhaps, have saved Hapley from his fate by entering into his delusion, and covering his face with gauze, as he prayed might be done. But, as I say, the doctor was a blockhead, and until the leg was healed Hapley was kept tied to his bed, and with the imaginary moth crawling over him. It never left him while he was awake and it grew to a monster in his dreams. While he was awake he longed for sleep, and from sleep he awoke screaming.

So now Hapley is spending the remainder of his days in a padded room, worried by a moth that no one else can see. The asylum doctor calls it hallucination; but Hapley, when he is in his easier mood, and can talk, says it is the ghost of Pawkins, and consequently a unique specimen and well worth the trouble of catching.

John Masefield wrote his now forgotten novel, Multitude and Solitude, *in 1907. It treats with some authenticity the story of two young men, trying primitive serology (which had come into prominence at that time – the issues in fact would have been highly topical) in Africa with the aim of curing trypanosomiasis, or sleeping sickness.*

"Well. If I've had sleeping sickness, how comes it that I'm here, talking to you? You say yourself the atoxyl was lost."

"Lionel," said Roger, "I injected you with a dead culture. After that, I shot a couple of koodoos (if they were koodoos), a cow and a fawn. The fawn had nagana or something. I took sera from them, and injected the sera into both of us. Great big doses in both cases. I injected the sera into seven poor devils in the village, and they all swelled up and died. It was awful, Lionel. What makes people swell up?"

"I don't know," said Lionel. "I suppose it might be anthrax. Was there fever?"

"Intense pain, very high fever, and death apparently from exhaustion. And you and I swelled up a little; and I made sure yesterday that we were both going to die too. I wrote letters and stuck them up on a bar inside there."

"Oh, so that was what the rod was for? I thought it was something funny. And now we are both cured?"

"Yes. My God, Lionel, I'm thankful to hear your voice again. You don't know what it's been."

They shook hands.

"You're a public benefactor," said Lionel. He looked hard at Roger. "I give you best," he added. "I thought you were a griff. But you've found a cure, it seems. Eh? Look at him. It's the first time he's realized it!"

"But," Roger stammered, "I've killed seven with it; that's not what I call a cure."

"Did you inject the seven with the dead culture first?" Lionel asked.

"No. Only myself and you."

"There you are," said Lionel. "You griffs make the discoveries, and haven't got the gumption to see them. My good Lord! It's as plain as measles. You inject the dead culture. That's the first step. That makes the trypanosomes agglutinize. Very well, then. You inject your serum when they are agglutinized; not before. When they are agglutinized, the serum destroys them, after raising queer symptoms. When they are not agglutinized the serum destroys you by the excess of what causes the queer symptoms. I don't understand those symptoms. They are so entirely unexpected. Did you examine the blood?"

"One cubic centimetre of the venous blood killed a guinea-pig in three hours."

"Yes, no doubt, But did you look at the blood microscopically?"

"No," said Roger, ashamed. "I looked at my sera for streptococci."

"You juggins!" said Lionel. "Yet you come out and land on a cure. Well, well! You're a lucky dog. Let's go in and look at our glands." Roger noticed that he walked with the totter of one newly risen from a violent attack of fever.

Four months later, the two men reached Shirikanga in a canoe of their own making. They were paddled by four survivors from the village. All the rest were dead, either of sleeping sickness or of the serum. Lionel had not discovered what it was in the serum which caused the fatal symptoms. It contained some quality which caused the streptococci, or pus-forming microbes, to increase; but, as far as he could discover, this quality was exerted only when the patient's blood contained virulent trypanosomes, or some other active toxin-producing micro-organisms in the unagglutinized condition. They cured four of the villagers. They might have saved more had they been able to begin the treatment earlier in the disease. They were not dissatisfied with their success.

162

Here is the earlier scene in the drawing-room, where they first meet and the discussion turns to tropical diseases, and yellow fever in particular.

"Belize," said Lionel. "My chief was in Belize. Was there any yellow fever there, when you were there?"

"There was one case," said Roger.

"Did you see it?"

"No," said Roger; "I didn't."

"I should like to see yellow fever," said Lionel simply. "I suppose there was a good deal of fuss directly this case occurred?"

"Yes," said Roger. "A gang came round at once. I think they put paraffin in the cisterns. They sealed the infected house with brown paper and fumigated it."

"And that stopped it?"

"Yes. There were no other cases."

"It's all due to a kind of mosquito," said Lionel. "The white-ribbed mosquito. He carries the organism. You put paraffin on all standing puddles and pools to prevent the mosquito's larvae from hatching out. My old chief did a lot of work in Havana, and the West Indies, stampin' out yellow fever. It has made the Panama Canal possible."

"Are you a doctor, then may I ask?" said Roger.

"No," said Lionel. "I do medical research work; but I don't know much about it. I never properly qualified. I'm interested in all that kind of thing."

"What medical research do you do? Would it bore you to tell me?"

"I've been out in Uganda, doing sleeping sickness."

"Have you?" said Roger. "That's very interesting. I've been reading a lot of books about sleeping sickness."

"Are you interested in that kind of thing?" Lionel asked.

"Yes."

"If you care to come round to my rooms some time I would shew you some relics. I live in Pump Court. I'm generally in all the morning, and between four and six in the evening. I could shew you some trypanosomes. They're the organisms."

"What are they like?" Roger asked.

"They're like little wriggly flattened membranes. Some of them have tails. They multiply by longitudinal division. They're unlike anything else. They've got a pretty bad name."

"And they cause the disease?"

"Yes. You know, of course, that they are spread by the tsetse fly? The tsetse fly sucks them out of an infected fish or mammal, and develops them, inside his body probably for some time, during which the organism probably changes a good deal. When the tsetse bites a man, the developed trypanosome gets down the proboscis into the blood. About a week after the bite, when the bite itself is cured, the man gets the ordinary trypanosome fever, which makes you pretty wretched, by the way."

"Have you had it?"

"Yes; rather. I have it now. It recurs at intervals."

"And how about sleeping sickness?"

"You get sleeping sickness when the trypanosome enters the cerebro-spinal fluid. You may not get it for six or seven years after the bite. On the other hand, you may get it almost at once."

"Then you may get it?" said Roger, startled, looking at the man with a respect which was half pity.

"I've got it," said Lionel.

"Got it? You?" said Roger. He stumbled in his speech. "But, forgive my speaking like this," he said; "is there a cure, then?"

"It's not certain that it's a permanent cure," said Lionel. "I've just started it. It's called atoxyl. Before I tried atoxyl I had another thing called trypanroth, made out of aniline dye. It has made my eyes red, you see? Dyed them. You can have 'em dyed blue, if you prefer. But red was good enough, I thought, Now I'm afraid I'm talking rather about myself."

163

Compare now Ronald Ross, the discoverer of the malaria parasite and its life-cycle (see also below, no. 207), in bitter mood.

THE ANNIVERSARY
(20th August, 1917)

Now twenty years ago
  This day we found the thing;
With science and with skill
  We found; then came the sting –
What we with endless labour won
  The thick world scorned;
Not worth a word to-day –
  Not worth remembering.

O Gorgeous Gardens, Lands
  Of beauty where the Sun
His lordly raiment trails
  All day with light enspun,
We found the death that lurk'd beneath
  Your purple leaves,
We found your secret foe,
  The million-murdering one;

And clapp'd our hands and thought
  Your teeming width would ring
With that great victory – more
  Than battling hosts can bring.
Ah, well – men laugh'd. The years have pass'd;
  The world is cold –
Some million lives a year,
  Not worth remembering!

Ascended from below
  Men still remain too small;
With belly-wisdom big
  They fight and bite and bawl,
These larval angels! – but when true
  Achievement comes –
A trifling doctor's matter –
  No consequence at all!

164

Alex Comfort's cheerful novel, Come out to Play, *has the discoverer of a human pheromone (the structural formula in the novel makes it steroidal in nature) trying it out on a sceptical colleague.*

And I worked the idea out a little from there. It would amuse the conference, give the Press a 'story' which would keep them from misreporting more serious scientific business (I could probably keep my name out on professional grounds) and save me having to think about a review article. This *jeu d'esprit* was accepted – it earned me my fare and two double-column headlines for the Conference, which pleased the commercial sponsors greatly, event hough they were – SEX MEN MEET: BRITON SLATES 'CHEMICAL WEAPONS' FOR LOVERS, and SMELL ME, LOVE ME: PERFUME OF FUTURE SPELLS DOOM OF MALE: that had been all, so far as I was concerned.

As Marcel talked I felt decidedly humble. Evidently, and not for the first time, a jocular coffee speculation in the common-room had turned out to be correct, and it had unloosed some of the 'strange repercussions' I'd flippantly catalogued to the scientific peasantry of San Francisco – in fact, I had experienced some of them already

– all in all it was a little alarming, but morally salutary.

Marcel was working for Francodor, the firm that has been trying for years to beat the natural essential oils with its synthetics. He was a dedicated man, for whom perfumes obviously had a deep unconscious meaning, Groddeck or no Groddeck – probably his mother was a broken flower, I thought; a tuberose at least.

Fired by that infernal lecture, he'd started with civetone, which was where I'd left off, and worked his way through a forest of big-ring cycloketones. Apart from muscone, which was already known, and things like exaltone, all of these had smelt, but none smelt exciting. The most promising one was a substance which came as an impurity in one of his syntheses. It appeared to be a double molecule – half was a cyclo-pentadecanone, the other half was heterocyclic. "I was my own experimental animal," said Marcel, "and this substance had an effect on me. It might, of course, be suggestion. Now on theoretical grounds, as you know" (he knows, I thought, that I don't but he's telling me), "the odour should be enhanced by acetylation. I accordingly made the derivative. It was quite, quite odourless – as odourless as sugar. Misery! I can't do any more that day, so I go home. But I've been too hasty – I can't smell it, but it is active as hell: dogs can smell it – they follow me – dogs penetrate into the laboratory! The dirty old concierge, not the present one – asks finally if I am in season and can he have one of the pups? I have to destroy the stuff quickly – the animal house is in uproar – it can be heard all over the building. Two months later a dog still comes in occasionally. Well, at least we can send this to Australia for dingo bait, and there is something there. Like Pasteur I now have to ask myself, what? After a great deal of thought I acetylate again, not here, on carbon two, but on carbon three, here." He was half into my plate. My glass went over. "Now the odour is back unequivocally. This time there is no doubt. The theory of Goggins is vindicated. Within two hours I have had to dismiss a frustrated secretary, poor woman, for importunity – on the way home people behave peculiarly to me. Victory, then, at least in part."

It suddenly crossed my mind that Marcel was off his head. Women made advances to him. People whispered because he smelt. It would do for a delusion; then I remembered myself, and Dulcinea, and the dog. There was the other possibility, that he was both right *and* crazy. I filed that for future reference.

"Was that the stuff which got out yesterday?" I asked him.

"No, I am not there yet. Courage, In order, I set about acetylating each carbon in turn. I go round the molecule clockwise."

He counted on his fingers in my food – I had to eat round the demonstration of stereochemistry.

"1-acetyl, inactive: 2-acetyl, attracts dogs; 3-acetyl, active; 4-acetyl, nothing; 5-acetyl, active to about the same extent as 3-acetyl, with which I think it is also contaminated – the melting-points are very close. But 6-acetyl is very active. It's mono-lactone also: I had got careless and inhaled a lot of it, and I practically lost interest in chemistry for a week. The sensation was pleasant, but I realize I must watch for dangers. I make next all the di-lactones. They are inactive. This is not the road. So what then? I ask myself. I start on the di-acetyl derivatives."

He looked rather alarmingly excited, I thought, and he'd a definite tremor of the fingers. I kept wondering, counting round the peas his sleeve had left on my plate, "junkie, maniac, sane French scientist, addict,psychopath, con-man, thief"?

"So far," said Marcel, "I have made two, Two only. The substance which you experienced was the 1, 3-di-acetyl compound. You'll forgive me saying this breakdown was a lucky accident. This was the second batch I've made. And you've been able to judge if it's effective."

"How much got out yesterday?" I asked. "I mean, you could avoid some of these crises if you made smaller amounts . . ."

He'd long since forgotten to eat his own meal, and I had lapped him by a course. He leaned over into my plate again.

"My friend, there was one milligram. Not all of that escaped. You see?"

I certainly did, and I whistled.

"But that isn't all," he continued. "We have a substance of quite formidable power. But everything suggests that the derivative which yet remains to be made, the 1,6-di-acetyl compound will be more active." I moved my plate out of the way of his dramatic sense.

"Enormously more active."

He was in the plate again with his sleeve nevertheless. I moved it farther. "Any idea by what factor?" I said.

"By three, by four, by five orders of magnitude."

The plate went off the end of the table with a crash. The handful of other diners looked round, and the waiter came to retrieve the pieces.

"In that case," I said, "you'd better be damned careful with it. In the first place, don't make it in the Institute, in the middle of Paris, Do it in a field . . ."

"My friend! I *am* careful. You don't need to warn me. I dare not make it at all. For three months I have been pausing and considering the implications of this work – the *wider* implications, you understand. I'm pretty well worn out with them. But I'm going to make it – that is not avoidable. The trouble is I've had nobody to consult. That's why I had to wait for your arrival. That's why I'm in a state of excitement. Nobody!"

165

H. G. Wells's didactic streak was never far from the surface. His description of the nature of transparency in The Invisible Man *is masterly.*

"Before we can do anything else," said Kemp, "I must understand a little more about this invisibility of yours." He had sat down, after one nervous glance out of the window, with the air of a man who has talking to do. His doubts of the sanity of the entire business flashed and vanished again as he looked across to where Griffin sat at the breakfast-table, – a headless, handless dressing-gown, wiping unseen lips on a miraculously held serviette.

"It's simple enough – and credible enough," said Griffin, putting the serviette aside and leaning the invisible head on an invisible hand.

"No doubt, to you, but –" Kemp laughed.

"Well, yes; to me it seemed wonderful at first, no doubt. But now, great God! – But we will do great things yet! I came on the stuff first at Chesilstowe."

"Chesilstowe?"

"I went there after I left London. You know I dropped medicine and took up physics? No? – well, I did, Light – fascinated me."

"Ah!"

"Optical density! The whole subject is a network of riddles – a network with solutions glimmering elusively through. And being but two-and-twenty and full of enthusiasm, I said, 'I will devote my life to this. This is worth while.' You know what fools we are at two-and-twenty?"

"Fools then or fools now," said Kemp.

"As though Knowing could be any satisfaction to a man!

"But I went to work – like a nigger. And I had hardly worked and thought about the matter six months before light came through one of the meshes suddenly – blindingly! I found a general principle of pigments and refraction, – a formula, a geometrical expression involving four dimensions. Fools, common men, even common mathematicians, do not know anything of what some general expression may mean to the student of molecular physics. In the books – the books that Tramp has hidden – there are marvels, miracles! But this was not a method, it was an idea that might lead to a method by which it would be possible, without changing any other property of matter, – except, in some instances, colours, – to lower the refractive index of a substance, solid or liquid, to that of air – so far as all practical purposes are concerned."

"Phew!" said Kemp. "That's odd! But still I don't see quite – I can understand that thereby you could spoil a valuable stone, but personal invisibility is a far cry."

"Precisely," said Griffin. "But consider: Visibility depends on the action of the visible bodies on light. Either a body absorbs light, or it reflects or refracts it, or does all these things. If it neither reflects nor refracts nor absorbs light, it cannot of itself be visible. You see an opaque red box, for instance, because the colour absorbs some of the light and reflects the rest, all the red part of the light, to you. If it did not absorb any particular part of the light, but reflected it all, then it would be a shining white box. Silver! A diamond box would neither absorb much of the light nor reflect much from the general surface, but just here and there where the surfaces were favourable the light would be reflected and refracted, so that you would get a brilliant appearance of flashing reflections and translucencies, – a sort of skeleton of light. A glass box would not be so brilliant, not so clearly visible, as a diamond box, because there would be less refraction and reflection. See that? From certain points of view you would see quite clearly through it. Some kinds of glass would be more visible than others, a box of flint glass would be brighter than a box of ordinary window glass. A box of very thin common glass would be hard to see in a bad light, because it would absorb hardly any light and refract and reflect very little. And if you put a sheet of common white glass in water, still more if you put it in some denser liquid than water, it would vanish almost altogether, because light passing from water to glass is only slightly refracted or reflected or indeed affected in any way. It is almost as invisible as a jet of coal gas or hydrogen is in air. And for precisely the same reason!"

"Yes," said Kemp, "that is pretty plain sailing."

"And here is another fact you will know to be true. If a sheet of glass is smashed, Kemp, and beaten into a powder, it becomes much more visible while it is in the air; it becomes at last an opaque white powder. This is because the powdering multiplies the surfaces of the glass at which refraction and reflection occur. In the sheet of glass there are only two surfaces; in the powder the light is reflected or refracted by each grain it passes through, and very little gets right through the powder. But if the white powdered glass is put into water, it forthwith vanishes. The powdered glass and water have much the same refractive index; that is, the light undergoes very little refraction or reflection in passing from one to the other.

"You make the glass invisible by putting it into a liquid of nearly the same refractive index; a transparent thing becomes invisible if it is put in any medium of almost the same refractive index. And if you will consider only a second, you will see also that the powder of glass might be made to vanish in air, if its refractive index could be made the same as that of air; for then there would be no refraction or reflection as the light passed from glass to air."

"Yes, yes," said Kemp. "But a man's not powdered glass!"

"No," said Griffin. "He's more transparent!"

"Nonsense!"

"That from a doctor! How one forgets! Have you already forgotten your physics, in ten years? Just think of all the things that are transparent and seem not to be so. Paper, for instance, is made up of transparent fibres, and it is white and opaque only for the same reason that a powder of glass is white and opaque. Oil white paper, fill up the interstices between the particles with oil so that there is no longer refraction or reflection except at the surfaces, and it becomes as transparent as glass. And not only paper, but cotton fibre, linen fibre, wool fibre, woody fibre, and *bone*, Kemp, *flesh*, Kemp, *hair*, Kemp, *nails*, and *nerves*, Kemp, in fact the whole fabric of a man except the red of his blood and the black pigment of hair, are all made up of transparent, colourless tissue. So little suffices to make us visible one to the other. For the most part the fibres of the living creature are no more opaque than water."

"Great Heavens!" cried Kemp. "Of course, of course! I was thinking only last night of the sea larvae and all jelly-fish!"

"*Now* you have me! And all that I knew and had in mind a year after I left London – six years ago. But I kept it to myself. I had to do my work under frightful disadvantages. Oliver, my professor, was a scientific bounder, a journalist by instinct, a thief of ideas, – he was always prying! And you know the knavish system of the scientific world, I simply would not publish, and let him share my credit. I went on working. I got nearer and nearer making my formula into an experiment, a reality. I told no living soul, because I meant to flash my work upon the world with crushing effect, – to become famous at a blow. I took up the question of pigments to fill up certain gaps. And suddenly, not by design but by accident, I made a discovery in physiology."

"Yes?"

"You know the red colouring matter of blood; it can be made white – colourless – and remain with all the functions it has now!"

Kemp gave a cry of incredulous amazement.

The Invisible Man rose and began pacing the little study. "You may well exclaim. I remember that night. It was late at night, – in the daytime one was bothered with the gaping, silly students, – and I worked then sometimes till dawn. It came suddenly, splendid and complete into my mind. I was alone; the laboratory was still, with the tall lights burning brightly and silently. In all my great moments I have been alone. 'One could make an animal – a tissue – transparent! One could make it invisible! All except the pigments. I could be invisible!' I said, suddenly realising what it meant to be an albino with such knowledge. It was overwhelming. I left the filtering I was doing, and went and stared out of the great window at the stars. 'I could be invisible!' I repeated.

"To do such a thing would be to transcend magic. And I beheld, unclouded by doubt, a magnificent vision of all that invisibility might mean to a man, – the mystery, the power, the freedom. Drawbacks I saw none. You have only to think! And I, a shabby, poverty-struck, hemmed-in demonstrator, teaching fools in a provincial college, might suddenly become – this. I ask you, Kemp, if *you* – Any one, I tell you, would have flung himself upon that research. And I worked three years, and every mountain of difficulty I toiled over showed another from its summit. The infinite details! And the exasperation, – a professor, a provincial professor, always prying. 'When are you going to publish this work of yours?' was his everlasting question. And the students, the cramped means! Three years I had of it –

"And after three years of secrecy and exasperation, I found that to complete it was impossible, – impossible."

"How?" asked Kemp.

"Money," said the Invisible Man, and went again to stare out of the window.

He turned round abruptly. "I robbed the old man – robbed my father.

"The money was not his, and he shot himself."

166

J. D. Watson's story, The Double Helix, *about the discovery of DNA which won the Nobel Prize for him and Crick, together with Wilkins, caused outrage by its candour. It was arguably the first of a new literary genre, revealing the scientist as a reporter. Watson makes it an exciting and engaging story. (The* dramatis personae *are Linus Pauling, the unseen rival in California, Peter, his son, Crick (Francis), Wilkins (Maurice), Rosalind Franklin (Rosy – which she was never called), Sir Lawrence Bragg, the long-suffering Head of the Cavendish laboratory, John Kendrew and Max Perutz).*

No further news emerged from Pasadena before Christmas. Our spirits slowly went up, for if Pauling had found a really exciting answer the secret could not be kept long. One of his graduate students must certainly know what his model looked like, and if there were obvious biological implications the rumour would have quickly reached us. Even if Linus was somewhere near the right structure, the odds seemed against his getting near the secret of gene replication. Also, the more we thought about DNA chemistry, the more unlikely seemed the possibility that even Linus could pick off the structure in total ignorance of the work at King's.

Maurice was told that Pauling was in his pasture when I passed through London on my way to Switzerland for a Christmas skiing holiday. I was hoping that the urgency created by Linus's assault on DNA might make him ask Francis and me for help. However, if Maurice thought that Linus had a chance to steal the prize, he didn't let on. Much more important was the news that Rosy's days at King's were numbered. She had told Maurice that she wanted soon to transfer to Bernal's lab at Birkbeck College. Moreover, to Maurice's surprise and relief, she would not take the DNA problem with her. In the next several months she was to conclude her stay by writing up her work for publication. Then, with Rosy at last out of his life, he would commence an all-out search for the structure.

Upon my return to Cambridge in mid January, I sought out Peter to learn what was in his recent letters from home. Except for one brief reference to DNA, all the news was family gossip. The one pertinent item, however, was not reassuring. A manuscript on DNA had been written, a copy of which would soon be sent to Peter. Again there was not a hint of what the model looked like. While waiting for the manuscript to arrive, I kept my nerves in check by writing up my ideas on bacterial sexuality. A quick visit to Cavalli in Milan, which occurred just after my skiing holiday in Zermatt, had convinced me that my speculations about how bacteria mated were likely to be right. Since I was afraid that Lederberg might soon see the same light, I was anxious to publish quickly a joint article with Bill Hayes. But this manuscript was not in final form when, in the first week of February, the Pauling paper crossed the Atlantic.

Two copies, in fact, were dispatched to Cambridge – one to Sir Lawrence, the other to Peter. Bragg's response upon receiving it was to put it aside. Not knowing that Peter would also get a copy, he hesitated to take the manuscript down to Max's office. There Francis would see it and set off on another wild-goose chase. Under the present timetable there were only eight months more of Francis's laugh to bear. That is, if his thesis was finished on schedule. Then for a year, if not more, with Crick in exile in Brooklyn, peace and serenity would prevail.

While Sir Lawrence was pondering whether to chance taking Crick's mind off his thesis, Francis and I were poring over the copy that Peter brought in after lunch. Peter's face betrayed something important as he entered the door, and my stomach sank in apprehension at learning that all was lost. Seeing that neither Francis nor I could bear any further suspense, he quickly told us that the model was a three-chain helix with the sugar-phosphate backbone in the centre. This sounded so suspiciously like our aborted effort of last year that immediately I wondered whether we might already have had the credit and glory of a great discovery if Bragg had not held us back. Giving Francis no chance to ask for the manuscript, I pulled it out of Peter's outside coat pocket and began reading. By spending less than a minute with the summary and the introduction, I was soon at the figures showing the locations of the essential atoms.

At once I felt something was not right. I could not pinpoint the mistake, however, until I looked at the illustrations for several minutes. Then I realized that the phosphate groups in Linus's model were not ionized, but that each group contained a bound hydrogen atom and so had no net charge. Pauling's nucleic acid in a sense was not an acid at all. Moreover, the uncharged phosphate groups were not incidental features. The hydrogens were part of the hydrogen bonds that held together the three intertwined chains. Without the hydrogen atoms, the chains would immediately fly apart and the structure vanish.

Everything I knew about nucleic-acid chemistry indicated that phosphate groups never contained bound hydrogen atoms. No one had ever questioned that DNA was a moderately strong acid. Thus, under physiological conditions, there would always be positively charged ions like sodium or magnesium lying nearby to neutralize the negatively charged phosphate groups. All our speculations about whether divalent ions held the chains together would have made no sense if there were hydrogen atoms firmly bound to the phosphates. Yet somehow Linus, unquestionably the world's most astute chemist, had come to the opposite conclusion.

When Francis was amazed equally by Pauling's unorthodox chemistry, I began to breathe slower. By then I knew we were still in the game. Neither of us, however, had the slightest clue to the steps that had led Linus to his blunder. If a student had made a similar mistake, he would be thought unfit to benefit from Cal Tech's chemistry faculty. Thus, we could not but initially worry whether Linus's model followed from a revolutionary re-evaluation of the acid-base properties of very large molecules. The tone of the manuscript, however, argued against any such advance in chemical theory. No reason existed to keep secret a first-rate theoretical breakthrough. Rather, if that had occurred Linus would have written two papers, the first describing his new theory, the second showing how it was used to solve the DNA structure.

The bloomer was too unbelievable to keep secret for more than a few minutes. I dashed over to Roy Markham's lab to spurt out the news and to receive further reassurance that Linus's chemistry was screwy. Markham predictably expressed pleasure that a giant had forgotten elementary college chemistry. He then could not refrain from revealing how one of Cambridge's great men had on occasion also forgotten his chemistry. Next I hopped over to the organic chemists', where again I heard the soothing words that DNA was an acid.

By teatime I was back in the Cavendish, where Francis was explaining to John and Max that no further time must be lost on this side of the Atlantic. When his mistake became known, Linus would not stop until he had captured the right structure. Now our immediate hope was that his chemical colleagues would be more than ever awed by his intellect and not probe the details of his model. But since the manuscript had already been dispatched to the *Proceedings of the National Academy*, by mid March at the latest Linus's paper would be spread around the world. Then it would be only a matter of days before the error would be discovered. We had anywhere up to six weeks before Linus again was in full-time pursuit of DNA.

Though Maurice had to be warned, we did not immediately ring him. The pace of Francis's words might cause Maurice to find a reason for terminating the conversation before all the implications of Pauling's folly could be hammered home. Since in several days I was to go up to London to see Bill Hayes, the sensible course was to bring the manuscript with me for Maurice's and Rosy's inspection.

Then, as the stimulation of the last several hours had made further work that day impossible, Francis and I went over to the Eagle. The moment its doors opened for the evening we were there to drink a toast to the Pauling failure. Instead of sherry, I let Francis buy me a whisky. Though the odds still appeared against us, Linus had not yet won his Nobel.

Maurice was busy when, just before four, I walked in with the news that the Pauling model was far off base. So I went down the corridor to Rosy's lab, hoping she would be about. Since the door was already ajar, I pushed it open to see her bending over a lighted box upon which lay an X-ray photograph she was measuring. Momentarily startled by my entry, she quickly regained her composure and, looking straight at my face, let her eyes tell me that uninvited guests should have the courtesy to knock.

I started to say that Maurice was busy, but before the insult was out I asked her whether she wanted to look at Peter's copy of his father's manuscript. Though I was curious how long she would take to spot the error, Rosy was not about to play games with me. I immediately explained where Linus had gone astray. In doing so, I could not refrain from pointing out the superficial resemblance between Pauling's three-chain helix and the model that Francis and I had shown her fifteen months earlier. The fact that Pauling's deductions about symmetry were no more inspired than our awkward efforts of the year before would, I thought, amuse her. The result was just the opposite. Instead, she became increasingly annoyed with my recurring references to helical structures. Coolly she pointed out that not a shred of evidence permitted Linus, or anyone else, to postulate a helical structure for DNA. Most of my words to her were superfluous, for she knew that Pauling was wrong the moment I mentioned a helix.

Interrupting her harangue, I asserted that the simplest form for any regular polymeric molecule was a helix. Knowing that she might counter with the fact that the sequence of bases was unlikely to be regular, I went on with the argument that, since DNA molecules form crystals, the nucleotide order must not affect the general structure. Rosy by then was hardly able to control her temper, and her voice rose as she told me that the stupidity of my remarks would be obvious if I would stop blubbering and look at her X-ray evidence.

I was more aware of her data than she realized. Several months earlier Maurice had told me the nature of her so-called antihelical results. Since Francis had assured me that they were a red herring, I decided to risk a full explosion. Without further hesitation I implied that she was incompetent in interpreting X-ray pictures. If only she would learn some theory, she would understand how her supposed antihelical features arose from the minor distortions needed to pack regular helices into a crystalline lattice.

Suddenly Rosy came from behind the lab bench that separated us and began moving towards me. Fearing that in her hot anger she might strike me, I grabbed up the Pauling manuscript and hastily retreated to the open door. My escape was blocked by

Maurice, who, searching for me, had just then stuck his head through. While Maurice and Rosy looked at each other over my slouching figure, I lamely told Maurice that the conversation between Rosy and me was over and that I had been about to look for him in the tea room. Simultaneously I was inching my body from between them, leaving Maurice face to face with Rosy. Then, when Maurice failed to disengage himself immediately, I feared that out of politeness he would ask Rosy to join us for tea. Rosy, however, removed Maurice from his uncertainty by turning round and firmly shutting the door.

Walking down the passage, I told Maurice how his unexpected appearance might have prevented Rosy from assaulting me. Slowly he assured me that this very well might have happened. Some months earlier she had made a similar lunge towards him. They had almost come to blows following an argument in his room. When he wanted to escape, Rosy had blocked the door and had moved out of the way only at the last moment. But then no third person was on hand.

My encounter with Rosy opened up Maurice to a degree that I had not seen before. Now that I need no longer merely imagine the emotional hell he had faced during the past two years, he could treat me almost as a fellow collaborator rather than as a distant acquaintance with whom close confidences inevitably led to painful misunderstandings. To my surprise, he revealed that with the help of his assistant Wilson he had quietly been duplicating some of Rosy's and Gosling's X-ray work. Thus there need not be a large time gap before Maurice's research efforts were in full swing. Then the even more important cat was let out of the bag: since the middle of the summer Rosy had had evidence for a new three-dimensional form of DNA. It occurred when the DNA molecules were surrounded by a large amount of water. When I asked what the pattern was like, Maurice went into the adjacent room to pick up a print of the new form they called the 'B' structure.

The instant I saw the picture my mouth fell open and my pulse began to race. The pattern was unbelievably simpler than those obtained previously ('A' form). Moreover, the black cross of reflections which dominated the picture could arise only from a helical structure. With the A form, the argument for a helix was never straightforward, and considerable ambiguity existed as to exactly which type of helical symmetry was present. With the B form, however, mere inspection of its X-ray pictures gave several of the vital helical parameters. Conceivably, after only a few minutes' calculations, the number of chains in the molecule could be fixed. Pressing Maurice for what they had done using the B photo, I learned that his colleague R. D. B. Fraser earlier had been doing some serious playing with three-chain models but that so far nothing exciting had come up. Though Maurice conceded that the evidence for a helix was now overwhelming – the Stokes-Cochran-Crick theory clearly indicated that a helix must exist – this was not to him of major significance. After all, he had previously thought a helix would emerge. The real problem was the absence of any structural hypothesis which would allow them to pack the bases regularly in the inside of the helix. Of course this presumed that Rosy had hit it right in wanting the bases in the centre and backbone outside. Though Maurice told me he was now quite convinced she was correct, I remained sceptical, for her evidence was still out of reach of Francis and me.

On our way to Soho for supper I returned to the problem of Linus, emphasizing that smiling too long over his mistake might be fatal. The position would be far safer if Pauling had been merely wrong instead of looking like a fool. Soon, if not already, he would be at it day and night. There was the further danger that if he put one of his assistants to taking DNA photographs, the B structure would also be discovered in Pasadena. Then, in a week at most, Linus would have the structure.

Maurice refused to get excited. My repeated refrain that DNA could fall at any moment sounded too suspiciously like Francis in one of his overwrought periods. For years Francis had been trying to tell him what was important, but the more dispassionately he considered his life, the more he knew he had been wise to follow up his own hunches. As the waiter peered over his shoulder, hoping we would finally order, Maurice made sure I understood that if we could all agree where science was going, everything would be solved and we would have no recourse but to be engineers or doctors.

With the food on the table I tried to fix our thoughts on the chain number, arguing that measuring the location of the innermost reflection on the first and second layer lines might immediately set us on the right track. But since Maurice's long-drawn-out reply never came to the point, I could not decide whether he was saying that no one at King's had measured the pertinent reflections or whether he wanted to eat his meal before it got cold. Reluctantly I ate, hoping that after coffee I might get more details if I walked him back to his flat. Our bottle of Chablis, however, diminished my desire for hard facts, and as we walked out of Soho and across Oxford Street, Maurice spoke only of his plans to get a less gloomy apartment in a quieter area.

Afterwards, in the cold, almost unheated train compartment, I sketched on the blank edge of my newspaper what I remembered of the B pattern. Then as the train jerked towards Cambridge, I tried to decide between two- and three-chain models. As far as I could tell, the reason the King's group did not like two chains was not foolproof. It depended upon the water content of the DNA samples, a value they admitted might be in great error. Thus by the time I had cycled back to college and climbed over the back gate, I had decided to build two-chain models. Francis would have to agree. Even though he was a physicist, he knew that important biological objects come in pairs.

167

The discovery of insulin had instant and dramatic consequences. Dying diabetics were brought back from the grave's brink. The relations between those involved in the discovery were marked by rancour and acrimony. The version that stuck was essentially that spread by Banting, a man in the grip of chronic paranoia. It was left to a recent historian, Michael Bliss, in The Discovery of Insulin, *to restore the balance and rehabilitate J. J. R. Macleod, the Professor at Toronto, with whom Banting shared the Nobel Prize. The extract illustrates the relationship that prevailed between the four main protagonists.*

One of the more remarkable personal confrontations in the history of science occurred sometime between January 17 and January 24 [1922]. There are no contemporary accounts of it, no references whatever by Collip, and only the two following accounts, neither of which should be considered totally reliable.

Banting wrote in 1940 as follows:

> The worst blow fell one evening toward the end of January. Collip had become less and less communicative and finally after a week's absence he came into our little room about five thirty one evening. He stopped inside the door and said "Well fellows I've got it."

I turned and said, "Fine, congratulations. How did you do it?"
Collip replied, "I have decided not to tell you."
His face was white as a sheet. He made as if to go. I grabbed him with one hand by the overcoat where it met in front and almost lifting him I sat him down hard on the chair. I do not remember all that was said but I remember telling him that it was a good job he was so much smaller – otherwise I would "knock hell out of him." He told us that he had talked it over with Macleod and that Macleod agreed with him that he should not tell us by what means he had purified the extract.

Best, not having read Banting's account, gave his version of the incident in a letter to Sir Henry Dale, written in 1954 and intended for the historical record:

One evening in January or February, 1922, while I was working alone in the Medical Building, Dr. J. B. Collip came into the small room where Banting and I had a dog cage and some chemical apparatus. He announced to me that he was leaving our group and that he intended to take out a patent in his own name on the improvement of our pancreatic extract. This seemed an extraordinary move to me, so I requested him to wait until Fred Banting appeared, and to make quite sure that he did I closed the door and sat in a chair which I placed against it. Before very long Banting returned to the Medical Building and came along the corridor to this little room. I explained to him what Collip had told me and Banting appeared to take it very quietly. I could, however, feel his temper rising and I will pass over the subsequent events. Banting was thoroughly angry and Collip was fortunate not to be seriously hurt. I was disturbed for fear Banting would do something which we would both tremendously regret later and I can remember restraining Banting with all the force at my command.

Except for a veiled but important reference in Banting's 1922 account, there are no other useful written records of this incident. Clark Noble once drew a cartoon, unfortunately now lost, of Banting sitting on Collip, choking him; he captioned it "The Discovery of Insulin."

The one surviving artifact of the fight is an agreement signed by Banting, Best, Collip, and Macleod, dated January 25, 1922, and entitled, "Memorandum in Reference to the Co-operation of the Connaught Anti-Toxin Laboratories in the Researches of Dr. Banting, Mr. Best and Dr. Collip – Under the General Direction of Professor J. J. R. Macleod to obtain an Extract of Pancreas Having a Specific Effect on the Blood Sugar Concentration." The two key conditions of Connaught's co-operation with the team were:

1. Dr. Banting, Mr. Best and Dr. Collip each agrees not to take any steps which will result in the process of obtaining an extract or extracts

of pancreas, being patented, prepared by any commercial firm with aid of any of the above or otherwise exploited during the period of co-operation with the Connaught Anti-Toxin Laboratories.
2. That no step involving any modification in policy concerning these researches be taken without preliminary joint conference between Dr. Banting, Mr. Best and Dr. Collip, and Professor Macleod and Professor Fitzgerald be held.

The rest of the document spelled out technical and financial details.

What had happened? What had Collip said to Banting to cause the attack? Why had he said it? There seems little doubt that Collip said three things that night in the lab to Banting and Best: first, he would not tell them how he had made his breakthrough; second, he had told Macleod, who had agreed that Collip did not have to tell Banting and Best; and third, he might go ahead and take out a patent on his process.

What was going on in Collip's mind and what Banting and Best said to him in the course of the conversation can only be speculated upon. He was probably tired – they were all probably tired – after days of hard work and extreme pressure. I presume that Collip and Macleod had little use for Banting's conduct in the past several weeks, particularly Banting's breaking of the spirit of the collaboration by himself and Best making the extract for the first clinical test. And, it appeared, Banting had appropriated some of Collip's improvements in making that extract. Banting had shown his distrust of them; now they had no reason to trust him. It was Collip's job to purify the extract, not Banting and Best's. Collip and Macleod may have decided that Banting was trying to take credit away from Collip – that if he knew the process for making the extract he would claim it as his own. They may have believed, after the misadventure of January 11, that Banting could not be trusted not to try to forestall the rest of the team by applying for a patent. Paranoia begat paranoia. So Collip and Macleod decided not to tell Banting and Best the secret of making an effective anti-diabetic extract.

Speculating further, the kind of things Collip likely said that night are these: "Why should I tell you? . . . It's my job, not yours, to get it ready . . . What do you want to know for? So you can run your own test again? . . . Stick to your job, I'll do mine . . . You'll know in good time when *we* see how it works . . . Don't worry, you'll get your share of credit for the work you've done . . . I'm not going to let you take credit for my work . . . you've already tried to do it once . . . I don't have to put up with your kind of nonsense . . . maybe I'll just go back to Alberta and patent my method. . . ."

From Stephen Hall's book about the chase after the insulin gene, Invisible Frontiers, *sixty years on from Banting and Macleod, and Best and Collip, here is a moment of high comedy, as the exigent and abrasive gang of Americans, going hell-for-leather for the finishing line, are tripped up by the unyielding bureaucracy of one of the stuffier British institutions, the Ministry of Defence research establishment at Porton Down on Salisbury Plain.*

"I remember being enormously impressed by their arrival . . ." Peter Greenaway would remark later. "They jumped out of the car, introduced themselves, and then began pulling out what appeared to be endless numbers of suitcases and trunks, all of which contained reagents necessary to do the work that they had planned." The Harvard researchers remember being impressed, too, but for entirely different reasons.

On their first day at Porton, the laboratory was closed (work was not conducted on weekends), so they could do little more than unpack their scientific gear and set it up in the appropriate spot. But all four were issued little pink identity cards, to be carried at all times, which – in impeccably cordial, restrained, and unalarmist tones – warned physicians to treat the bearer with caution and care: "In cases of P.U.O. [fever of undetermined origin], especially when associated with respiratory or CNS [central nervous system] symptomatology, the possibility of an accidental laboratory infection cannot be excluded and you are asked to contact as soon as possible . . ." As if that did not give sufficient pause, each of the scientists was asked by Greenaway to sign a "code of practice," which outlined work procedures in the containment facility; one of the conditions imposed by the British authorities, Greenaway points out, was that "any breach of MRE's safety code for the facility should immediately lead to their becoming *persona non grata.*"

Because sensitive military work was going on at the laboratory, Greenaway was required to accompany the Harvard researchers at all times on the premises, and although this role seems to have begun as a cordial chaperoning, it did not take long before it warmed into friendly comradery. He was a tall, thin, personable fellow who immediately won flattering reviews from his American guests for everything from courtesy and scientific knowledge to skills of diplomacy. "He didn't watch the clock and he didn't wear a tie," Villa-Komaroff recalls approvingly, as if those were the most telltale traits of a good biologist. It was not merely that he "baby-sat" the Americans, as Efstratiadis put it; he also proved to be a helpful and unflappable ambassador who moved easily between the two very different scientific cultures. Just how different was revealed on a daily, if not hourly, basis. "Work for normal people at Porton started at eight-forty and finished at five P.M.," Greenaway dryly observes. "The fact that the American group wanted to work after hours was considered really strange."

Time was tight, however; and the Gilbert group had a human insulin gene they wanted to plug into *E. coli.* The members of "Operation Lollipop" – as Villa-Komaroff unofficially christened the enterprise – knew exactly what they wanted to do and immediately launched into the work. On the Sunday after their arrival, according to an informal ledger kept by Villa-Komaroff, the group prepared an overnight culture of HB101 – the Herb Boyer strain of *E. coli.* The following morning, on September 4, they made their first attempt at transformation. Villa-Komaroff used one-tenth of her cloning material – pBR322 plasmid into which Efstratiadis's insulin gene had been spliced – and smeared the transformed bacteria out on plates. On Tuesday, she performed a second massive transformation of *E. coli* using another tenth of the DNA. Then they waited. The nice thing about bacteria was that you didn't have to wait long.

Unfortunately, bureaucratic and logistical interruptions constantly intruded. For example, on their first full day at the lab, they were obliged to get fitted for gas masks, which in Villa-Komaroff's opinion may have been the most harrowing aspect of the work – certainly more frightening than exposure to any recombinant organisms.

"This little man gave us a gray raincoat, sort of gave us a look, and then found a mask that he *thought* would fit," she recalls. "Then he waved an ammonium salt around, asked if we smelled anything. Everybody said no. So then he said, 'Well, shake your head and nod up and down.' We did that. And then he put us all in a little room, along with a guy from Porton who was not sure his mask was intact and wanted to check it. So there we were, standing there with these gas masks and these long gray British raincoats, in this glass room, and the man drops these two pellets into this little cup and this *gas* rises!

"He's having us nod our heads up and down and side to side, and we're standing there taking deep, huffing breaths. The guy who was checking his mask – it turned out his was *indeed* not intact. He started gasping and choking and scratching at the door, and they let him out. When we got out ten minutes later, the guy was still sitting there, tears streaming from his eyes." The test substance, of course, had been tear gas. The scientists did not have to wear the gas masks continuously,but they did have to put them on during such routine bench operations as running the centrifuges or transferring material from test tube to test tube with the use of pipettes. The fear was that either operation could create "aerosols," invisible droplets of material that could be accidentally inhaled or ingested by a researcher. The precaution seemed excessive for work involving insulin and *E. coli*, but such standard P4 protocols were mandatory for experiments involving other research pathogens at Porton, which at the time included the deadly hemorrhagic fever viruses (Lassa fever, for one) and the botulism bacterium, *Clostridium botulinum*.

Then there was the lab. Considering the general casualness of today's molecular biology laboratory, and given the relative freedom with which Genentech's tennis-shoed, T-shirted researchers bounced through their insulin work in 1978, the extraordinarily restrictive conditions at Porton made the Harvard team a potato-sack entry in the insulin race. Indeed, the futility of the entire regulatory picture could be seen in pathetic perspective from Porton's P4 lab – while Genentech's scientists worked with their synthetic human material under P2 conditions and Axel Ullrich was cloning the human gene in a P3 lab in France, the Harvard researchers were wearing gas masks in England. In his inimitably straightforward fashion, Argiris Efstratiadis says, "The working conditions were the pits, okay?"

Merely *entering* the P4 lab was an ordeal. After removing all clothing, each researcher donned government-issue white boxer shorts, black rubber boots, blue pajamalike garments, a tan hospital-style gown open in the back, two pairs of gloves, and a blue plastic hat resembling a shower cap. Everything then passed through a quick formaldehyde wash. Everything. All the gear, all the bottles, all the glassware, all the equipment. All the scientific recipes, written down on paper, had to pass through the wash; so the researchers slipped the instructions, one sheet at a time, inside plastic Ziploc bags, hoping that formaldehyde would not leak in and turn the paper into a brown, crinkly, parchmentlike mess. Any document exposed to lab air would ultimately have to be destroyed, so the Harvard group could not even bring in their lab notebooks to make entries. After stepping through a basin of formaldehyde, the workers descended a short flight of steps into the P4 lab itself. The same hygienic rigmarole, including a shower, had to be repeated whenever anyone left the lab.

At the bottom of the steps, through a door opening to the right, was the warm room, which maintained a steady 37°C (98.6°F) temperature. Here bacteria replicated and the biologists dozed. To the left was the main laboratory area. Under a bank of windows stood three glove boxes – sealed boxes into which an experimenter could stick his or her hands to handle materials, pour reagents together, pipette material, and perform other operations deemed potentially dangerous in the open air. After each day's work, the glove boxes too were flushed out with formaldehyde.

At the far end of the room stood a metal table and metal countertops (the easier to wipe up spills), three huge autoclaves, and a door that led to the kitchen, where media were prepared and glassware cleaned. The autoclaves, where everything from utensils to growth media is "cooked" under pressure to ensure sterilization, swallowed all manner of jetsam in the Porton lab: not just the usual lab refuse, but also waste water from the showers that each researcher had to take before leaving the lab and even paperback books, from science fiction to *All Things Bright and Beautiful*, that the Harvard workers read in the laboratory to kill time between experiments. Into the autoclave they would go, and then sterile and sopped and ruined, into the trash.

As if the physical containment procedures were not sufficiently troublesome, the highly routinized way of doing things at Porton came as a rude shock to researchers accustomed to the twenty-four-hour-a-day,open-access, come-as-you-are atmosphere back at Harvard. "The people were pretty friendly, to the degree they could be friendly," Efstratiadis says. His reaction was partially conditioned by the epic journey he took one day to procure a rotor, which looks somewhat like an imploded bowling ball but is in fact the circular object with holes that holds test tubes in a centrifuge.

"They were so funny, those British people," Efstratiadis recalls. "I mean, they said person X had a rotor for the centrifuge, so I walked a *mile* – it's an *enormous* building, okay? – to find that, and

I said, 'Can I take your rotor?'

"The guy said 'What time do you want it?' I said, 'At five o'clock in the afternoon.' And he said, 'Fine. Sign for the rotor.'

"And so I did, and I started going, and he said, 'Hey! Where are you going? You didn't sign for the key!' I said, '*What* key?' He said, 'The key that opens the locker where the rotor is. But then you have to sign for *another* key that opens the door to go to where the locker is.' They had three keys or something, [and] you had to sign for each and every one of them." In the midst of this military milieu, it is difficult to imagine four less militarily inclined individuals than the Harvard researchers.

Even Wally Gilbert, he of Olympian scientific repute, found himself relegated to the role of bottle washer and microbial saucier. Porton provided no kitchen staff to prepare culture plates; in fact, the Harvard group did not even have access to kitchen-scale autoclaves for such routine operations as preparing growth media. As Villa-Komaroff recalls, "Wally ended up making the plates. We couldn't use the autoclave because it was very hierarchical. Only autoclave people could use the autoclaves, and they turned them off at five o'clock. So he would make media in hundred-milliliter bottles in a pressure cooker. Usually you make media by the liter, three liters at a whack. So he'd make one hundred milliliters [a bit more than three fluid ounces] at a time. One hundred milliliters makes three plates. We were going through *lots* of plates. He was," Villa-Komaroff adds, "very good about that."

The work began well. They looked at the plates from Monday's cloning run and discovered a number of colonies growing up through the antibiotics in the media. A good sign: the insulin-bearing plasmid had made its way into some of the bugs. A sense of excitement began to build. Within a day or two, the positive clones could be grown into colonies and tested by hybridizing probes designed to zero in on insulin. That would nail down one step: cloning of the human gene. Positive colonies from that procedure would then be turned over to Efstratiadis for analysis and to Broome for a check to see if any of the bugs were reading that gene and making human insulin. The analysis would take a few days, but on that Tuesday, September 5, Lydia Villa-Komaroff made a one-word entry in her ledger: "Whoopee." Things were going exactly according to plan.

Despite, it should be added, continuing logistical problems.

On the first full day, the Americans made a disastrous excursion to the officers' mess hall, which was open only for an hour during midday – not the kind of flexible schedule preferred by scientists working with bugs on a microbiological, not military, schedule. Male patrons were obliged to wear ties. Efstratiadis claims to have owned only one tie at the time, black, to wear to funerals (which, he says, "I didn't care to bring with me"). But Gilbert, with characteristic color panache, had prepared for such a contingency by packing two ties, one green and the other bright orange. Both Gilbert and Efstratiadis shared a certain disdain for formality in convention and dress. (One Harvard graduate student tells an amusing story about the time Gilbert went to an exclusive, highly formal restaurant in Boston with a purple tie tucked in his pocket; sure enough, the restaurant pointed out that ties were required, and so Gilbert took great pleasure putting on this loud purple cravat over his customary orange turtleneck. "And really," he is reported to have said, "it only made *them* look silly that I had to wear this tie.")

The Harvard group did not realize, however, that only recently had women been allowed access to the officers' mess, and only if properly attired in dress or skirt. When Villa-Komaroff and Broome strode in, wearing the standard-issue slacks and shirts of biology's front-line troops, and the unshaven Efstratiadis bounced in on sneakers, it was the Boston Tea Party all over again. The place experienced the kind of silent uproar only the British can manage. "The presence of distinguished American guests in flagrant violation of mess rules caused obvious embarrassment to the gentlemen scientists reclining in their comfy armchairs for a quick nap after dinner," observes Peter Greenaway. "I believe I am right in saying that this was the only time the American group went to the officers' mess."

Next, they tried picnics. When the morning work seemed fairly well along and a natural break could be anticipated, one of the scientists would shower out of the lab and drive back to the Pheasant Inn in Salisbury, where they all stayed, to pick up a prepared picnic lunch. But this seemingly innocuous plan, too, ran afoul of the authorities. "The picnics were initially taken on the grass directly in front of the building," Greenaway recalls. "However, complaints were made to the director concerning the bad image that this presented, particularly when we were drinking beer directly from the bottles." They were tactfully asked to remove themselves from the front of the main building and to picnic behind the cycle sheds.

169

The world, it seems, holds scientists cheap – nowhere more so than in England. Lord Todd (see above, no. 1) in A Time to Remember: the Autobiography of a Chemist, *has an agreeable reminiscence, going back to 1941. He was engaged in chemical defence work and had occasion to travel to Porton to watch a demonstration. The ambience of Porton was evidently little different from the way Gilbert and his young playmates found it some forty years later. Todd had taken a long time getting there from Manchester by train, because of the vicissitudes of war-time travel. Long before his arrival he had run out of cigarettes, and he was desperate, for he was a heavy smoker and cigarettes were not easily come by. But then, after a morning of demonstrations on Salisbury Plain, he was taken to the officers' mess for lunch:*

After a wash I proceeded to the bar where – believe it or not – there was a white-coated barman who was not only serving drinks but also cigarettes. I hastened forward and rather timidly said 'Can I have some cigarettes?'

'What's your rank?' was the slightly unexpected reply.

'I am afraid I haven't got one', I answered.

'Nonsense – everyone who comes here has a rank.'

'I'm sorry but I just don't have one.'

'Now that puts me in a spot,' said the barman, 'for orders about cigarettes in this camp are clear – twenty for officers and ten for other ranks. Tell me what exactly are you?'

Now I really wanted those cigarettes so I drew myself up and said 'I am the Professor of Chemistry at Manchester University.' The barman contemplated me for about thirty seconds and then said 'I'll give you five.'

Since that day I have had few illusions about the importance of professors!

Now a rumination by John Updike. Somatotrophin was one of the first products of the infant recombinant DNA industry, and, one could say, H. G. Wells's Food of the Gods.

ODE TO GROWTH

Like an awl-tip breaking ice
the green shoot cleaves the gray spring air.
The young boy finds his school-pants cuffs
too high above his shoes when fall returns.
The pencilled marks on the bathroom doorframe climb.
The cells replicate,
somatotrophin
comes bubbling down the bloodstream, a busybody
with instructions for the fingernails,
another set for the epiderm,
a third for the budding mammae,
all hot from the hypothalamus
and admitting of no editing,
lest dwarves result, or cretins, or neoplasms.
In spineless crustaceans
the machinery of molting is controlled
by phasing signals from nervous ganglia
located, often, in the eyestalks, where these exist.
In plants
a family of auxins,
shuttling up and down,
inhibit or encourage cell elongation
as eventual shapeliness demands,
and veto lateral budding while apical growth proceeds,
and even determine abscission –
the falling of leaves.
For death and surrender
are part of growth's package. . . .

171

From Olga Metchnikoff's Life of Elie Metchnikoff, *a reminder of the restorative properties that a new idea, coming unbidden from the void, can have.*

He said to himself: "Why live? My private life is ended; my eyes are going; when I am blind I can no longer work, then why live?" Seeing no issue to his situation, he absorbed the morphia. He did not know that too strong a dose, by provoking vomiting, eliminates the poison. Such was the case with him. He fell into a sort of torpor, of extraordinary comfort and absolute rest; in spite of this comatose state he remained conscious and felt no fear of death. When he became himself again, it was with a feeling of dismay. He said to himself that only a grave illness could save him, either by ending in death or by awaking the vital instinct in him. In order to attain his object,he took a very hot bath and then exposed himself to cold. As he was coming back by the Rhone bridge, he suddenly saw a cloud of winged insects flying around the flame of a lantern. They were Phryganidae, but in the distance he took them for Ephemeridae, and the sight of them suggested the following reflection: "How can the theory of natural selection be applied to these insects? They do not feed and only live a few hours; they are therefore not subject to the struggle for existence, they do not have time to adapt themselves to surrounding conditions."

His thoughts turned towards Science; he was saved; the link with life was re-established.

172

And lastly, the headiness of victory – from Nicholas Wade's The Nobel Prize Duel. *The quarry has finally been run to earth after years of increasingly desperate pursuit.*

Matsuo decided to synthesize for himself the more likely of his two possible candidates for LRF. He worked round the clock for ten days, eating and sleeping in the laboratory. When the structure was synthesized, he gave it to Arimura for bioassay and turned to synthesizing the second candidate.

Arimura had developed an assay based on the radioimmunoassay technique invented by Solomon Berson and Rosalyn Yalow. Unlike the old erratic Parlow assay and the other assays with which Guillemin and Schally had wrestled for so long, the new radioimmunoassays were exquisitely sensitive and precise. They enabled the releasing factors and other peptide hormones to be measured directly and in exact amounts. By 1970 the technique was bringing about a revolution in endocrinology, for which Yalow was later to win the Nobel Prize, Berson by then being dead.

Since Arimura had had the foresight to develop a radioimmunoassay that would test for LRF, everything was prepared when Matsuo asked him to test the synthetic peptide for LRF-like activity. The day set for the experiment was Sunday, April 25. Just in case the test should be positive, which would mean that Matsuo's peptide was identical with the real LRF and that the structure had been solved, Arimura asked his wife to prepare a champagne dinner. On his way into work, a hubcap fell off his car. Since this had never happened before, it seemed a bad omen. It seemed an even worse omen when, a few minutes after he had replaced the hubcap, it fell off again. Arimura usually paid no attention to omens, but this caused some uneasiness. He entered the laboratory, which was still and empty, quite in contrast to his inner feelings. He placed the test tubes in the radioactivity counter, which would signal whether or not Matsuo's peptide was biologically active.

The machine registered a sudden signal of high activity. Matsuo had synthesized LRF. Arimura was the first person in the world to know the structure of the hormone that governed the breeding behavior of pigs, and probably of man and all other mammals as well. He immediately rang Matsuo's house to break the good news. Matsuo's wife refused to wake him. It was 9 A.M. and Matsuo had only just gone to bed, having been up all night synthesizing the second possible structure. Arimura felt sure Matsuo wouldn't mind being waked up and insisted on speaking to him. "I found activity," Arimura told him. It was a moment that Matsuo would never forget. "At last you've got it!" he told himself. He visualized the Japanese characters of the report he would write describing his discovery.

For Matsuo, it was a brilliant achievement. He had arrived in New Orleans on December 31 to tackle a problem which many scientists deemed impossible, and he had solved it by April 25, less than four months later. Schally had provided 250 micrograms of LRF, but Matsuo used only 50 micrograms to determine the structure. Through novel and daring techniques he had accomplished a tour de force.

Making the discovery was one thing; making it known was another. The first problem was the authorship of the paper describing the structure of LRF. Matsuo felt that his name should be first. Baba, who had worked for a year longer on the structure and had done all the groundwork, considered that if he weren't the first author, then Schally, not Matsuo, should be.

Schally took a detached view of the dispute. "What did it matter to me? It's my lab – I got the glory anyway," he observes, giving utterance to a cardinal principle of the organization of American science.

The author list was decided as Matsuo, Baba, R. M. G. Nair (Schally's mass spectroscopist), Arimura, and Schally. The paper was sent to the journal *BBBC* in early May but was scheduled to appear only in late June.

Schally wanted to save the public announcement of the structure of LRF for the Endocrine Society meeting to be held June 24 to 26, 1971, in San Francisco. Most of the researchers in the field would be there and the announcement would have considerable impact. But a smaller meeting of the specialty was due to take place in New York a month earlier, at the end of May. Max Amoss, the member of Guillemin's team responsible for doing assays of LRF, was on the list of speakers. Schally had heard rumors that both Guillemin and McCann had solved the structure of LRF. He instructed Arimura and Matsuo to attend the New York meeting but to reveal the structure of LRF only if one of the other groups did so.

Matsuo was unhappy with this arrangement. The New York conference was an important meeting and would serve as a perfectly adequate forum for the announcement of their discovery, regardless of what the other groups had to declare. Schally was adamant that no disclosure should be made, except in the emergency of Guillemin's reporting success. Matsuo could not understand the reason for Schally's behavior.

At the New York conference Amoss spoke first. He reported on behalf of Guillemin's group that LRF had nine amino acids, and that no one had yet worked out their sequence. Arimura gave the next lecture. He did not report the structure of LRF but, at Schally's instruction, he included among his slides a chart comparing the activity of natural and synthetic LRF, which quickly made plain that he knew what the structure was.

There was a gasp from the audience, followed by a barrage of questions from scientists demanding to know what the structure was. Arimura declined to divulge it, explaining that Schally wished to announce the structure himself and would do so at the Endocrine Society meeting next month. In Arimura's personal opinion, Schally had been searching for LRF for years and deserved the opportunity to declare his achievement at a time and place of his choosing.

Science and War

It was in the First World War that scientists were first widely called upon to apply their skills to killing the enemy. Organic and inorganic chemists on both sides dedicated themselves with greater or lesser enthusiasm to the development of poison gases and also of course to the means of protecting their own armies against the enemy's gases. There appear to have been more scruples on the German side, and many scientists refused the call. The leading proponent of gas warfare on the German side was Fritz Haber, who had already incurred his country's gratitude in 1914 by devising a method for synthesising nitric acid from ammonia. This made possible the manufacture of high explosives, despite the Atlantic blockade, which was restricting the supplies of 'Chile saltpetre'. Haber believed that effective exploitation of gas could drastically shorten the war. His wife, also a chemist, objected and pleaded with her husband to abandon this work. Haber insisted that his patriotic duty came before all else, and his wife committed suicide the evening before he left for the front. The Professor in André Malraux's novel, The Walnut Trees of Altenburg, *is Haber.*

The Professor's voice was summing up the advantages and disadvantages of phosgene, and my father was conscious of the depth of the Slav world extending as far as the Pacific. In the trees there were still a few birds; other were returning to their nests, with the hoarse, anxious cheep of the swallows carried from afar as darkness fell. The Professor threw his extinguished cigarette out of the window, which he shut, and came back beaming with joy:

'The wind's still perfect, still perfect! Anyway I'm less frightened of a change of wind than of a sudden dampness.'

He had already lit another cigarette and started pacing up and down again.

'But we're still at the level of pre-history in chemical warfare! Ethyldichloride sulphate is perhaps the best fighting gas of all. A caustic product, blistering and poisonous at the same time! Particularly dangerous, mind you, since the victim is not affected at the actual moment of poisoning: it begins to take effect several hours afterwards . . .'

He stopped talking; his hand was still moving, the sausage in it beating time to humanity's funeral march.

'And even then . . .'

He kept them in a state of suspense that seemed to give him physical pleasure, then suddenly appeared to forget them.

'Eight centigrammes vaporised . . . a cubic metre of air,' he murmured, 'fatal casualties in under half an hour. Effective – it's marvellous – up to . . . up to . . .'

He could not get through his mental calculation. He took out a fountain pen, looked for something on the table; Max snatched up the photograph of the dogs; his father seized the one of his beloved house and turned it over, then began to scribble down some figures. It was difficult to see. The Captain lit a paraffin-lamp. The Professor went on with the calculations he had started on the back of the snapshot and got up. He looked at the officers as though they were the enemy – eyes half-shut, jaw slightly tilted:

'Effective up to one part in fourteen million parts of air.'

And waving his photograph like a proof:

'Perhaps chemistry is the final weapon, the superior weapon, which will give the people who use it properly – who master it! – world-wide supremacy. Perhaps even the Empire of the World!'

The lamp-light had suddenly made the window black and opaque. Beyond it, my father was still conscious of Russia and its sunflowers blooming in the night as far as Mongolia.

'Isn't it likely that the enemy Intelligence Service could soon get hold of our formulae?' asked the Captain.

'But in less than six months we shall have used six different kinds of gas! You see, between the production of the various kinds of gas and the means of protection against them, the same race will go on as was started thousands of years ago, since . . . the first manufacturer of the bludgeon, between the lance and the breastplate, the bullet and armour-plating. Only, and this is the most important side of the question . . .'

He started pacing up and down again, creating the same suspense he had created once already:

'. . . all the time that struggle has been going on, it's never been armour-plating that has scored the winning run.'

There was a short silence.

'Professor, you did say, didn't you,' asked Wurtz, 'that the latest discoveries are capable of poisoning the enemy troops without their knowing it?'

'Sufficiently so for the fighting spirit of the evacuated troops to be reduced to zero against a fresh attack.'

Wurtz smiled uneasily, like an overgrown child caught in the wrong. The Professor lit another cigarette, stopped pacing up and down, and said nothing. He knew what the Captain was thinking. He waited. Wurtz shook his head, looking resigned but not in agreement.

'In this war that's being waged against us by the whole world,' the Professor said at last, almost gently, 'Germany has no choice.

'Victory will save the lives of hundreds of thousands of our soldiers. And the most effective ways of achieving it are the best ones.'

Once again he brandished the photograph of his house:

'One part in fourteen million of air!'

And armour-plating is always less strong than a shell, my father reflected.

'I admit that,' said the Captain. 'But . . . can you tell me why we are despised?'

He began to trace stars on the table-cloth with the point of his knife.

'If you look at it objectively,' said the Professor in a voice of authority, 'gas constitutes the most humane method of warfare. For gas, mind you, announces its presence! The opaque cornea first goes blue, the breath starts to come in hisses, the pupil – it's really very odd! – becomes almost black. In a word, the enemy is forewarned. Now, if I think I still have a chance, even a faint chance, I'm brave: but if I know in my heart of hearts that I haven't, then courage is no use. Nothing to be done about it!'

To the Captain, these two men were enemies. Men of words and figures, 'intellectuals', who wanted to do away with courage.

174

Haber narrowly escaped trial as a war criminal in 1919, and then dedicated himself to a project for extracting gold from sea-water to pay off the German war debt. Haber was Jewish by birth and his country repaid him cruelly for his services. He had forsaken his heritage and been baptised, but it availed him nothing. One of his associates observed that before the age of thirty-five Haber was too young for a professorial chair, after forty-five he was too old, and in between he was a Jew. With the rise of Hitler, Haber was hounded out of his country, Nobel Prize notwithstanding. He arrived in Cambridge, a broken man. Here, from Chaim Weizmann's autobiography, Trial and Error, *is his description of Haber before and after his fall, one of the tragic figures of his time.*

Haber's anti-Zionist prejudices must have been wearing off, perhaps under the influence of developments in Germany. I found him, somewhat to my surprise, extremely affable. He even invited me to visit him at his research institute, which had the high-sounding name of *Kaiser Wilhelm Forschungs Institut*, in Dahlem, which I did towards the end of 1932, on one of my visits to Berlin.

It was a magnificent collection of laboratories, superbly equipped, and many-sided in its programme, and Haber was enthroned as dictator. He guided me through building after building, and after the long tour of inspection invited me to lunch with him at his villa in Dahlem. He was not only hospitable; he was actually interested in my work in Palestine. Frequently, in the course of our conversation on technical matters, he would throw in the words: 'Well, Dr. Weizmann, you might try to introduce that in Palestine.' He repeated several times that one of the greatest factors in the development of Palestine might be found in technical botany. This is a combination of plant physiology, genetics and kindred sciences, which was represented in Dahlem both by great laboratories and by first-class men conducting them. I was comparing in my mind those mighty institutions which served the agriculture of Germany with our little Agricultural Experiment Station at Rehovoth, and hoping that the new Institute which I contemplated might help to fill some of the gaps in our reconstruction.

I left Dahlem heavy-hearted and filled with forebodings, which I remember communicating to my wife on my return to London.

Not long afterwards, I received a telephone call at my home in London from Haber. He was in the city, staying at the Russell Hotel. He had had to leave Berlin precipitately, stripped of everything – position, fortune, honours – and take refuge in London, a sick man, suffering from angina pectoris, not quite penniless, but with very small reserves. I went to him at once, and found him broken, muddled, moving about in a mental and moral vacuum.

I made a feeble attempt to comfort him, but the truth is that I could scarcely look him in the eyes. I of course invited him to the house, and he visited us repeatedly. He told me that Cambridge was prepared to provide him with a laboratory, but he did not think he could really settle down. The shock had been too great. He had occupied too high a position in Germany; his fall was therefore all the harder to bear.

It must have been particularly bitter for him to realize that his baptism, and the baptism of his family, had not protected him. It was difficult for me to speak to him; I was ashamed for myself, ashamed for this cruel world, which allowed such things to happen, and ashamed for the error in which he had lived and worked throughout all his life. And yet it was an error which was common enough; there were many Jews with his outlook – though not with his genius – who had regarded Zionists as dreamers or, worse, as kill-joys, or even as maniacs, who were endangering the positions they had fought through to after many years.

I began to talk to him then about coming out to us in Palestine, but did not press the matter. I wanted him first to take a rest, recover from his shock and treat his illness in a suitable climate.

He went south, and that summer (1933), following my hasty visit to America, we met again in Switzerland. I was staying in Zermatt, at the foot of the Matterhorn, and Haber was somewhere in the Rhône valley and came over to see us. We dined together that evening. I found him a little improved, somewhat settled and past the shock. The surroundings in the Rhône valley had had a beneficent effect on him.

During the dinner, at which my wife and my son Michael were also present, Haber suddenly burst into an eloquent tirade. The reason was the following: the eighteenth Zionist Congress was then being held in Prague. I had refused to attend, not wishing to be involved in any political struggle. During the dinner repeated calls came from Prague, and frantic requests that I leave Zermatt at once and betake myself to the Congress. I persisted in my refusal, and though I said nothing to Haber about these frequent interruptions, except to mention that they came from Prague, he guessed their purport from something he had read in the papers, and he said to me, with the utmost earnestness:

'Dr. Weizmann, I was one of the mightiest men in Germany. I was more than a great army commander, more than a captain of industry. I was the founder of industries; my work was essential for the economic and military expansion of Germany. All doors were open to me. But the position which I occupied then, glamorous as it may have seemed, is as nothing compared with yours. You are not creating out of plenty – you are creating out of nothing, in a land which lacks everything; you are trying to restore a derelict people to a sense of dignity. And you are, I think, succeeding. At the end of my life I find myself a bankrupt. When I am gone and forgotten your work will stand, a shining monument, in the long history of our people. Do not ignore the call now; go to Prague, even at the risk that you will suffer grievous disappointment there.'

I remember watching my young son, as he listened to Haber, who spoke a halting English which his asthma made the more difficult to follow. Michael was literally blue to the lips, so painfully was he affected, so eager was he for me to take Haber's advice, even though it meant my leaving him in the middle of his holiday.

I did not go to Prague, much to Haber's disappointment. But I made use of the opportunity to press upon him our invitation to come out to Palestine and work with us. I said: 'The climate will be good for you. You will find a modern laboratory, able assistants. You will work in peace and honour. It will be a return home for you – your journey's end.'

He accepted with enthusiasm, and asked only that he be allowed to spend another month or two in a sanatorium. On this we agreed – and in due course he set out for Palestine, was taken suddenly ill in Basle, and died there. Willstätter came from Munich to bury him. Some ten years later Willstätter too died in Switzerland, like Haber, an exile from Germany.

175

Rubber was one of the critical raw-materials in the Second World War. Vicki Baum's novel The Weeping Wood, *tells something of this story.*

'What do you want from me, Kim,' said George Tyler an hour later to his old friend Senator Kimball B. Bland, 'a lecture on synthetic rubber so you can shine when you make your report to the President? Why don't you ask Randy Warrens; he's the synthetic man. I know only what I read in the papers. And do you know what I read in the papers? That you fellows are losing an unholy amount of time by dillydallying and bickering and listening to this one and to that one and by letting everyone give his opinion and push his little private scheme and by simply not being able to make up your damned minds once and for all. Now, Kim, you've got to get one thing clear in your mind: there are as many brands of synthetic rubber as there are brands of bread in the average market; the trouble with us is that, as a nation, we are too ingenious. If we had one single brand of synthetic rubber with the Government's blessing, like the Germans or the Russians, we would have started production a long time ago. Now listen and try not to get confused and I'll try to make it as simple as a nursery book. Rubber – I mean natural rubber – is essentially a hydrocarbon; that much was known as early as 1826. Ever since then the chemists have figured that if they could break it up, analyse it, and piece it together again, they would have synthetic rubber. But rubber is not that simple, no, sir. In 1860 a certain Williams broke it down all right and got a substance he called isoprene – and isoprene it has remained to this day. For more than sixty years chemists fooled around with this isoprene; they cooked up all sorts of rubbery, sticky messes, but none of them had the properties of real rubber and none of them could be used. That is, during the first World War the Germans produced something like a hundred and fifty tons a month of a thing they call Weichgummi; it took six months to coagulate in their vats and was terrible, and they dropped it like a hot potato as soon as the war was over. But the Germans, being what they are, did not give up, and finally they approached it from a different angle, and by and by they got this Buna-S on which most of our synthetic rubber programme is to be based. It might seem a little ironical that we hope to win this war by running it on a German invention, but that's how it is. Now watch for this, Kim: Buna-S is a so-called polymer – as is natural rubber – and its raw materials are butadiene and styrene. Butadiene and styrene, in turn, can be made from almost any hydrocarbon. The Germans make it from coal and limestone, because that's what they've got. We, naturally, would make it from oil, because we have plenty of that. The Russians made it from grain in the beginning but are turning more and more to oil too. Theoretically you could make it from potatoes, from sugar, even from whisky if you felt like it. That is, you can always turn any starchy substance into alcohol, and you can, by some further process, finally turn alcohol into butadiene. If your farmers throw a spanner into the works and clamour to be let in on the breadbasket of a synthetic-rubber programme by which butadiene is made from grain and you want to listen to them, that's up to you. You're a politician, and I'm a rubber man, and I say make it from oil, and I say every day that's lost by quibbling and looking for private advantages means playing Hitler's and Hirohito's game. Now take a note of this: Buna is so much in the foreground because it is the one synthetic rubber that makes satisfactory tyres. Firestone's are experimenting with the new tyres now. Goodyear makes something called Chemigum, and Goodrich makes something called Ameripol;

basically they are made by the same process. There is also Perbunan, which is very good but very expensive and rather inelastic, and for the time being we will forget about it.

'This is one family of synthetics. And now comes the next one, which belongs to the family of Neoprene and is something entirely different. It has most of the properties of rubber but is quite different chemically. It was first discovered by Father Julius Nieuwland, in Notre Dame, as a by-product of his studies on acetylene. Du Pont acquired this patent, and Carothers improved it; in fact, Du Pont has been manufacturing several thousand tons of Neoprene every year for quite some time. Because it is especially resistant to oils and organic solvents it is used for hose, and because of its resistance to ozone and sunrays it is very important for airplanes. But it isn't good to make tyres from, because it hasn't got the necessary elasticity. The same holds true for another group, and I don't want to confuse you by too many chemical details. It's enough if you know that there are Koroseal and Thiokol, both good for raincoats and in bulletproof gasoline tanks and various other goods – but not for tyres. The thing which might eventually become usable for tyres is this stuff the papers are making such a noise about – Standard Oil's Butyl and USA Oil's Butanex. These again are American inventions and have great possibilities, and don't tell me that I'm plugging for the oil companies, because it isn't true. No, I have no reason whatsoever to pull Randolph Warrens' chestnuts out of the fire – ' George Tyler said, fingering the letter in his wallet with his left hand. 'I'm just telling you the little I know. The great trouble with these new synthetics is, however, that they are still in the experimental stage, and you can't build a programme on something that isn't ready yet. Between you and me and the lamp-post, that is where we are guilty. We've done some research – but not enough. We've made experiments, but not enough. If we had let our chemists experiment sufficiently in past years, we would have a cheap finished product today and we wouldn't be in the jam we're in now. If you ask me, it all boils down to something I heard my friend Hoover say at a banquet, as early as 1926, when he told us that if there wasn't more money, more brains, more men put to research efforts in the United States, we were going to lose out in the world contest. I think we all felt that he was right. At any rate, the research effort in this country has grown tremendously since that day, and you can take this from an old production man like me: We're yet going to show the world a surprise or two in this war, so help us God –'

176

The development of radar was the physicists' first critical intervention in the Second World War. The source of the radiation, the cavity magnetron, was invented in M. L. Oliphant's department in Birmingham by Randall and Boot. It appears to have been a serendipitous discovery, and the American physicists found it difficult to analyse just how it worked. A well-known episode is here related by Daniel Kevles in his history of American physics, The Physicists.

With Lawrence's help, DuBridge recruited a scientific staff, and soon some three dozen physicists were tuning up experimental equipment in MIT's massive buildings and fitting out a new rooftop microwave laboratory constructed of wood and covered with gray-green tarpaper. Their average age in the mid-thirties – forty-three-year-old I. I. Rabi was one of the old men of the enterprise – the group included some of the best nuclear physicists in the country. They knew something about high-frequency radiation from their work on cyclotrons, but the magnetron confounded even them at first.

'It's simple,' Rabi told the theorists who were seated around a table staring at the disassembled parts of the tube. 'It's just a kind of whistle.'

'Okay, Rabi,' Edward U. Condon asked, 'how does a whistle work?'

Rabi was at a loss for a satisfactory explanation.

177

Following the discovery of nuclear fission and the realisation that a chain reaction was a realistic possibility, a hysteria gripped the European physicists. To deny the Germans the materials, particularly Norwegian heavy water, that would be needed to produce a fission bomb, became a pressing preoccupation. Otto Frisch, the nephew of Lise Meitner and associate of Otto Hahn (for the story of how he and his aunt perceived the significance of Hahn's results, see no. 115), was in Birmingham, exploring methods of achieving a chain reaction. This anecdote from his memoirs, What Little I Remember, *illustrates the overheated atmosphere of the time.*

The report which Peierls and I had sent to (Sir) Henry Tizard on Oliphant's advice had triggered the formation of a committee, with (Sir) George Thomson as chairman, which was given the code name 'Maud Committee'. The reason for that name was a telegram which had arrived from Niels Bohr, ending with the mysterious words 'AND TELL MAUD RAY KENT'. We were all convinced that this was a code, possibly an anagram, warning us of something or other. We tried to arrange the letters in different ways and came out with mis-spelt solutions like 'Radium taken', presumably by the Nazis, and 'U and D may react', meant to point out that one could get a chain reaction by using uranium in combination with heavy water, a compound of oxygen and the heavy hydrogen isotope called deuterium, abbreviated D. The mystery was not cleared up until after the war when we learned that Maud Ray used to be a governess in Bohr's house and lived in Kent.

178

Tizard was one of the principal protagonists in C. P. Snow's account of intrigue and jealousy in Whitehall during the war, Science and Government.

We had never had the conventional English faith in strategic bombing, partly on military and partly on human grounds. But now it came to the point it was not Lindemann's ruthlessness that worried us most, it was his calculations.

The paper went to Tizard. He studied the statistics. He came to the conclusion, quite impregnably, that Lindemann's estimate of the number of houses that could possibly be destroyed was five time too high.

The paper went to Blackett. Independently he studied the statistics. He came to the conclusion, also quite impregnably, that Lindemann's estimate was six times too high.

Everyone agreed that, if the amount of possible destruction was as low as that calculated by Tizard and Blackett, the bombing offensive was not worth concentrating on. We should have to find a different strategy, both for production and for the use of élite troops. It fell to Tizard to argue this case, to put forward the view that the bombing strategy would not work.

I do not think that, in secret politics, I have ever seen a

minority view so unpopular. Bombing had become a matter of faith. I sometimes used to wonder whether my administrative colleagues, who were clever and detached and normally the least likely group of men to be swept away by any faith, would have acquiesced in this one, as on the whole they did, if they had had even an elementary knowledge of statistics. In private we made the bitter jokes of a losing side. 'There are the Fermi-Dirac statistics,' we said. 'The Einstein-Bose statistics. And the new Cherwell nonquantitative statistics.' And we told stories of a man who added up two and two and made four. 'He is not to be trusted,' the Air Ministry then said. 'He has been talking to Tizard and Blackett.'

The Air Ministry fell in behind the Lindemann paper. The minority view was not only defeated, but squashed. The atmosphere was more hysterical than is usual in English official life; it had the faint but just perceptible smell of a witch hunt. Tizard was actually called a defeatist. Strategic bombing, according to the Lindemann policy, was put into action with every effort the country could make.

The ultimate result is well known. Tizard had calculated that Lindemann's estimate was five times too high. Blackett had put it at six times too high. The bombing survey after the war revealed that it had been ten times too high.

After the war Tizard only once said 'I told you so.' He gave just one lecture on the theory and practice of aerial bombing. 'No one thinks now that it would have been possible to defeat Germany by bombing alone. The actual effort in manpower and resources that was expended on bombing Germany was greater than the value in manpower of the damage caused.'

During the war, however, after he had lost that second conflict with Lindemann, he went through a painful time. It was not easy, for a man as tough and brave as men are made, and a good deal prouder than most of us, to be called a defeatist. It was even less easy to be shut out of scientific deliberations, or to be invited to them on condition that he did not volunteer an opinion unless asked. It is astonishing in retrospect that he should have been offered such humiliations. I do not think that there has been a comparable example in England this century.

However, the Establishment in England has a knack of looking after its own. At the end of 1942 he was elected to the presidency of Magdalen College, Oxford. This is a very honourable position, which most official Englishmen would accept with gratitude. So did Tizard. There are no continuous diary entries at this period, although now he had plenty of time. For once his vitality seems to have flagged.

I think there is little doubt that, sitting in the Lodgings at Magdalen during the last thirty months of war, he often thought of Whitehall with feelings both of outrage and regret. Here he was, in one of the most splendid of honorific jobs, but his powers were rusting – powers that were uniquely fitted for this war. He knew, more accurately than most men, what he was capable of. He believed, both in his dignified exile in Oxford and to the end of his life, that if he had been granted a fair share of the scientific direction between 1940 and 1943, the war might have ended a bit earlier and with less cost.

179

This world, in which adversaries manoeuvred for advantage at green baize tables, is captured with remarkable immediacy in Nigel Balchin's novel of science in the Second World War, The Small Back Room. *(The scientists in those days were known as back-room boys, or boffins.)*

Gladwin wiped the sweat off his forehead and said, 'Gentlemen, I have called this meeting on the Minister's instructions to clear up the position about the Reeves gun. The Minister has his own views about the Reeves. But before making any final decision, he wants to be sure that the various points of view of you gentlemen are fully understood by us. There have been extensive demonstrations and experiments. The question now is, what have those experiments shown and what are we going to do about it?' He gave his forehead a final mop and said in a loud crow, 'That's all I want to say. Now you talk and I'll listen.'

There was a pause. Then Gladwin said, 'I believe Professor Mair has been interested in these experiments. Perhaps he'll give us his conclusions?'

I looked at the Old Man a bit anxiously. The chances were that he would up and do one of his enthusiastic acts, which were always bad salesmanship. But he just went on filling his pipe and said slowly, 'Well, in a few words, Mr. Chairman, I should say that the Reeves gun is one of the most promising developments I've seen – from some points of view. But that's only an opinion, and for my part I should prefer to hear other people before being too dogmatic.'

Gladwin nodded. There was silence again. Then the brigadier said, 'Well, frankly, Mr. Chairman, we don't like the Reeves.'

'You don't, eh?' said Gladwin. As he must have known they'd been fighting it tooth and nail for weeks, I thought he did pretty well to sound surprised.

'No. It has a lot of snags from the user point of view, and we don't think it has sufficient advantages to offset them.'

'Well, of course that's the question,' squeaked Gladwin. 'What's the balance of advantage and disadvantage? Everything has disadvantages.' He sank back and mopped his forehead as though thinking that up had exhausted him. Nobody seemed to know the next line.

Easton had been sitting staring blankly at the window. He turned and looked through the chairman and said, 'It might possibly be of interest to the meeting, Mr. Chairman, to hear the views of the National Scientific Advisory Council, of which I have the honour to be chairman.'

'Please,' said Gladwin in a loud squawk.

'The National Scientific Advisory Council . . .' said Easton.

'Of which he has the honour to be chairman,' Waring muttered in my ear.

'. . . is the body officially deputed by the Cabinet to offer advice on all major scientific issues. The Council appointed a special sub-committee to examine and report on the Reeves gun. Dr. Brine, who is with us here to-day, acted as convener of this sub-committee, which was given the fullest facilities for the examination of this weapon. I suggest that Dr. Brine should give the views of his sub-committee.'

'Please,' said Gladwin a bit pathetically, mopping away. I had the feeling that unless somebody really gave some views pretty soon, he would burst into tears.

The blue-chinned man stroked the blue part and said, 'Well, Mr. Chairman, my colleagues and I approached this matter purely as scientists.' He said it as though everybody else had approached it as income tax inspectors or jobbing gardeners. 'And our conclusion was that scientifically speaking it was not a sound conception. Not at all a sound conception. In fact I'll go further and say that no scientist could feel happy about many of the principles involved.' He sat back and looked pointedly at Mair. I glanced at the Old Man. His bottom lip was beginning to stick out – that meant he was bloody angry. The blue-chinned chap was obviously trying to be rude.

I scribbled on a bit of paper, 'Ask if the sub-committee saw the gun fired' and passed it to Mair.

'I'm interested to hear that, Mr. Chairman,' said the brigadier. 'Because in our unscientific way that's what we thought.'

'Just what scientific principles did *you* think were unsound?' said Mair quickly.

The brigadier hesitated and old Holland chipped in. 'We're talking about different sorts of principles,' he said. 'We didn't like it because it had user snags. Dr. Brine is talking about principles of – of physics or something like that.'

'Well, then, may we take Dr. Brine's statement first?' said Mair, unfolding my note and glancing at it. 'What didn't the sub-committee like?' Before Brine could answer he added, 'Perhaps first he could tell us who were his colleagues?'

Brine said, 'The sub-committee consisted of Professor Char, Dr. Goulder, Dr. Pease and myself.'

Mair smiled. 'One crystallographer, one vital statistician, one embryologist and one – let's see . . .?' he looked inquiringly at Brine with a charming smile. It couldn't have been more beautifully done if he'd slapped his face.

Easton went very red and said icily, 'If I might answer for Dr. Brine, he is, of course, one of the best-known organic chemists in the country.'

'Right,' said Mair. 'Now we've got that straight.' He leant back and added casually, 'By the way – you did see the gun firing, I take it?'

Brine hesitated. Then he said, a bit feebly, 'We were not actually present at the trials.'

'But you've seen the gun *fire*? You didn't just look at it as a piece of furniture?'

'No,' said Brine defiantly. 'We didn't see it firing.'

Mair looked at him as though he were astounded. Then he said, 'Oh . . . Well, well – never mind. Can we get back to what the committee didn't like?'

It was very pretty. The wretched bloke Brine started off, but it was obvious that no one was going to accept him as evidence after that. What he said was quite all right though not very profound. But he was batting on a ruined wicket. Easton weighed in once or twice, but he was so darned pompous that he only made things worse. So far the Old Man had the meeting cold.

At last old Holland looked up and said, 'Mr. Chairman, might we consider this from another aspect? There seems to be some difference of opinion on the scientific side, and I for one am not competent to know who's right or wrong. But our objections to the Reeves aren't particularly scientific. Professor Mair may be quite right in saying that it's a grand idea, scientifically, but what we're interested in is whether it's a good *gun* as it stands.'

This was just the line I wanted to avoid. I glanced at Mair, hoping he'd stall. But by that time he had his tail well up and he went in head first.

'Well now,' he said cheerfully, 'that's surely a matter of fact. We've had experiments and trials. What do they say?'

The brigadier nearly saved him by shaking his head doubtfully and saying, 'Of course trials are one thing, and service in the field is another. Far too many weapons are put out without proper user consultation. That's the trouble.'

This was a grand red herring, because it woke up Gladwin's boys. They recognised the opening notes of one of the eternal rows with the army and started to rally round.

Styles, one of Gladwin's senior people, said, 'One of *our* difficulties, Mr. Chairman, is that we can never get at the facts on which these user opinions are based.'

'Hear, hear!' said another of the crowd.

'Our trials may not be very good,' said Styles. 'But at least they're an advance on looking at the thing, firing it twice and deciding that you don't like the noise it makes.'

The brigadier went rather red and I thought we were safe. They'd quarrel happily now for an hour about whether the army knew what it wanted or not. But old Holland chipped in again and said, 'Mr. Chairman, this is an old argument, and it won't get us far. Could we get back to Professor Mair's question – what do the results of the trials show?' He looked across at me and said, 'I'm not a scientist or a statistician. But my reading of the figures I've seen suggests that, in practice, we don't *get* these advantages that have been talked about. I may be wrong.'

'I think Colonel Holland's taking altogether too gloomy a view,' said Mair, still in a high good humour. 'Perhaps, Mr. Chairman, you'd allow Mr. Rice of my staff, to give us the facts. Mr. Rice has done all the statistical work on the trials.'

'Please,' squawked Gladwin.

Everybody looked at me. I took the papers out of my bag. I knew the stuff by heart. But it gave me time to think. My hands were shaking slightly. Looking at the figures it suddenly struck me that they probably wouldn't mean anything to the meeting, because nobody but some of Gladwin's boys would know the comparative figures for other weapons. I just started to read out the figures rather quickly.

It worked quite well. Gladwin leant back and mopped himself in a bored way, and Brine and the brigadier had that nice vacant look of people who are plodding on through the snow but have lost their way pretty thoroughly. I could see that Gladwin's crowd were a bit doubtful about what to do. They didn't like us, but if they shot at the thing they would be backing the army, which was dead against their principles.

When I finished there was a silence. Then old Holland suddenly said:

'That's fine, Mr. Chairman. Now may I ask Mr. Rice what it all adds up to?'

I said, 'I hardly think that's for me to say. I was merely giving the results of the trials.'

'And on these results you think that the Reeves is a first rate weapon?'

I hesitated for a moment and then said, 'I think Professor Mair's already given the view of the section.'

'And you share that view?' said Holland quietly.

'Oh come!' said Waring. 'That's scarcely a fair question, is it?

'Why not?' said Holland.

'Well, Mr. Chairman,' said Waring. 'I suggest that if Colonel Holland expressed a view, he'd hardly expect us to ask one of his junior officers if he agreed.'

'Quite,' said Gladwin. 'I don't think you can ask Rice to argue with his chief, Holland.'

Old Holland was still looking at me. He sat quiet for a moment and then he started to pat gently on the blotting paper in front of him.

'Mr. Chairman,' he said quietly, 'I want to be quite frank. We don't like this gun. We're told that those figures show that we're wrong. Professor Mair suggested that his expert should give us the facts. If Mr. Rice, who carried out this work, feels that his facts prove Professor Mair's case, I've no more to say. But surely I'm entitled to ask him what his figures mean?' He gave his blotter a sharp pat. 'After all, this is an important matter. We aren't debating, or defending a point of view. We're trying to get at the facts.' He was still staring at me with his rather pale, washed-out blue eyes. His voice was very quiet. 'If the Reeves gun is accepted, sooner or later men have got to fight with it. If there is anything wrong – if we've been too optimistic or anything has been glossed over . . .' He shrugged his shoulders. '*They'll* be the sufferers. We shan't.'

There was a rather uncomfortable silence. This was the first time anybody had said anything as though he really meant it.

'Well, well, Mr. Chairman,' said Mair cheerfully. 'Nobody wants to hide anything. If Colonel Holland would like Mr. Rice's views, I have no objection at all.' He leaned back and gazed at me inquiringly.

'Well, Mr. Rice?' said Holland. He turned the light blue eyes on me again.

I hesitated, and then said slowly, 'I agree with Professor Mair that the idea is excellent . . .'

'And the weapon?'said Holland.

'I don't think it's right yet.'

There was a very faint rustle of interest.

'Would you be happy to see it accepted in its present form?' said Holland mercilessly. 'On these figures?'

I could feel Waring's angry eyes on me. My throat was very dry.

I said, 'No, I shouldn't.'

Holland sat back in his chair. 'Thank you,' he said quietly. 'That's all I wanted to know.'

Easton said, 'You agree, in fact, with the view of our sub-committee?'

'I don't know how your sub-committee arrived at its view,' I said. 'My opinion is simply based on the figures.'

'I think it's rather important that that should be realised, Mr. Chairman,' said Waring. 'Mr. Rice is a technician and what he has given is purely a technician's view.'

'Quite,' said Gladwin, mopping his forehead very hard indeed. 'The position's quite understood.'

I looked round the table. It was understood all right.

Gladwin shut the meeting down as soon as he decently could and said something vague about reporting to the Minister.

As soon as we were outside afterwards I said, 'I'm sorry, sir.'

'Oh, that's all right,' said Mair cheerfully. 'I thought it was quite a good meeting. Made Easton and his sub-committee look pretty silly anyhow.'

'Well, it didn't leave us looking very clever ourselves, did it?' said Waring bitterly.

'Why not?' said Mair in surprise. It was clear that he hadn't noticed anything wrong.

'Well, Christ . . . !' said Waring irritably. 'If you're going to say one thing and Rice is going to say the exact opposite . . .'

I said, 'Well, what the hell was I to do? I warned you before hand that we were on dangerous ground.'

'You could have stalled,' said Waring savagely. 'As it is, you've given old Holland and that other red-tabbed stooge exactly what they wanted.'

'I simply told the truth. I'm a physicist, not a salesman.'

180

And here for comparison is another fragment of Snow's Science and Government, *depicting one of the ruthless power struggles that surged around the committee tables.*

Almost from the moment that Lindemann took his seat in the committee room, the meetings did not know half an hour's harmony or work undisturbed. I must say, as one with a taste for certain aspects of human behaviour, I should have dearly like to be there. The faces themselves would have been a nice picture. Lindemann, Hill, and Blackett were all very tall men of distinguished physical presence – Blackett sculptured and handsome, Hill ruddy and English, Lindemann pallid, heavy, Central European. Blackett and Hill would be dressed casually, like academics. Tizard and Lindemann, who were both conventional in such things, would be wearing black coats and striped trousers, and both would come to the meetings in bowler hats. At the table Blackett and Hill, neither of them specially patient men nor overfond of listening to nonsense, sat with incredulity through diatribes by Lindemann, scornful, contemptuous, barely audible, directed against any decision that Tizard had made, was making, or ever would make. Tizard sat it out for some time. He could be irritable, but he had great resources of temperament, and he knew that this was too serious a time to let the irritability flash. He also knew, from the first speech that

Lindemann made in committee, that the friendship of years was smashed.

There must have been hidden resentments and rancours, which we are now never likely to know and which had been latent long before this. No doubt Lindemann, who was a passionate man, with the canalised passion of the repressed, felt that he ought to have been doing Tizard's job. No doubt he felt, because no one ever had more absolute belief in his own conclusions, that he would have done Tizard's job much better, and that his specifics for air defence were the right ones, and the only right ones. No doubt he felt, with his fanatical patriotism, that Tizard and his accomplices, these Blacketts, these Hills, were a menace to the country and ought to be swept away.

It may have been – there are some who were close to these events who have told me so – that all his judgments at these meetings were due to his hatred of Tizard, which had burst out as uncontrollably as love. That is, whatever Tizard wanted and supported, Lindemann would have felt unshakably was certain to be wrong and would have opposed. The other view is that Lindemann's scientific, as well as his emotional, temperament came in: it was not only hatred for Tizard, it was also his habit of getting self-blindingly attached to his own gadgety ideas that led him on. Whatever the motive was, he kept making his case to the committee in his own characteristic tone of grinding certainty. It was an unjustifiable case.

The issue in principle was very simple. Radar was not yet proved to work: but Tizard and the others, as I have said, were certain that it was the only hope. None of them was committed to any special gadget. That was not the cast of their minds. There was only a limited amount of time, of people, of resources. Therefore the first priority must be given to radar – not only to making the equipment, but to making arrangements, well in advance even of the first test, for its operational use. (It was in fact in the operational use of radar, rather than in the equipment, that England got a slight tactical lead.)

Lindemann would not have any of this. Radar was not proved. He demanded that it should be put much lower on the priority list and research on other devices given the highest priority. He had two pet devices of his own. One was the use of infrared detection. This seemed wildly impracticable then, to any of the others and to anyone who heard the idea. It seems even more wildly impracticable now. The other putative device was the dropping, in front of hostile aircraft, of parachute bombs and parachute mines. Mines in various forms had a singular fascination for Lindemann. You will find Heath Robinson-like inspirations about them – aerial mines, fluvial mines, and so on – all over the Churchillian minutes from 1939–1942. They keep coming in as a final irritation to a hard-pressed man in Tizard's records of his conversations with Churchill. All these mine inspirations originated from Lindemann. None of them was ever any practical good at all.

For twelve months Lindemann ground on with his feud on the committee. He was tireless. He was ready at each meeting to begin again from the beginning. He was quite unsoftened, quite impregnable to doubt. Only a very unusual man, and one of abnormal emotional resistance and energy, could sit with men so able and not be affected in the slightest regard.

They themselves were not affected so far as choice was concerned. Tizard went ahead with the radar decisions and they let Lindemann register his disagreement. But gradually they got worn down. Neither Blackett nor Hill was phlegmatic enough to endure this monomaniac tension for ever. In July 1936, when the committee were preparing a report, Lindemann abused Tizard in his usual form, over the invariable issue of too much priority for radar, but in terms so savage that the secretaries had to be sent out of the room.

At that point Blackett and Hill had had enough of it. They resigned and did not try to give an emollient excuse for doing so. Whether this was done after discussion with Tizard is not clear. No discussion was really necessary. They all believed that this friction was doing too much harm. They were all experienced enough to know that, with Churchill still out of office, they could make their own terms.

Within a short time the committee was reappointed. Tizard was still chairman, Blackett and Hill were still members. Lindemann, however, was not. He was replaced by E. V. Appleton, the greatest living English expert on the propagation of radio waves. Radar itself was an application of Appleton's fundamental work. The announcement of his name meant, in the taciturn eloquence of official statements, a clear victory for radar and for Tizard. The radar stations and the radar organisation were ready, not perfect but working, in time for the Battle of Britain. This had a major, and perhaps a decisive, effect.

This cautionary story of the first Lindemann-Tizard collision seems to me to contain a number of lessons, some of them not obvious. But there is one, at the same time so obvious and so ironic, that I shall mention it now. It is simply that the results of closed politics can run precisely contrary to the results of open politics. That is an occupational feature of the way in which closed politics works and the way in which secret choices are made. Probably not more than a hundred people had any information whatever about Tizard's first radar decision; not more than twenty people took any effective part in it, and at the point of choice not more than five or six.

181

C. M. Bowra, known in Oxford as a consummate academic politician, used to say that scientists were treacherous allies on committees, for they were apt to change their minds in response to arguments. This is not the attribute most conspicuous in contemporary accounts of science policy-making during the war. What communicates itself instead is the extent to which the players' passions were engaged in defence of their theories. R. V. Jones in Most Secret War *speculated on what it was all about.*

I concluded with a section on German policy in which I tried to answer the bewilderment still existing in Whitehall about why the Germans had developed such a weapon at all. I suspected that Hitler had been carried away by the romance of the rocket, just as our own politicians had been carried away by its threat: for some psychological reason they seemed far more frightened by one ton of explosive delivered by rocket than by five tons delivered by air-craft:

A rational approach brought us nearest the truth regarding the technique of the Rocket. When, however, we try to understand the policy behind it, we are forced to abandon rationality, and instead to enter a fantasy where romance has replaced economy.

The Germans have produced a weapon which, at the cost of years of intense research, throws perhaps a one or two ton warhead into the London area for the expenditure of an elaborate radio controlled carcase consuming eight or so tons of fuel. Their own Flying Bomb achieves the same order of result far more cheaply. Why, then have they made the Rocket?

The answer is simple: no weapon yet produced has a comparable romantic appeal. Here is a 13 ton missile which traces out a flaming ascent to heights hitherto beyond the reach of man, and hurls itself 200 miles across the stratosphere at unparalleled speed to descend – with luck – on a defenceless target. One of the greatest realizations of human power is the ability to destroy at a distance, and the Nazeus would call down his thunderbolts on all who displease him. Perhaps we may be permitted to express a slight envy of his ability, if not to destroy his victims, at least to raise one of the biggest scares in history by virtue of the inverted romance with which those victims regard the Rocket.

182

At Los Alamos, meanwhile, the atomic bomb was taking shape. It was a remarkable triumph of team-work, in which a large proportion of the world's foremost physicists and chemists participated. The physical dangers to which some of these exposed themselves are well conveyed by Otto Frisch in What Little I Remember.

Everything happened exactly as it should. When the core was dropped through the hole we got a large burst of neutrons and a temperature rise of several degrees in that very short split second during which the chain reaction proceeded as a sort of stifled explosion. We worked under great pressure because the material had to be returned by a certain date to be made into metal and assembled to form a real atom bomb. During those hectic weeks I worked about seventeen hours a day and slept from dawn till mid-morning.

Other experiments with this atom bomb material were less flamboyant but in fact more dangerous. One man was killed by a runaway reaction while I was there, and later my successor, Louis Slotin – very likeable and popular – was the victim of the second accident. The danger was psychological: assembling a mass of uranium-235 was something we completely understood, and as long as we hadn't reached the critical amount – when the chain reaction began to grow spontaneously – the assembly was completely harmless. But critical conditions could be reached very suddenly as the result of a minor mistake.

Most of the assemblies we made were meant to help us find out exactly how much material would be needed for the bomb; they had neutron-reflecting material round them and were assembled from small bricks, with the reflector in the form of somewhat bigger bricks made from non-fissile material. We had made a strict rule that nobody was to work all by himself, and that nobody should ever hold a piece of material in such a way that if it dropped it might cause the assembly to become critical. The first man to be killed, Harry Daghlian, had broken both rules. He was so eager that he wanted to do one more assembly after everybody else had gone, and a large chunk of heavy metal slipped out of his fingers, and fell on the almost completed assembly; even as he instantly swept it aside with a blow of his muscular arm he saw a brief blue aura of ionized air around the assembly. He felt nothing else but became sick as the ambulance (he had reported by phone) took him to the hospital; two weeks later, his blood count way down, he died from some trivial infection which his body could no longer fight. Louis Slotin, I was told, lived only nine days after a pencil he had placed under a piece of reflector slipped out.

How did an experienced and cautious physicist like Louis Slotin make such a stupid and fatal mistake? Did he really think a pencil was a safe support for a critical piece of material? Or did something deep down in his mind tempt him to play atomic roulette? We shall never know.

There was one occasion when I myself very nearly became victim of a similar accident. We were building an unusual assembly, with no reflecting material around it; just the reacting compound of uranium-235, because this was a good way to test the reliability of our calculations. For obvious reason we called it the Lady Godiva assembly. I did follow all the rules. I had a student helping me. He was standing by the neutron-counting equipment, and we both watched the little red signal lamps blinking faster and faster and the meter clattering with increasing speed. Suddenly to my surprise the meter stopped; I looked up and saw that the student had unplugged it. Immediately I leant forward and called out 'Do put the meter back, I am just about to go critical', and at that moment, out of the corner of my eye I saw that the little red lamps had stopped flickering. They appeared to be glowing continuously. The flicker had speeded up so much that it could no longer be perceived.

Hastily I took off some of the blocks of the uranium compound which I had just added, and the lamps slowed down again to a visible flicker. It was clear to me what had happened: by leaning forward I had reflected some neutrons back into Lady Godiva and thus caused her to become critical. I hadn't felt anything, but after we had completed the experiment with appropriate care, I took a few of the blocks and checked their radioactivity with a counter. Sure enough, the activity was many times larger than what should have accumulated if that little incident hadn't occurred. From that I could calculate that during the two seconds while I was leaning over the assembly the reaction had been increasing, not explosively but at a very fast rate, by something like a factor of a hundred every second. Actually the dose of radiation I had received was quite harmless; but if I had hesitated for another two seconds before removing the material (or if I hadn't noticed that the signal lamps were no long flickering!) the dose would have been fatal.

183

The German atom bomb programme never got properly off the starting-line. This, in the light of historical study, was clearly not because of any reluctance on the part of the German physicists to work for the victory of an evil cause. Hahn and especially von Laue (who was ostracised for it) were noble exceptions. The Dutch-American physicist, Sam Goudsmit, was placed in charge by General Groves (the organiser of the Manhattan Project) of a mission (given the transparent code-name of Alsos – the Greek for grove) to investigate the state of progress of atomic physics in Germany, as soon as the invasion of Europe allowed. Goudsmit chronicled his experiences in Alsos. *After the defeat of Germany in 1945 a group of leading physicists was brought to England and kept under guard in a country mansion, where their conversation was picked up by hidden microphones and recorded. Here Goudsmit describes their reactions to the news of the Hiroshima bomb.*

It was on August 6, 1945, at dinner time, that the interned German physicists first heard the news of Hiroshima. Their initial reaction was one of utter incredulity. Impossible, they said. After all, they themselves had been working on the uranium problem for several years and they had proved that an atomic bomb was too difficult to achieve in such short order. Then how could the Americans do it? It was preposterous.

'It can't be an atomic bomb,' one of their number said. 'It's probably propaganda, just as it was in Germany. They may have some new explosive or an extra large bomb they call "atomic," but it's certainly not what we would have called an atomic bomb. It has nothing whatever to do with the uranium problem.'

That being settled, the German scientists were able to finish their dinner in peace and even partially digest it. But at nine o'clock came the detailed news broadcast, the same one, very likely, that I heard at Frankfurt.

The impact on the ten scientists was shattering. Their whole world collapsed. At one stroke, all their self-confidence was gone and the belief in their own scientific superiority gave way to an intense feeling of despair and futility. Then all their work of the last six years had been wasted; their hope for a bright future for German science was only so much illusion.

Only one of the group was not affected, at least not personally, and that was von Laue. He had been merely an onlooker, never a participant in the physicists' dream of power, the atomic bomb. For the atomic bomb meant power not only for the country that could solve this problem, but power as well for the physicists and their science, which would now be recognized as indispensable. Von Laue was probably the only man among the ten interned on that English estate who realized thoroughly the world-shaking effects this Hiroshima bomb would produce everywhere, in ever-widening circles.

Von Laue, at any rate, took the news calmly. Not so the others. There were bitter words as the question arose why they, the superior Germans, had not succeeded. Some of the younger men angrily reproached their elders for not having had more vision, for having failed Germany in her hour of need.

Gerlach was the most violently upset of all. He acted like a defeated general. He, the 'Reichsmarschal' for nuclear physics, had not succeeded in his assignment. He took the remarks of the younger men as a personal criticism and was profoundly depressed for several days. His colleagues and friends had to reason with him and comfort him to bring him back to his senses.

The rest got over their hysteria. They spent hours discussing the science of the bomb and tried to figure out its mechanism. But the radio for all its details, had not given enough, and the German scientists still believed that what we had dropped on Hiroshima was a complete uranium pile. No wonder they were bewildered. For us, or anyone, to have dropped a complete uranium pile would indeed have been a super achievement of modern engineering; as yet, no plane has ever been built that could do that. Even if there were such a plane, a uranium pile could never be a bomb. It could only be a fizz. But as yet, these Germans experts failed to realize this basic fact.

Failing to understand it, they began to carp at details. Why wasn't Otto Hahn, the discoverer of fission, given credit on the radio and in the newspapers? Why did the papers erroneously state that Lise Meitner had discovered the splitting of uranium and stress that she was Jewish? But especially, and above all, why didn't Germany succeed in making the bomb? They looked for all sorts of excuses and rationalizations. The Nazi government, they said, would never have backed the scientists the way the Allies did. Apparently, they had conveniently forgotten the active interest the Gestapo and other government agencies had taken in their uranium work. They forgot that they themselves had not been very convinced of their own chances of success; had the conviction been there, they would probably have had all the government support they could use.

But even if we had had more government support, some of them argued, we would never have given Hitler such a devastating weapon. This may have been true of von Laue and Hahn, but it is extremely doubtful if it would apply to the rest of them. At any rate, intentions in the long run are fruitless; there are always enough co-workers who fail to see the consequences of their handiwork as clearly as the few great minds with vision and ethical scruples. After all, the atom bomb would have made Germany strong, and even if a von Laue gave his life to prevent Hitler's having the bomb, the sacrifice would have been in vain.

184

After Hiroshima and Nagasaki, Oppenheimer, who had directed the Manhattan Project, and many of his colleagues felt that science would never be the same again, that scientists had, in Oppenheimer's words, known sin. The conflict and torment to which physicists had now become subject was the stuff of drama. The Swiss playwright, Friedrich Dürrenmatt, exploited it in his remarkable play, Die Physiker *(The Physicists), which is set in a lunatic asylum, where Möbius, who has solved the problems left unsolved by Einstein, has sought refuge. He is stalked by two physicists, representatives of the Eastern and Western powers, who pretend to the delusion of being Newton and Einstein.*

EINSTEIN. Now we can only get out of the asylum if we act together.

MÖBIUS. But I don't want to run away.

EINSTEIN. Möbius –

MÖBIUS. I can't see the slightest reason for it. On the contrary. I am satisfied with my fate.

(*Silence*)

NEWTON. But I am not satisfied with it, a rather decisive consideration, don't you think? All respect to your personal feelings, but you are a genius and therefore common property. You have penetrated into new realms of physics. But you don't hold a monopoly over science. You have the duty to unlock the door to us as well, the non-geniuses. Come with me and in one year we shall put you in a frock-coat, carry you off to Stockholm and you shall have the Nobel Prize.

MÖBIUS. Your Secret Service is very selfless.

NEWTON. I admit, Möbius, that it is particularly animated by the suspicion that you may have solved the problem of gravitation.

MÖBIUS. Correct.

(*Silence*)

EINSTEIN. You say it with such composure?

MÖBIUS. How should I say it?

EINSTEIN. My Secret Service thought the unified theory of elementary particles might –

MÖBIUS. I can set their minds at rest there too. The Unified Field theory is discovered.

(*Newton dabs the sweat from his forehead with his napkin*)

EINSTEIN. And the System of All Possible Discoveries, Möbius?

MÖBIUS. That also. I set it up out of curiosity, as a practical corollary to my theoretical discoveries. Should I feign innocence? Our thoughts have their consequences. It was my duty to study the consequences that would flow from my field theory and from my theory of gravitation. The results are devastating. New, unimaginable energies would be liberated and a technology would become possible that defies the imagination, if my results were to fall into the hands of mankind.

EINSTEIN. That can scarcely be avoided.

NEWTON. The question is only who gets it first.

(*Möbius laughs*)

MÖBIUS. You want that good fortune for your Secret Service, Kilton, and for the military command that stands behind it?

NEWTON. Why not? Any high command that brings the greatest physicist of all time back into the community of physicists is blessed for me. It's a matter of the freedom of our science and nothing else.

Who guarantees this freedom is immaterial. I serve any system that leaves me alone. I know everyone talks these days about the responsibility of physicists. We are suddenly confronted by fear and we become morally aware. That is nonsense. We have to do pioneering work and nothing more. Whether mankind finds out how to follow the trail that we blaze is its affair and not ours.

EINSTEIN. Granted. We have to do pioneering work. That is my view too. But we can't exclude responsibility. We hand enormous power to mankind. That entitles us to make conditions. We have to become power-politicians because we are physicists. We must decide for whose benefit we apply our science, and I have decided. You are a wretched aesthete, Kilton. Why don't you join us if you are only concerned with the freedom of science? We also have been unable for a long time now to regulate our scientists. We too need results. Our political system also has to eat out of science's hand.

NEWTON. Both our political systems, Eisler, have to eat above all out of Möbius's hand now.

185

And at the start of the denouement, which precedes the above interchange:

NEWTON. . . . The warders compel me to act. This very day.
MÖBIUS. That's your affair.
NEWTON. Not altogether. A confession Möbius: I am not insane.
MÖBIUS. But of course not, Sir Isaac.
NEWTON. I am not Sir Isaac Newton.
MÖBIUS. I know. Albert Einstein.
NEWTON. Rubbish. Nor Herbert Georg Beutler, as they think here. My name is Kilton, my boy.

(*Möbius, startled, stares at him*)

MÖBIUS. Alec Jasper Kilton?
NEWTON. The same.
MÖBIUS. The founder of the Theory of Correspondence?
NEWTON. He.

(*Möbius approaches the table*)

MÖBIUS. You have insinuated yourself here?
NEWTON. By playing the lunatic.

186

Edward Teller, supported by the Polish mathematician, Stanislas Ulam, whom we have already encountered, was the true begetter of the hydrogen bomb ('the Super') programme. Both had been members of Oppenheimer's team at Los Alamos. Many physicists never forgave Teller for what they saw as his betrayal of Oppenheimer during the McCarthy hearings. The events took their emotional toll, as this extract from Nuel Pharr Davis's Lawrence and Oppenheimer *suggests.*

Even before the war ended, Ulam had been caught up in the conferences with Teller and the others about the Super, but in 1946 his participation had been interrupted. He showed symptoms of an unusual illness. 'When he was well, I would guess you'd put his I.Q. at about 180,' says a Mesa associate with medical connections. 'First it went down to average; he couldn't solve problems any better than I can. It kept on going down, and he seemed to have difficulty articulating his thoughts. Then lower still, and he couldn't talk at all. They took him to a hospital where he had to be fed and cared for like a baby. In spite of what people say about a brain tumor, there was absolutely no physical cause. After a while he recovered and came to the Mesa, perfectly sound, just the same as before, except that he could hold his own a little better with Teller. I don't think they liked each other too well. They got on each other's nerves. In my opinion what Teller would consider a favorable atmosphere for work would be one where everybody revolved around him.'

If what had made Ulam ill in 1946 was recoil from a glimpse of the missing idea, he was still able to keep it buried in his psyche in 1949 and for two years more. 'I don't believe he would ever have come out with it if Teller hadn't got him angry,' says Frederick Seitz. Meanwhile in the bright, shortening days of early October in 1949, Ulam kept talk about the Super centered on the unsolved problem of hydrodynamics. Excited by the White House announcement of the Red bomb, Teller fretted at Ulam's cool dispassion and wondered whether the Mesa had been the right place to come to.

187

A kind of Edward Teller of an earlier time and another place was the noted French chemist, Marcellin Berthelot. He was Professor at the

Collège de France, and during the war with Prussia he became an advisor to the Government on the defence of Paris; a Scientific Defence Committee was formed with Berthelot as its President. He supervised the manufacture of explosives (which had been an early interest of his) and guns, and became in due course President of the Committee of Explosives and a Senator. In Emile Zola's novel, Paris, *Berthelot appears as Bertheroy, the worldly old professor.*

Just as the daylight was departing, and he was sitting at his large table near the window, again checking and classifying the documents and plans connected with his invention, he was surprised to see his old master and friend Bertheroy enter the work room. The illustrious chemist called on him in this fashion at long intervals, and Guillaume felt the honour thus conferred on him by the old man to whom eminence had brought so many titles, offices and decorations. Moreover, Bertheroy, with his position as an official *savant* and member of the Institute, showed some courage in thus venturing to call on one whom so-called respectable folks regarded with contumely. On this occasion, however, Guillaume at once understood that it was some feeling of curiosity that had brought him. And so he was greatly embarrassed, for he hardly dared to remove his papers and plans which were lying on the table.

'Oh don't be frightened', gaily exclaimed Bertheroy, who, despite his careless and abrupt ways, was really very shrewd. 'I haven't come to pry into your secrets. Leave your papers there, I promise you that I won't read anything.'

Then in all frankness, he turned the conversation to the subject of explosives, which he was still studying, he said, with passionate interest. He had made some new discoveries which he did not conceal. Incidentally, too, he spoke of the opinion he had given in Salvat's affair. His dream was to discover some explosive of great power, which one might attempt to domesticate and reduce to complete obedience. And with a smile he pointedly concluded: 'I don't know where that madman found the formula of his powder. But if you should ever discover it, remember that the future perhaps lies in the employment of explosives as motive power.'

We consider now how innocents can get caught up in the toils of war. First a story from the autobiography of the physicist, cosmologist and theoretical biologist, George Gamow, My World Line.

Here is a story told to me by one of my friends who was at that time a young professor of physics in Odessa. His name was Igor Tamm (Nobel Prize laureate in Physics, 1958). Once when he arrived in a neighboring village, at the period when Odessa was occupied by the Reds, and was negotiating with a villager as to how many chickens he could get for half a dozen silver spoons, the village was captured by one of the Makhno bands, who were roaming the country, harassing the Reds. Seeing his city clothes (or what was left of them), the capturers brought him to the Ataman, a bearded fellow in a tall black fur hat with machine-gun cartridge ribbons crossed on his broad chest and a couple of hand grenades hanging on the belt.

'You son-of-a-bitch, you Communistic agitator, undermining our Mother Ukraine! The punishment is death.'

'But no,' answered Tamm. 'I am a professor at the University of Odessa and have come here only to get some food.'

'Rubbish!' retorted the leader. 'What kind of professor are you?'

'I teach mathematics.'

'Mathematics?' said the Ataman. 'All right! Then give me an estimate of the error one makes by cutting off Maclaurin's series at the nth term. Do this, and you will go free. Fail, and you will be shot!'

Tamm could not believe his ears, since this problem belongs to a rather special branch of higher mathematics. With a shaking hand, and under the muzzle of the gun, he managed to work out the solution and handed it to the Ataman.

'Correct!' said the Ataman. 'Now I see that you really are a professor. Go home!'

Who was this man? No one will ever know. If he was not killed later on, he may well be lecturing now on higher mathematics in some Ukrainian university.

The school of brilliant young mathematicians that arose miraculously in Poland between the wars was abruptly annihilated in 1939. Jews and non-Jews fled (if they were lucky), went underground or in most cases joined the army and were killed in the brief campaign or vanished into Russia and were never seen again. Here is a tribute to just one of his friends by Marc Kac in Enigmas of Chance.

In less than a year the world exploded and much of my part of it was consumed by flames. Millions, including my parents and my brother, were murdered by the Germans and many disappeared without a trace in the vastness of the Soviet Union. Wladek Hetper was one of them. In 1939, before the outbreak of the war, he received his *veniam legendi*. When hostilities started, as a reserve officer he was called to active duty and sent to the Eastern Front. I heard nothing more of him until, long after the war, an article in a London-based Polish literary journal printed excerpts from the diary of a Feliks Lachman who had spent time with Wladek in a Soviet prisoners' camp in 1940. He wrote a short, moving account of their friendship, which was interrupted by Lachman's sentencing and deportation to Siberia. Wladek, according to this account, was in poor health, which, considering what a superb physical specimen he had been, leads me to conjecture that he was suffering from malnutrition. On a visit to London I contacted Lachman and some time later (April 1974) he sent me an excerpt from his diary:

1940, December, Starobielsk   The Fourth Dimension

My acquaintance with Wladek the mathematician lasted for not longer than five weeks. Like the homeless swallow in the Norse saga, he flew into my life out of darkness, and into darkness he went away. I never met him again, neither do I know where his bones lie now. He was thirty-one when I met him, a pessimist, though a deeply religious man. He played chess marvelously and could solve difficult mathematical problems in his memory. We spent many a long and thrilling evening discussing Bertrand Russell, the principles of topology and entertaining one another with physical-mathematical puzzles and intelligence tests. We succeeded in reconstructing Cardano's solution of the cubic equation. And then, head first, we plunged into multi-dimensional worlds.

This was the last glimpse I had of my friend in a cruelly distorted mirror image of our times together in Lwów. We, too, had talked of Russell and topology and had entertained each other with all sorts of puzzles. And cubic equations! Another shadow of the past? I'll never know.

190

The chemist Primo Levi (see above, no. 145) was sent to Auschwitz. Even in the extermination camp there were chemical laboratories – in the Buna rubber factory. Levi survived the nightmare by being selected for laboratory work.

Kommando 98, called the Chemical Kommando, should have been a squad of skilled workers.

The day on which its formation was officially announced a meagre group of fifteen Häftlinge gathered in the grey of dawn around the new Kapo in the roll-call square.

This was the first disillusion: he was a 'green triangle', a professional delinquent, the *Arbeitsdienst* had not thought it necessary for the Kapo of the Chemical Kommando to be a chemist. It was pointless wasting one's breath asking him questions; he would not have replied, or else he would have replied with kicks and shouts. On the other hand, his not very robust appearance and his smaller than average stature were reassuring.

He made a short speech in the foul German of the barracks, and the disillusion was confirmed. So these were the chemists: well, he was Alex, and if they thought they were entering paradise, they were mistaken. In the first place, until the day production began, Kommando 98 would be no more than an ordinary transport-Kommando attached to the magnesium chloride warehouse. Secondly, if they imagined, being *Intelligenten*, intellectuals, that they could make a fool of him, Alex, a *Reichsdeutscher*, well, *Herrgottsakrament*, he would show them, he would . . . (and with his fist clenched and index finger extended he cut across the air with the menacing gesture of the Germans); and finally, they should not imagine that they would fool anyone, if they had applied for the position without any qualifications – an examination, yes gentlemen, in the very near future; a chemistry examination, before the triumvirate of the Polymerization Department: Doktor Hagen, Doktor Probst and Doktor Ingenieur Pannwitz.

And with this, *meine Herren*, enough time had been lost, Kommandos 96 and 97 had already started, forward march, and to begin with, whosoever failed to walk in line and step would have to deal with him.

He was a Kapo like all the other Kapos.

Leaving the camp, in front of the musical band and the SS counting-post we march in rows of five, beret in hand, arms hanging down our sides and neck rigid; speaking is forbidden. Then we change to threes and it is possible to exchange a few words amidst the clatter of ten thousand pairs of wooden shoes.

Who are my new comrades? Next to me walks Alberto; he is in his third year at university, and once again we have managed to stay together. The third person on my left I have never seen; he seems very young, is as pale as wax and has the number of the Dutch. The three backs in front of me are also new. It is dangerous to turn around, I might lose step or stumble; but I try for a moment, and see the face of Iss Clausner.

So long as one walks there is no time to think, one has to take care not to step on the shoes of the fellow hobbling in front, and not let them be stepped on by the fellow behind; every now and again there is a hole to be walked over, an oily puddle to be avoided. I know where we are, I have already come here with my preceding Kommando, it is the H-Strasse, the road of the stores, I tell Alberto, we are really going to the magnesium chloride warehouse, at least that was not a lie.

We have arrived, we climb down into a large damp cellar, full of draughts; this is the headquarters of the Kommando, the Bude as it is called here. The Kapo divides us into three squads: four to unload the sacks from the wagon, seven to carry them down, four to pile them up in the deposit. We form the last squad, I, Alberto, Iss and the Dutchman.

At last we can speak, and to each one of us what Alex said seems a madman's dream.

With these empty faces of ours, with these sheared craniums, with these shameful clothes, to take a chemical examination. And obviously it will be in German; and we will have to go in front of some blond Aryan doctor hoping that we do not have to blow our noses, because perhaps he will not know that we do not have handkerchiefs, and it will certainly not be possible to explain to him. And we will have our old comrade hunger with us, and we will hardly be able to stand still on our feet, and he will certainly smell our odour, to which we are by now accustomed, but which persecuted us during the first days, the odour of turnips and cabbages, raw, cooked and digested.

Exactly so, Clausner confirms. But have the Germans such great need of chemists? Or is it a new trick, a new machine '*pour faire chier les Juifs*'? Are they aware of the grotesque and absurd test asked of us, of us who are no longer alive, of us who have already gone half-crazy in the dreary expectation of nothing?

Clausner shows me the bottom of his bowl. Where others have carved their numbers, and Alberto and I our names, Clausner has written: '*Ne pas chercher à comprendre.*'

Although we do not think for more than a few minutes a day, and then in a strangely detached and external manner, we well know that we will end in selections. I know that I am not made of the stuff of those who resist, I am too civilized, I still think too much, I use myself up at work. And now I also know that I can save myself if I become a Specialist, and that I will become a Specialist if I pass a chemistry examination.

Today, at this very moment as I sit writing at a table, I myself am not convinced that these things really happened.

Three days passed, three of those usual immemorable days, so long while they are passing, and so short afterwards, and we were already all tired of believing in the chemistry examination.

The Kommando was reduced to twelve men: three had disappeared in the customary manner of down there, perhaps into the hut next door, perhaps cancelled from this world. Of the twelve, five were not chemists; all five had immediately requested permission from Alex to return to their former Kommandos. They were given a few kicks, but unexpectedly, and by who knows whose authority, it was decided that they should remain as auxiliaries to the Chemical Kommando.

Down came Alex into the magnesium chloride yard and called us seven out to go and face the examination. We go like seven awkward chicks behind the hen, following Alex up the steps of the *Polimerisations-Büro*. We are in the lobby, there is a brass-plate on the door with the three famous names. Alex knocks respectfully, takes off his beret and enters. We can hear a quiet voice; Alex comes out again. '*Ruhe, jetzt. Warten,*' wait in silence.

We are satisfied with this. When one waits time moves smoothly without need to intervene and drive it forward, while when one works, every minute moves painfully and has to be laboriously driven away. We are always happy to wait; we are capable of waiting for hours with the complete obtuse inertia of spiders in old webs.

Alex is nervous, he walks up and down and we move out of his way each time. We too, each in our own way, are uneasy; only Mendi is not. Mendi is a rabbi; he comes from sub-Carpathian Russia, from that confusion of peoples where everyone speaks at least three languages, and Mendi speaks seven. He knows a great number of things; besides being a rabbi, he is a militant Zionist, a comparative philologist, he has been a partisan and a lawyer; he is not a chemist, but he wants to try all the same, he is a stubborn, courageous, keen little man.

Balla has a pencil and we all crowd around him. We are not sure if we still know how to write, we want to try.

Kohlenwasserstoffe, Massenwirkungsgesetz. The German names of compounds and laws float back into my memory. I feel grateful towards my brain: I have not paid much attention to it, but it still serves me so well.

Here is Alex. I am a chemist. What have I to do with this man Alex? He plants his feet in front of me, he roughly adjusts the collar of my jacket, he takes out my beret and slaps it on my head, then he steps backwards, eyes the result with a disgusted air, and turns his back, muttering: '*Was für ein Muselmann Zugang.*' What a messy recruit!

The door opens. The three doctors have decided that six candidates will be examined in the morning. The seventh will not. I am the seventh, I have the highest entry number, I have to return to work. Alex will only come to fetch me in the afternoon. What ill-luck, I cannot even talk to the others to hear what questions they are asking.

This time it really is my turn. Alex looks at me blackly on the doorstep; he feels himself in some way responsible for my miserable appearance. He dislikes me because I am Italian, because I am Jewish and because of all of us, I am the one furthest from his sergeants' mess ideal of virility. By analogy, without understanding anything, and proud of this very ignorance, he shows a profound disbelief in my chances for the examination.

We have entered. There is only Doktor Pannwitz; Alex, beret in hand, speaks to him in an undertone: '. . . an Italian, has been here only three months, already half kaputt . . . *Er sagt er ist Chemiker . . .*' But he, Alex, apparently has his reservations on the subject.

Alex is briefly dismissed and put aside, and I feel like Oedipus in front of the Sphinx. My ideas are clear, and I am aware even at this moment that the position at stake is important; yet I feel a mad desire to disappear, not to take the test.

Pannwitz is tall, thin, blond; he has eyes, hair and nose as all Germans ought to have them, and sits formidably behind a complicated writing-table. I, Häftling 174517, stand in his office, which is a real office, shining, clean and ordered, and I feel that I would leave a dirty stain whatever I touched.

When he finished writing, he raised his eyes and looked at me.

From that day I have thought about Doktor Pannwitz many times and in many ways. I have asked myself how he really functioned as a man; how he filled his time, outside of the Polymerization and the Indo-Germanic conscience; above all when I was once more a free man, I wanted to meet him again, not from a spirit of revenge, but merely from a personal curiosity about the human soul.

Because that look was not one between two men; and if I had known how completely to explain the nature of that look, which came as if across the glass window of an aquarium between two beings who live in different worlds, I would also have explained the essence of the great insanity of the third Germany.

One felt in that moment, in an immediate manner, what we all thought and said of the Germans. The brain which governed those blue eyes and those manicured hands said: 'This something in front of me belongs to a species which it is obviously opportune to suppress. In this particular case, one has to first make sure that it does not contain some utilizable element.' And in my head, like seeds in any empty pumpkin: 'Blue eyes and fair hair are essentially wicked. No communication possible. I am a specialist in mine chemistry. I am a specialist in organic syntheses. I am a specialist . . .'

And the interrogation began, while in the corner that third zoological specimen, Alex, yawned and chewed noisily.

'*Wo sind Sie geboren?*' He addresses me as *Sie*, the polite form of address: Doktor Ingenieur Pannwitz has no sense of humour. Curse him, he is not making the slightest effort to speak a slightly more comprehensible German.

I took my degree at Turin in 1941, *summa cum laude* – and while I say it I have the definite sensation of not being believed, of not even believing it myself; it is enough to look at my dirty hands covered with sores, my convict's trousers encrusted with mud. Yet I am he, the B.Sc. of Turin, in fact, at this particular moment it is impossible to doubt my identity with him, as my reservoir of knowledge of organic chemistry, even after so long an inertia, responds at request with unexpected docility. And even more, this sense of lucid elation, this excitement which I feel warm in my veins, I recognize it, it is the fever of examinations, *my* fever of *my* examinations, that spontaneous mobilization of all my logical faculties and all my knowledge, which my friends at university so envied me.

The examination is going well. As I gradually realize it, I seem to grow in stature. He is asking me now on what subject I wrote my degree thesis. I have to make a violent effort to recall that sequence of memories, so deeply buried away: it is as if I was trying to remember the events of a previous incarnation.

Something protects me. My poor old 'Measurements of dielectrical constants' are of particular interest to this blond Aryan who lives so safely: he asks me if I know English, he shows me Gatterman's book, and even this is absurd and impossible, that down here, on the other side of the barbed wire, a Gatterman should exist, exactly similar to the one I studied in Italy in my fourth year, at home.

Now it is over: the excitement which sustained me for the whole of the test suddenly gives way and, dull and flat, I stare at the fair skin of his hand writing down my fate on the white page in incomprehensible symbols.

'*Los, ab!*' Alex enters the scene again, I am once more under his jurisdiction. He salutes Pannwitz, clicking his heels, and in return receives a faint nod of the eyelids. For a moment I grope around for a suitable formula of leave-taking: but in vain. I know how to say to eat, to work, to steal, to die in German; I also know how to say sulphuric acid, atmospheric pressure, and short-wave generator, but I do not know how to address a person of importance.

Here we are again on the steps. Alex flies down the stairs: he has leather shoes because he is not a Jew, he is as light on his feet as the devils of Malabolge. At the bottom he turns and looks at me sourly as I walk down hesitantly and noisily in my two enormous unpaired wooden shoes, clinging on to the rail like an old man.

It seems to have gone well, but I would be crazy to rely on it. I already know the Lager well enough to realize that one should never anticipate, especially optimistically. What is certain is that I have spent a day without working, so that tonight I will have a little less hunger, and this is a concrete advantage, not to be taken away.

To re-enter Bude, one has to cross a space cluttered up with piles of cross-beams and metal frames. The steel cable of a crane cuts across the road, and Alex catches hold of it to climb over: *Donnerwetter*, he looks at his hand black with thick grease. In the meanwhile I have joined him. Without hatred and without sneering, Alex wipes his hand on my shoulder, both the palm and the back of the hand, to clean it; he would be amazed, the poor brute Alex, if someone told him that today, on the basis of this action, I judge him and Pannwitz and the innumerable others like him, big and small, in Auschwitz and everywhere.

191

The Cold War: June Goodfield's immunological spy story, Courier to Peking, *contains many recognisable characters, including one who is obviously P. B. Medawar.*

'Let me try again. Suppose you wanted to send a simple message, using a sophisticated but foolproof technique. You could use the immunological principle. You could make a different combination of numbers stand for whatever message you wanted to pass – like those servicemen's cables in wartime. Nine antigens, say, tabulated, could stand – in various combinations – for more than five hundred different messages. All you'd have to do would be to have your courier injected with the relevant ones – and God knows what would be easy enough.

'Why a routine vaccination visit would provide you with a fine opportunity. Off the man goes and his body immediately begins making the antibodies, and they don't fade. Once he gets to the other end you prick his finger – jab him by mistake, somehow. You need only a couple of drops of blood, and very quickly and accurately you can tell which combination of your nine antigens was present in the original injection.

'Of course, if you want to be really sophisticated, then you can do a quantitative test. It's known as Farr's technique. But that's a reasonably expert job, which can't be done at the kitchen sink. However any amateur can do the Ouchterlony test, which simply depends on diffusing the substance through gelatine and looking for a reaction line.'

Trust him to rub it in, thought Edds grimly at the words 'any amateur.' By now he knew. They were in trouble.

In a crisis, Edds always reacted the same way. He became cooler and more detached; his mind worked more and more clearly. Unfortunately, the same was true of the other two. Whereas, by now, the strain might have caused other people to break, all three were still in command of themselves. Only the pitch of tension continued to mount, higher and higher, until it was almost screeching. But externally no one gave any sign.

As usual, Edds's mind had been working on two levels at once. At the beginning he had wondered whether Tanyard was only acting in his usual, self-important fashion, taking it for granted that his country's intelligence agency would find it useful to get his own first-hand impressions of China and Chinese science. But, he realized that the inconceivable had happened. The technique, with its three strands separately fed into the situation, should have been foolproof even to a trained immunologist like Tanyard: so something, somewhere, must have aroused his suspicions. And unless he was very careful it could be embarrassing. But whatever had happened his best line was total incomprehension, since he knew that Tanyard had no evidence that could actually stand up under cross-examination. So with a show of impatience, he shuffled his papers and said, in a dismissive tone, 'Well Andy, we're both of us busy men. What you have been saying could well be valuable. I'd be glad to have one of our technical people drop in to see you sometime in the next few weeks, if you could spare an hour or so . . .'

Tanyard acted as though he hadn't heard a word. He was still concentrating on Groover. The man wanted to know, didn't he? All right, then he'd tell him in a manner that would hurt.

'The sweetness of the method is its simplicity,' he reiterated. 'It's a code that would be almost impossible to crack. If the other side wants to break it, they have to know who the courier is – whose blood to go for. Then they'd have to know the nine antigens whose antibodies you had to look for. And thirdly, they'd have to know what message the combinations stood for. It would be quite remarkable if any outsider, any counterspy, if you will, could pick them all up, for you'd surely feed them all in separately. It's as near one hundred percent secure as it could be.'

Groover followed all the way, as Tanyard knew he would.

The Pathology of Science

Here are collected some examples of obsessives – the men with staring eyes – of people blinkered by dogma, political or religious, of plagiarists, fakers and cheats, and of the incompetent or merely unlucky.

We begin with Martin Gardner, the best of all writers about perversions of science, giving us a rundown on some categories of crank in Fads and Fallacies in the Name of Science.

The modern pseudo-scientist . . . stands entirely outside the closely integrated channels through which new ideas are introduced and evaluated. He works in isolation. He does not send his findings to the recognized journals, or if he does, they are rejected for reasons which in the vast majority of cases are excellent. In most cases the crank is not well enough informed to write a paper with even a surface resemblance to a significant study. As a consequence, he finds himself excluded from the journals and societies, and almost universally ignored by the competent workers in his field. In fact, the reputable scientist does not even know of the crank's existence unless his work is given widespread publicity through non-academic channels, or unless the scientist makes a hobby of collecting crank literature. The eccentric is forced, therefore, to tread a lonely way. He speaks before organizations he himself has founded, contributes to journals he himself may edit, and – until recently – publishes books only when he or his followers can raise sufficient funds to have them printed privately.

A second characteristic of the pseudo-scientist, which greatly strengthens his isolation, is a tendency toward paranoia. This is a mental condition (to quote a recent textbook) 'marked by chronic, systematized, gradually developing delusions, without hallucinations, and with little tendency toward deterioration, remission or recovery.' There is wide disagreement among psychiatrists about the causes of paranoia. Even if this were not so, it obviously is not within the scope of this book to discuss the possible origins of paranoid traits in individual cases. It is easy to understand, however, that a strong sense of personal greatness must be involved whenever a crank stands in solitary bitter opposition to every recognized authority in his field.

If the self-styled scientist is rationalizing strong religious convictions, as often is the case, his paranoid drives may be reduced to a minimum. The desire to bolster religious beliefs with science can be a powerful motive. For example, in the case of George McCready Price, the greatest of modern opponents of evolution, it is clear that his devout faith in Seventh Day Adventism is a sufficient explanation for his curious geological views. But even in such cases, an element of paranoia is nearly always present. Otherwise the pseudo-scientist would lack the stamina to fight a vigorous, single-handed battle against such overwhelming odds. If the crank is insincere – interested only in making money, playing hoax, or both – then obviously paranoia need not enter his make-up. However, very few cases of this sort will be considered.

There are five ways in which the sincere pseudo-scientist's paranoid tendencies are likely to be exhibited.

(1) He considers himself a genius.

(2) He regards his colleagues, without exception, as ignorant blockheads. Everyone is out of step except himself. Frequently he insults his opponents by accusing them of stupidity, dishonesty, or other base motives. If they ignore him, he takes this to mean his arguments are unanswerable. If they retaliate in kind, this strengthens his delusion that he is battling scoundrels.

Consider the following quotation: 'To me truth is precious. . . . I should rather be right and stand alone than to run with the multitude and be wrong. . . . The holding of the views herein set forth has already won for me the scorn and contempt and ridicule of some of my fellow men. I am looked upon as being odd, strange, peculiar . . . But truth is truth and though all the world reject it and turn against me, I will cling to truth still.'

These sentences are from the preface of a booklet, published in 1931, by Charles Silvester de Ford, of Fairfield, Washington, in which he proves the earth is flat. Sooner or later, almost every pseudo-scientist expresses similar sentiments.

(3) He believes himself unjustly persecuted and discriminated against. The recognized societies refuse to let him lecture. The journals reject his papers and either ignore his books or assign them to 'enemies' for review. It is all part of a dastardly plot. It never occurs to the crank that this opposition may be due to error in his work. It springs solely, he is convinced, from blind prejudice on the part of the established hierarchy – the high priests of science who fear to have their orthodoxy overthrown.

Vicious slanders and unprovoked attacks, he usually insists, are constantly being made against him. He likens himself to Bruno, Galileo, Copernicus, Pasteur, and other great men who were unjustly persecuted for their heresies. If he has had no formal training in the field in which he works, he will attribute this persecution to a scientific masonry, unwilling to admit into its inner sanctums anyone who has not gone through the proper initiation rituals. He repeatedly calls your attention to important scientific discoveries made by laymen.

(4) He has strong compulsions to focus his attacks on the greatest scientists and the best-established theories. When Newton was the outstanding name in physics, eccentric works in that science were violently anti-Newton. Today, with Einstein the father-symbol of authority, a crank theory of physics is likely to attack Einstein in the name of Newton. This same defiance can be seen in a tendency to assert the diametrical opposite of well-established beliefs. Mathematicians prove the angle cannot be trisected. So the crank trisects it. A perpetual motion machine cannot be built. He builds one. There are many eccentric theories in which the 'pull' of gravity is replaced by a 'push.' Germs do not cause disease, some modern cranks insist. Disease produces the germs. Glasses do not help the eyes, said Dr. Bates. They make them worse. In our next chapter we shall learn how Cyrus Teed literally turned the entire cosmos inside-out, compressing it within the confines of a hollow earth, inhabited only on the inside.

(5) He often has a tendency to write in a complex jargon, in many cases making use of terms and phrases he himself has coined. Schizophrenics sometimes talk in what psychiatrists call 'neologisms' – words which have meaning to the patient, but sound like Jabberwocky to everyone else. Many of the classics of crackpot science exhibit a neologistic tendency.

When the crank's I.Q. is low, as in the case of the late Wilbur Glenn Voliva who thought the earth shaped like a pancake, he rarely achieves much of a following. But if he is a brilliant thinker, he is capable of developing incredibly complex theories. He will be able to defend them in books of vast erudition, with profound observations, and often liberal portions of sound science. His rhetoric may be enormously persuasive. All the parts of his world usually fit together beautifully, like a jig-saw puzzle. It is impossible to get the best of him in any type of argument. He has anticipated all your objections. He counters them with unexpected answers of great ingenuity. Even of the subject on the shape of the earth, a layman may find himself powerless in a debate with a flat-earther. George Bernard Shaw, in *Everybody's Political What's What*, gives an hilarious description of a meeting at which a flat-earth speaker completely silenced all opponents who raised objections from the floor. 'Opposition such as no atheist could have provoked assailed him'; writes Shaw, 'and he, having heard their arguments hundreds of times, played skittles with them, lashing the meeting into a spluttering fury as he answered easily what it considered unanswerable.'

193

Such monomaniacs have all too often (especially in medicine) been unloosed on the world. The proponents of eugenics in the first half of this century are a good example: the conviction that prevailed in some such circles, for example, that criminal behaviour was genetically determined, led in several states of the USA to the sterilisation of numerous criminals and even of their families (who had committed no offence). Here is another example from Denis Hamilton's The Monkey Gland Affair.

Dr Stanley's experience of gland-grafting at San Quentin prison appeared in the journal *Endocrinology*, where he reported impressive results in 643 inmates who have been given grafts or testicular extracts for a surprising range of problems. In addition, it was reported that thirteen physicians had received this type of treatment from Dr Stanley. According to Stanley, his ageing recipients were rejuvenated, and gland-grafting also improved acne and asthma. Stanley's report seemed particularly credible, since his work did not suffer the same disadvantages as that of other gland transplanters, namely that follow-up of their patients was usually poor and was accompanied by changes in diet and habits. Instead, Stanley's follow-up of his incarcerated patients was inevitably good. The prison environment was unchanging; in particular, the grafted patients had no change of food. While Voronoff wisely encouraged rest and temperance for his weary, rich, private patients, which helped the alleged rejuvenation effect, such abstinence has always been the lot of the lifer. But there was unspoken pressure on the prisoners to produce the expected result: parole or payments could be the reward of such experiments.

And Martin Gardner again on one of the most egregious twentieth-century cranks, who retains a following to this day.

It would be out of place to describe here at any greater length Reich's early contributions to psychiatric theory. Many of them are complex and technical, and in order to be understood, would require a mastery of the elaborate and cumbersome Reichian terminology. What has been said, however, should give a faint indication of the importance of the topics which Reich tackled courageously during the German phase of his career.

From this point onward, you may take your choice of one of three possible interpretations of Reich's development. (1) He became the world's greatest biophysicist. (2) He deteriorated from a competent psychiatrist into a self-deluded crank. (3) He merely switched to fields in which his former incompetence became more visible. Critics who favor the last view point out that psychoanalysis is still in such a confused pioneer state that writings by incompetent theorists are easily camouflaged by technical jargon and a sprinkling of sound ideas borrowed from others. When Reich turned to biology, physics, and astronomy – where there is a solid core of verifiable knowledge – his eccentric thinking became easier to detect.

Whatever the correct explanation may be, there is no doubt about the great turning point in Reich's career. It came in the late thirties when he discovered, in Norway, the existence of 'orgone energy.' Freud had earlier expressed the hope that some day his theory of the libido, or sexual energy, might be given a biological basis. Reich is convinced that his discovery of orgone energy fulfilled this hope – a discovery which he ranks in importance with the Copernican Revolution. Since coming to America, he has considered himself less a psychiatrist than a biophysicist, probing deeper into the mysteries of orgone energy, and applying this strange new knowledge to the treatment of bodily and mental ailments.

Exactly what is orgone energy? According to Reich it is a non-electro-magnetic force which permeates all of nature. It is the *élan vital* or life force, of Bergson, made practically accessible and usable. It is blue in color. To quote from one of Reich's booklets, 'Blue is the specific color of orgone energy within and without the organism. Classical physics tries to explain the blueness of the sky by the scattering of the blue and of the spectral color series in the gaseous atmosphere. However, it is a fact that blue is the color seen in all functions which are related to the cosmic or atmospheric or organismic orgone energy.' Protoplasm, says Reich, is blue with orgone energy, and loses its blueness when the cell dies. Orgone also causes the blue of oceans and deep lakes, and the blue coloration of certain frogs when they are sexually excited. 'The color of luminating, decaying wood is blue; so are the luminating tail ends of glowworms, St. Elmo's fire, and the aurora borealis. The lumination in evacuated tubes charged with orgone energy is blue.' (The latter has been photographed on color film and forms the cover photo of the booklet from which the above quotations are taken.)

The so-called 'heat waves' you often see shimmering above roads and mountain tops, are not heat at all, Reich declares, but orgone energy. These waves do not ascend. They move from west to east, at a speed faster that the earth's rotation. They cause the twinkling of stars. All phenomena which orthodox physicists attribute to 'static electricity' are produced by orgone energy – e.g., electric disturbances during sunspot activity, lightning, radio interference, and all other forms of static discharges. 'Cloud formations and thunderstorms,' he writes, '– phenomena which to date have remained unexplained – depend on changes in the concentration of atmospheric orgone.' That is why thunder clouds and hurricanes are deeply blue. 'One of the hurricanes which was personally experienced by the writer [Reich] in 1944 was of a deep blue-black color.' In an article in the *Orgone Energy Bulletin*, July 1951, Reich reports on some experiments made by himself which prove that dowsing rods operate by orgone energy!

In the human body, orgone is the basis of sexual energy. It is the *id* of Freud in a bio-energetic, concrete form. During coitus it becomes concentrated in the sexual parts. During orgasm, it flows back again through the entire body. By breathing, the body charges its red blood cells with orgone energy. Under the microscope, Reich has detected the 'blue glimmer' of red corpuscles as they absorb orgone. In 1947, he measured orgone energy with a Geiger counter. A recent film produced by his associates demonstrates how motors may some day be run by orgone energy.

The unit of living matter, Reich tells us, is not the cell but something much smaller which he calls the 'bion' or 'energy vesicle.' It consists of a membrane surrounding a liquid, and pulsates continually with orgone energy. This pulsation is the dance of life – the basic convulsive rhythm of love which finds its highest expression in the pulsation of the 'orgasm formula.' Bions propagate like bacteria. In fact, Reich's critics suspect, what he calls bions really *are* bacteria.

According to Reich, bions are constantly being formed in nature by the disintegration of both organic and inorganic substances. The bions first group themselves into clumps, then they organize into protozoa! In Reich's book, *The Cancer Biopathy*, 1948, are a series of photomicrographs showing various types of single-cell animals, such as amoebae and paramecia, in the process of formation from aggregates of bions.

Needless to say, no 'orthodox' biologist has been able to duplicate these revolutionary experiments. The opinion of bacteriologists who have troubled to look at Reich's photographs is that his protozoa found their way into his cultures from the air, or were already present on the disintegrating material in the form of dormant cysts. Reich is aware of these objections, of course, and vigorously denies that protozoa could have gotten into his cultures in any other than the way he describes.

In 1940, Reich invented a therapeutic box. Technically called an Orgone Energy Accumulator, it consists of a structure resembling a short phone booth, made of sheet iron on the inside, and organic material (wood or celotex) on the outside. Later, three to twenty-layer accumulators were made of alternate layers of steel wool and rock wool. The theory is that orgone energy is attracted by the organic substance on the outside, and is passed on to the metal which then radiates it inward. Since the metal reflects orgone, the box soon acquires an abnormally high concentration of the energy. In Reich's laboratory in Maine, he has a large 'orgone room' lined with sheet iron. When all the lights are turned out, he claims, the room takes on a blue-gray luminescence.

According to Dr. Theodore P. Wolfe, Reich's former translator, 'The Orgone Energy Accumulator is the most important single discovery in the history of medicine, bar none.' In 1951, Reich issued a booklet (there is no author's name on the title page) called *The Orgone Energy Accumulator*, which is the best available reference on the accumulator's construction and medical use. Most of the following material is taken from this work.

Orgone accumulators can be bought, but the Foundation holds rights to their medical use, and rents them to patients on a monthly basis, the charge varying with ability to pay. By sitting inside, lightly clothed, you charge your body with orgone energy. At first you feel a prickling, warm sensation, accompanied by reddening of the face and a rise in body temperature. There is a feeling that the body is 'glowing.' After you have absorbed as much orgone as your system demands, you begin to feel a slight dizziness and nausea. When this happens, you step out of the accumulator, breathe some fresh air, and the overcharge symptoms quickly vanish. 'Under no circumstances,' Reich's booklet reads, 'should one sit in the accumulator for hours, or, as some people do, go to sleep in it. This can cause serious damage (severe vomiting, etc.). It is better, if necessary, to use the accumulator several times a day at shorter intervals than to prolong one sitting unnecessarily. At this stage of research, no accumulator over 3-layers should be used without medical supervision.'

For people who are bedridden, Reich has developed an 'orgone energy accumulator blanket.' This is a curved structure which can be placed on the bed, over a reclining figure, while a set of flat layers goes beneath the mattress. There are also tiny orgone boxes called 'shooters,' for application to local areas. A flexible iron cable, from which the inner wires have been removed, carries the energy from the box to the part of the body being irradiated. If the body area is larger than the end of the cable, a funnel is attached. 'Only *metal* (iron) funnels can be used,' the booklet warns, 'funnels made of plastic are ineffective.'

It is Reich's belief that the natural healing process of a wound is greatly accelerated by applying the shooter. 'Even severe pain will be stopped soon after the accident if orgone energy is applied locally through the shooter,' the booklet states. 'In severe cases of burns, experience has revealed the amazing fact that no blisters appear, and that the initial redness slowly disappears. The wounds heal in a matter of a few hours; severe ones need a day or two. Only chronic advanced degenerating processes require weeks and months of daily irradiation. But here, too, severe lesions, as for instance *ulcus varicosus*, will yield to orgone energy irradiation.'

In addition to speeding up healing, the energy also sterilizes a wound. 'Microscopic observation shows that, for example, bacteria in the vagina will be immobilized after only one minute of irradiation through an inserted glass pipe filled with steel wool . . . *Do not mix orgone irradiation with other, chemical applications. Orgone energy is a very strong energy*. We do not know as yet what such a mixture can do.' (Italics his.)

Edmund Gosse wrote his classic memoir of his father, Phillip, and of his own early life, Father and Son, *in 1907. Phillip Gosse was a Victorian zoologist of some standing; that he took his profession seriously is clear from an entry in his Diary which read: 'E. delivered of a son. Green swallow received from Jamaica.' The discoveries of Darwin (whom he admired) and Lyell presented him with some agonising problems. As an austere Christian fundamentalist he felt it necessary to reconcile the fossil record with the creation. In his book,* Omphalos *(the navel) he treated, among many other intellectual conundrums, the difficult question of whether Adam could have possessed such an organ, like his descendants born of woman. Here Edmund looks back on his father's ordeal by ridicule.*

My Father had never admired Sir Charles Lyell. I think that the famous 'Lord Chancellor manner' of the geologist intimidated him, and we undervalue the intelligence of those whose conversation puts us at a disadvantage. For Darwin and Hooker, on the other hand, he had a profound esteem, and I know not whether this had anything to do with the fact that he chose, for his impetuous experiment in reaction, the field of geology, rather than that of zoology or botany. Lyell had been threatening to publish a book on the geological history of Man, which was to be a bomb-shell flung into the camp of the catastrophists. My Father, after long reflection, prepared a theory of his own, which, as he fondly hoped, would take the wind out of Lyell's sails, and justify geology to godly readers of 'Genesis.' It was, very briefly, that there had been no gradual modification of the surface of the earth, or slow development of organic forms, but that when the catastrophic act of creation took place, the world presented, instantly, the structural appearance of a planet on which life had long existed.

The theory, coarsely enough, and to my Father's great indignation, was defined by a hasty press as being this – that God hid the fossils in the rocks in order to tempt geologists into infidelity. In truth, it was the logical and inevitable conclusion of accepting, literally, the doctrine of a sudden act of creation; it emphasised the fact that any breach in the circular course of nature could be conceived only on the supposition that the object created bore false witness to past processes, which had never taken place. For instance, Adam would certainly possess hair and teeth and bones in a condition which it must have taken many years to accomplish, yet he was created full-grown yesterday. He would certainly – though Sir Thomas Browne denied it – display an *omphalos*, yet no umbilical cord had ever attached him to a mother.

Never was a book cast upon the waters with greater anticipations of success than was this curious, this obstinate, this fanatical volume. My Father lived in a fever of suspense, waiting for the tremendous issue. This 'Omphalos' of his, he thought, was to bring all the turmoil of scientific speculation to a close, fling geology into the arms of Scripture, and make the lion eat grass with the lamb. It was not surprising, he admitted, that there had been experienced an ever-increasing discord between the facts which geology brings to light and the direct statements of the early chapters of 'Genesis.' Nobody was to blame for that. My Father, and my Father alone, possessed the secret of the enigma; he alone held the key which could smoothly open the lock of geological mystery. He offered it, with a glowing gesture, to atheists and Christians alike. This was to be the universal panacea; this the system of intellectual therapeutics which could not but heal all the maladies of the age. But, alas! atheists and Christians alike looked at it and laughed, and threw it away.

In the course of that dismal winter, as the post began to bring in private letters, few and chilly, and public reviews, many and scornful, my Father looked in vain for the approval of the churches, and in vain for the acquiescence of the scientific societies, and in vain for the gratitude of those 'thousands of thinking persons,' which he had rashly assured himself of receiving. As his reconciliation of Scripture statements and geological deductions was welcomed nowhere; as Darwin continued silent, and the youthful Huxley was scornful, and even Charles Kingsley, from whom my Father had expected the most instant appreciation, wrote that he could not 'give up the painful and slow conclusion of five and twenty years' study of geology, and believe that God has written on the rocks one enormous and superfluous lie,' – as all this happened or failed to happen, a gloom, cold and dismal, descended upon our morning tea cups. It was what the poets mean by an 'inspissated' gloom; it thickened day by day, as hope and self-confidence evaporated in thin clouds of disappointment. My Father was not prepared for such a fate. He had been the spoiled darling of the public, the constant favourite of the press, and now, like the dark angels of old,

so huge a rout
Encumbered him with ruin.

He could not recover from amazement at having offended everybody by an enterprise which had been undertaken in the cause of universal reconciliation.

196

Geology and paleontology evidently troubled many honest citizens, including Flaubert's fervid pair, Bouvard and Pécuchet.

All the same, the end of the world, remote as it might be, made them despondent, and they walked side by side silently over the pebbles.

The cliff, perpendicular, all white, streaked in black here and there by bands of flint, ran off towards the horizon, like the curve of a rampart twelve miles long. A bitter cold east wind was blowing. The sky was grey, the sea greenish and looking almost swollen. From the top of the rocks birds flew up, wheeled, quickly came back to their holes. Sometimes a stone came loose, and bounced from place to place before coming down to them.

Pécuchet continued thinking aloud:

'Unless the earth is annihilated by a cataclysm! No one knows the length of our period. The central fire has only to go out of control.'

'But it is diminishing?'

'That has not stopped eruptions producing the island of Julia, Monte Nuovo, and many others too.' Bouvard remembered reading these details in Bertrand.

'But such cataclysms do not occur in Europe?'

'I'm very sorry, but look at the Lisbon one. As for our part of the world, there are a large number of coal and iron pyrite mines which could easily decompose and form volcanic outlets. In any case volcanoes always erupt near the sea.'

Bouvard looked out over the waves, and thought he could make out a distant plume of smoke going up towards the sky.

'Since the island of Julia,' replied Pécuchet, 'has disappeared, perhaps formations produced by the same cause will go the same way. One of the islands of the Archipelago is as important as Normandy, even Europe.'

Bouvard imagined Europe swallowed up in an abyss.

'Suppose,' said Pécuchet, 'that an earthquake takes place under the Channel; the water rushes into the Atlantic; the coasts of France and England totter on their bases, lean over, join up and presto! Everything between is crushed!'

Instead of answering Bouvard began to walk so fast that he was soon a hundred yards ahead of Pécuchet. Being alone he was disturbed by the idea of a cataclysm. He had not eaten since the morning; there was a throbbing in his temples. Suddenly the soil seemed to shudder and the top of the cliff above his head seemed to lean over. At that moment a shower of gravel rolled down from above.

Pécuchet saw him in headlong flight, understood his terror and cried from afar:

'Stop! Stop! The period is not over!'

Trying to catch him up he took enormous bounds, with his tourist's staff, bellowing the while: 'The period is not over! The period is not over!'

Bouvard, quite beside himself, kept running. The multi-branched umbrella fell, the skirts of his coat went flying, the haversack bounced on his back. He looked like a winged tortoise galloping among the rocks; one of the larger ones hid him from sight.

Pécuchet arrived there out of breath, saw no one, then turned back to reach the fields by a little valley which Bouvard had no doubt taken.

The narrow path was cut out in large steps up the cliff, the width of two men, gleaming like polished alabaster.

When he was fifty feet up Pécuchet wanted to come down. As the tide was on the flood he began scrambling.

At the second turning, when he saw the void, he froze with fear. As he came nearer the third his legs went limp. The layers of air vibrated round him, he was seized by cramp in the pit of his stomach; he sat down, with his eyes closed, no longer conscious of anything but the heartbeats choking him; then he threw away his tourist staff, and started climbing up again on hands and knees. But the three hammers fastened to his belt stuck into his stomach; the pebbles stuffed into his pockets banged his ribs; the peak of his cap blinded him; the wind became twice as strong. Finally he reached level ground and there found Bouvard, who had come up further along, by a less difficult valley.

A cart picked them up. They forgot about Étretat.

Next evening, at Le Havre, as they were waiting for the boat, they saw at the bottom of the newspaper a serial article entitled: 'On Teaching Geology'.

This article, full of facts, set out the question as it was understood at the time:

There had never been a complete global cataclysm, but a given species does not always last the same length of time, and becomes more quickly extinct in one place than another. Formations of the same age contain different fossils, just as widely distant formations contain similar ones. Ferns of past times are identical with those of the present. Many contemporary zoophytes are to be found in the most ancient strata. In brief, present modifications explain previous upheavals. The same causes always operate, there are no jumps in nature, and periods, Brongniart asserts, are after all only abstractions.

Up till then Cuvier had appeared to them in the radiance of a halo, at the height of an incontrovertible science. This was now sapped. Creation no longer showed the same discipline, and their respect for this great man decreased.

Through biographies and extracts they learned about the doctrines of Lamarck and Geoffroy de Saint-Hilaire.

All this went against received ideas, the authority of the Church. Bouyard had a sense of relief as though a yoke had been broken.

197

One whose beliefs were proof against the encroachments of science and rationalism was G. K. Chesterton. The works of man had to be put in their place. 'A man looking at a hippopotamus', he once observed, 'may sometimes be tempted to regard a hippopotamus as an enormous mistake; but he is also bound to confess that a fortunate inferiority prevents him personally from making such mistakes'. Here is Thomas Hardy's 'Epitaph for G. K. Chesterton'.

EPITAPH FOR G. K. CHESTERTON

Here lies nipped in his narrow cyst
The literary contortionist
Who'd prove and never turn a hair

That Darwin's theories were a snare
He'd hold as true with tongue in jowl,
That Nature's geocentric rule
. . . true and right
And if one with him could not see
He'd shout his choice word 'Blasphemy'.

198

A celebrated dissertation by Mark Twain on scientific reasoning comes from Life on the Mississippi *and goes as follows:*

Now, if I wanted to be one of those ponderous scientific people, and 'let on' to prove what had occurred in the remote past by what had occurred in a given time in the recent past, or what will occur in the far future by what has occurred in late years, what an opportunity is here! Geology never had such a chance, nor such exact data to argue from! Nor 'development of species', either! Glacial epochs are great things, but they are vague – vague. Please observe.

In the space of one hundred and seventy-six years the Lower Mississippi has shortened itself two hundred and forty-two miles. This is an average of a trifle over one mile and a third per year. Therefore, any calm person, who is not blind or idiotic, can see that in the Old Oolitic Silurian Period, just a million years ago next November, the Lower Mississippi River was upward of one million three hundred thousand miles long, and stuck out over the Gulf of Mexico like a fishing-rod. And by the same token any person can see that seven hundred and forty-two years from now the Lower Mississippi will be only a mile and three-quarters long, and Cairo and New Orleans will have joined their streets together, and be plodding comfortably along under a single mayor and a mutual board of aldermen. There is something fascinating about science. One gets such wholesale returns of conjecture out of such a trifling investment of fact.

H. L. Mencken regarded the (unrecorded) inventor of the thermostat as one of mankind's greatest benefactors. 'I wouldn't swap him for a dozen Marconis, a regiment of Bells, or a whole army corps of Edisons. Edison's life-work, like his garrulous and nonsensical talk, has been mainly a curse to humanity: he has greatly augmented its stock of damned nuisances.' As to Alexander Graham Bell, Thomas McMahon pictures him in his story, 'Bell and Langley', in a not very different light – a man prepared to follow his flickering intellectual star wherever it leads.

Years earlier, when Charles J. Guiteau had shot President Garfield, Bell had offered his services to the President's physicians, who had not been able to find the bullet. As Garfield lay near death, Bell experimented with techniques for finding metal in the body. He placed an electric light, along with a bullet, in the mouth of an assistant, and observed that the bullet cast a shadow on the man's cheek. He proposed that President Garfield should swallow an even more powerful electric light bulb connected to a source of electricity through wires leading from the mouth. The physicians protested that the President was too feeble to swallow weak tea, let alone a light bulb. Bell went back to his laboratory and returned two days later with a modified version of the Hughes Induction Balance. By this time, President Garfield had fallen into a state of septic delirium. None of his physicians had taken the trouble to wash their hands before probing his wounds, with the consequence that extensive infections now inflamed his body.

Bell set up the induction balance by the President's bed. He put an earphone to his ear and asked for silence. Slowly he moved the coil over the surface of the body. A newspaper man standing among the onlookers misinterpreted what he was seeing – later he wrote that Bell had been listening for the bullet with a telephone. In the course of seven hours' work, Bell located not one bullet but twelve. The bullets were arranged in a rectangular pattern over the President's thorax and abdomen. Bell drew a little circle on the skin around the location of each bullet, using a grease pencil. The physicians were dumbfounded, since they knew only one shot had been fired.

Much later, after the President's death, it became clear that the rectangular pattern sensed by the induction balance had been due to the steel springs in the President's bed.

Lysenko was a man possessed of invincible ignorance, but a sure instinct for power and self-advancement. He destroyed Russian biology and was responsible for the exile, imprisonment or death of many of his country's best scientists. Studies of Lysenko and his milieu *abound. Here is a portrayal of Lysenko in fiction: he is Avdiyev in Dudintsev's well-known novel,* Not by Bread Alone.

'Let's go, you'll have plenty of opportunity to enjoy the sight of your opponent later!' Nievrayev said, taking a folder from the table. He straightened his tie, passed a comb through his thin brush of hair, and they entered the same long hall, in which, a fortnight before, Lopatkin had written his application to the Minister.

The conference was to be held on the fourth floor, in Drozdov's office. At twelve o'clock those who were to take part in it were gathered in Drozdov's waiting-room. There were eight people there whom Lopatkin did not know. Some of them wore white tunics with white shoulder-straps; these were engineers, others thin light-coloured summer suits; these were the scientists. Nievrayev changed into another man as soon as he entered the waiting-room. His jacket was now buttoned up, and seemed to have become stiffer, not only hindering his movements but even making it difficult for him to turn his head. Flushed with the effort, he went up to each of those present, with a waddling gait and a severe expression, shook hands with everyone, and then entered Drozdov's office, without so much as a glance at Lopatkin.

He soon came out again and said:

'Come in, comrades.'

They all gathered at the door, entered the office, sat down on chairs placed along the wall, on which copies of Lopatkin's project were already pinned up. Drozdov was sitting at his desk, wearing a tunic of pale gold Shantung silk. Next to him sat Academician Florinski, his back humped, his hands on his stick, nodding every now and then, though no one had spoken to him. On the other side of the desk Professor Avdiyev was lounging in an armchair, shaking his yellow-grey curls. He was smoking, blowing the smoke towards the ceiling and knocking the ash off his cigarette into Drozdov's cast-iron ash-tray. Avdiyev was a huge man with a broad pink face and powerful pink neck covered with yellow freckles. His eyes surprised Lopatkin; they were like round, pale-blue, lack-lustre pebbles, which yet seemed to contain a wild gaicty. Avdiyev's voice, too, was strange; it resembled that of a woman with a violent cold in the head.

'Comrade Lopatkin, make your report to the conference!' Drozdov said.

'Why a report, everyone has already seen the project?' Avdiyev asked in a toneless voice, turning round and making his chair creak loudly. 'Hasn't everyone see it?'

'We have seen it. We know,' several voices replied.

'What is your opinion of it?' Drozdov asked.

'The Institute maintains its former opinion regarding the necessity of a scientific development of the main questions connected with the essential peculiarities of this scheme,' Avidyev said all in one breath, still sitting in his chair. He spoke only to Drozdov and the shorthand writer. 'However, taking into account our daily troubles, so to speak, and the already pressing need for such a machine, we consider possible the construction . . . mm . . . of an experimental prototype of the variant now proposed by the comrade inventor. The machine merits attention and testing on a level with the one now being made in Muzga . . . although the device now being made by order of the Ministry – I have in mind Uriupin's and Maxiutenko's device – is promising us a successful solution to the problem.'

'Peter Innokentievic, I believe you wanted to . . .' Drozdov said to Academician Florinski, then checked himself suddenly. 'I beg your pardon, Vassili Zakharovich, have you finished?' he asked Avdiyev.

'There's not much more to say,' Avidyev said hoarsely, wriggling his powerful back and taking a fresh cigarette out of his case. 'With one thing and another, we shall obtain our pipes now.' He turned and looked at Lopatkin, holding the cigarette between his strong teeth.

Before he began to speak Academician Florinski nodded repeatedly and tightened his hold on the stick on which he was leaning.

201

Zhores Medvedev, in his account, The Rise and Fall of T. D. Lysenko, *describes his manner with his colleagues and how he was lionised during his heyday.*

Lysenko's cult in these years was blown up to fabulous proportions. He is apparently the only biologist in history to whom the epithet 'great' was applied in his lifetime. His portraits hung in all scientific institutions. Art stores sold busts and bas-reliefs of Lysenko (these art works were still available in 1961, at triple discount). In some cities, monuments were erected to him. The State Chorus had in its repertory a hymn honoring Lysenko. In songbooks one could find folk doggerel along the lines:

Merrily play on, accordion,
With my girl friend let me sing
Of the eternal glory of Academician Lysenko.

He walks the Michurin path
With firm tread;
He protects us from being duped
by Mendelist-Morganists.

I remember well an interesting episode – Lysenko's first lecture at the TAA. In that academy, after the LAAAS session, the well-known economist Nemchinov was replaced as director by Stoletov, who had previously worked with Lysenko in the Institute of Genetics. As already noted, a number of outstanding scientists, such as Zhebrak and Paramonov, were dismissed. Together with all his other posts, Lysenko became professor of genetics and breeding, although before that he had never done any teaching.

And now this was to be his first lecture directed to the students. The compliant leadership of the Academy summoned to the lecture the whole staff, which occupied most of the seats, while the students, crowding the hallways, listened to a loudspeaker. The whole street was crowded with personal cars of Ministry executives, including that of the Minister of Agriculture himself. And now the illustrious LAAAS president arrives in his personal ZIS car. An especially summoned brass band begins to play a triumphal march, under the sounds of which Lysenko proceeds through the hailing rows to the rostrum to begin his first lecture. Seeing gray-haired scientists in the front rows of the audience, Lysenko exclaims with exaltation: 'Aha! You came to relearn?' I remember little of the content of the lecture – only the assertion that a horse is alive only in interaction with the environment; without interaction it is no longer a horse but a cadaver of a horse; that, when different birds are fed hairy caterpillars, cuckoos hatch from their eggs; that a new cell is not formed from a previously existing one, but near one; that the living body always wants to eat; etc, etc.

202

Daniil Granin's novel, The Geneticist, *is 'faction' – a life of Nikolai Timofeyev-Ressovsky (see above, no. 50), or Ur, as he was called. At a time when the mere mention of the word Drosophila, let alone that of Mendelian inheritance of genetic characteristics (coupled in the Orwellian newspeak of the Terror with the name of the American geneticist, Thomas Hunt Morgan as Morganism-Mendelism) would brand a man as a bourgeois and traitor, Timofeyev-Ressovsky found ways of continuing with his work on genetics, indeed on Drosophila.*

In August of 1948 there occurred the famous meeting of the Academy of Agricultural Sciences, as a result of which all opponents of Lysenko were denounced root and branch, branded and reviled, and many were forced to abandon their work. Biologists who did not share Lysenko's opinions had to relinquish their teaching and were dismissed. The wave did not strike Timofeyev's laboratory in the Urals until a good year later. The order went out that the Drosophila should be destroyed and all Morganism-Mendelism utterly forgotten. Now Ur's clarity of vision found expression. Uralez asked to see him and said: 'You, Nikolai Vladimirovich, are unaccustomed to our circumstances, so we shall have to handle matters differently for you. Go on with your genetics as before, but make sure that from now on no report or proposal that you sign as an expert makes any mention of genetics or of Drosophila.' 'Betrayal then.' 'But why? We have to drink from the river that we swim in.'

The little knowledge that Ur had been able to impart to him was enough to allow Uralez to form his own opinion of the absurdity of Lysenko's doctrine. He was able to dissociate genetics from Lysenkoism, to grasp the value of real science and to take what was in the prevailing circumstances – and especially for one in his position – a rather risky decision.

'We did a lot of experiments with Drosophila,' Ur related afterwards, 'but we were allowed to publish nothing. Publications issued from the Americans about their radioactivity programme, but not a line from us; we were the first, ahead of the Americans, to study the complexes that were responsible for eliminating radioisotopes from the human body. We also occupied ourselves with the removal of isotopic waste from water. Dozens of papers were published about this water problem alone.'

During all these years it was as if Ur had not existed. Where he was, whether he had survived the war, what had become of him were questions to which biologists abroad and at home had no answer, such were the circumstances of his research. Then one time he needed Drosophila cultures – before Lelka brought him some back from Berlin. Lieutenant Shvanyov was sent to the genetics laboratory of the Academy of Sciences in Moscow – this was before August 1948; after that there were no Drosophila to be had. So that he would know which cultures to bring, Ur made him a list and even indicated which laboratory would have which culture. He did not of course sign or even initial it. But the list sufficed to tell the geneticists in Moscow who it was, sitting in the Urals, working away. Some people, with whom he had corresponded when he was in Germany, recognised his writing.

In an instant the word spread: Kolyusha was alive.

203

The distortion of science in Russia during Stalin's reign penetrated, though not as pervasively, into physics. Mark Azbel is a theoretician, who details in his book, Refusenik, *how the historical travesties of the period impinged on the teaching of physics.*

While physics was liberated, fantastic obstacles to teaching and learning in other fields were still being erected, and the ridiculous fight for 'primacy' in Soviet science began. Everything that was done had to be attributed to Russian genius – the invention of airplanes, automobiles, engines of all kinds, telephones, locomotives – everything. Two of the most prominent professors at Kharkov were condemned by a special decision of the Ukrainian Central Committee because one of them, a physicist, had mentioned

Einstein in a lecture, and the other, a mathematician, had mentioned Newton!

The concept of Russian superiority in science died hard. Even after Stalin's death when I was teaching in an evening school, I was trying once to give my pupils an idea of Newton's First Law, the Second, the Third, and the Law of Gravity (by this time it was again permissible to mention Newton), and one of my pupils asked me, absolutely sincerely: 'Well, Mark Yakovlevitch, does this mean that Newton made a contribution as important as that of the Russian "*samo-uchki*"?' (The samo-uchki are the illiterate self-taught scientists and inventors, Russian 'Edisons,' 'Fords,' even 'Einsteins,' who were credited with all the modern technology and science.)!

I heard a story about that time from Professor Povzner, who taught a course at the Military Academy for Engineers. He walked into class one day, ready to start his lecture with a routine little spiel about Russian primacy in mathematics, and then settle down to a serious session of really teaching mathematics. But to his alarm, the minute he got up in front of the class he saw that among the audience was the general, the chief of the Academy. He pulled himself up short and decided that he had better devote the whole lecture to the subject of early Russian genius in mathematics. Luckily, he was a very talented man, good at thinking on his feet, so on the spur of the moment he invented a wonderful lecture on Russian mathematics in the twelfth century. He engaged in flights of fancy for the entire hour, stopping only five minutes before the end to ask, as was customary, 'Are there any questions?' He saw that one of the students had raised his hand.

'Yes?'

'This is so interesting, about medieval Russian mathematics. Could you tell us, please, where we could get more information about it – what the reference books would be? I would like to learn more.' Having no time to think, the professor immediately answered: 'Well, that's impossible! All the archives were burned during the Tatar invasion!'

When the class was over, the general came up to the lecturer and said, 'So, Professor . . . all the archives were burned?' Only then did poor Povzner realize what he had said. The unspoken question hung in the air: If all this evidence of Russian primacy in this science was burned, how in the world did the professor *himself* know the history of pre-invasion mathematics? He was ready to panic when, unexpectedly, the general smiled at him sympathetically, turned around, and left. This high-ranking commander was a clever and decent person; otherwise Professor Povzner would have been in deep trouble.

204

We turn now to fraud. The illusionist, Uri Geller, made fools of several apparently competent scientists, who should have known better. The episode, rich in humour, nevertheless came as a considerable shock to other scientists, who had not imagined that their colleagues (such as the Professor of Mathematics in the passage below) could be so credulous. Martin Gardner, in his collection. Science: Good, Bad and Bogus, *has fun with the story.*

In 1973, when Taylor appeared on a BBC television show with Uri Geller, he was so stunned by Geller's magic that he became an instant convert to the reality of ESP and PK. Geller did his familiar trick of duplicating a drawing in a sealed envelope. 'No methods known to science can explain his revelation of that drawing,' wrote Taylor with his usual dogmatism. The professor's jaw dropped even lower when Geller broke a fork by stroking it. 'This bending of metal is demonstrably reproducible,' Taylor later declared, 'happening almost wherever Geller wills. Furthermore, it can apparently be transmitted to other places – even hundreds of miles away.'

'I felt,' said Taylor in his most often quoted statement, 'as if the whole framework with which I viewed the world had suddenly been destroyed. I seemed very naked and vulnerable, surrounded by a hostile and incomprehensible universe. It was many days before I was able to come to terms with this sensation.'

Although Taylor was supremely ignorant of conjuring methods, and made not the slightest effort to enlighten himself, he at once set to work testing young children who had developed a talent for metal bending after seeing Geller on television. Taylor's controls were unbelievably inadequate. Children, for example, would put paper clips in their pockets and later take one out twisted. Nevertheless Taylor was persuaded that hundreds of youngsters in England had the mind power to deform metal objects. Curiously, Taylor never actually *saw* anything bend. One minute a spoon would be straight, later it would be found twisted. Taylor named this the 'shyness effect.' Metal rods were put inside sealed plastic tubes and children were allowed to take them home. They came back with the tubes still sealed and the rods bent. One boy startled Taylor by materializing an English five-pound note inside a tube.

So certain was Taylor that his IQ, combined with his knowledge of physics, gave him the ability to detect any kind of fraud that he rushed into print a big book called *Superminds* (published here by Viking in 1975). It will surely go down in the literature of pseudoscience as one of the funniest, most gullible books ever to be written by a reputable scientist. It is even funnier than Professor Johann Zöllner's *Transcendental Physics*, inspired by the psychic conjuring of the American medium Henry Slade. Taylor's book is crammed with photographs of grinning children holding up cutlery they have supposedly bent by PK, tables and persons floating in the air during old Spiritualist seances, glowing ectoplasmic ghosts, psychic surgeons operating in the Philippines, Rosemary Brown displaying a musical composition dictated to her by the spirit of Frederic Chopin, and numerous other wonders.

Not the least peculiar aspect of Taylor's volume was his argument that all paranormal feats, including religious miracles, are explainable by electromagnetism. "The Geller effect is a case in point. Will it ever turn out that the miracles of Jesus Christ also dissolve in scientific speculation . . . . This book has presented the case that for one modern "miracle," the Geller effect, there *is* a rational, scientific explanation. This explanation is also claimed to allow us to understand other apparently miraculous phenomena – ghosts, poltergeists, mediumship, and psychic healing. What, then, of other miracles? Can they too be explained by these newly discovered powers of the human body and mind, and the properties of matter broadly described in the book?'

After writing *Superminds*, of which let us hope he is now superashamed, Taylor slowly began to learn a few kindergarten principles of deception. When the Amazing Randi visited England in 1975, Taylor refused to see him, but Randi managed to call on him anyway, disguised as a photographer-reporter. You'll find a hilarious account of this in Chapter 10 of Randi's Ballantine paperback, *The Magic of Uri Geller.* Taylor proved to be easier to flimflam than a small child, and his 'sealed' tubes turned out to be so crudely sealed that Randi had no trouble uncorking one and corking it again while Taylor wasn't looking. Randi even managed to bend an aluminum bar when Taylor's attention was distracted, scratch on it 'Bent by

Randi,' and replace it among Taylor's psychic artifacts without Taylor noticing.

Another crushing blow to Taylor's naive faith in Geller was a test of the 'shyness effect' by two scientists at Bath University. They allowed six metal-bending children to do their thing in a room with an observer who was told to relax vigilance after a short time. All sorts of bending at once took place. None was observed by the observer, but the action was secretly being videotaped through a one-way mirror. The film showed, as the disappointed researchers wrote it up for *Nature* (vol.257, Sept.4, 1975, p.8): '*A* put the rod under her foot to bend it; *B, E* and *F* used two hands to bend the spoon . . . while *D* tried to hide his hands under a table to bend a spoon.'

Slowly, as more evidence piled up that Geller was a charlatan and that the 'Geller effect' never occurs under controlled conditions, Taylor began to have nagging doubts. After several years of silence, he suddenly announced his backsliding. Of course he didn't call it that. Instead he and a colleague at Kings College wrote a technical article for *Nature*, 'Can Electromagnetism Account for Extrasensory Phenomena?' (vol.276, Nov.2, 1978, pp.64-67; also *Skeptical Inquirer*, Spring 1979, p.3.)

205

C. P. Snow's novel, The Affair, *deals with a fraud and the misfortunes that can rebound on innocent parties, too close to the crime.*

'This does seem to be your thesis, doesn't it, Dr. Howard?' said Dawson-Hill, handing it to him.

'But of course.'

Dawson-Hill asked how many copies there were in existence.

The answer was, three more. In the Fellowship competition, the college asked for two copies. He had used the remaining two for other applications.

'This is the show copy, though?'

'You can call it that.'

'Then this –' There was a slip of paper protruding from the thesis and Dawson-Hill opened it at that page – 'might be your star print?'

It was the positive which everyone in that room knew. It was pasted in, with a figure 2 below it and no other rubric at all. It stood out, concentric rings of black and grey, like a target for a small-scale archery competition.

'It's a print, all right.'

'And this print is a fraud?'

I wished Howard would answer a straight question fast. Instead he hesitated, and only at last said, 'Yes.'

'That doesn't need proving, does it?' said Dawson-Hill. 'All the scientific opinion agrees that the drawing-pin hole is expanded? Isn't that true?'

'I suppose so.'

'That is, this print had been expanded, to make it look like something it wasn't?'

'I suppose so.'

'What about your other prints?'

'Which other prints?'

'You can't misunderstand me, Dr. Howard. The prints in the other copies of your thesis?'

'I think I re-photographed them from this one.'

'You *think*.'

'I must have done.'

'And this one, this fake one, came from a negative which you've never produced? Where is it, do you know?'

'Of course I don't know.'

For once articulate, Howard explained that the whole point of what he had said before lunch was that he *couldn't* know. He had not seen the negative; Palairet must have made the print and the measurements and put them in with a set of other positives.

At that, Nightingale broke in.

'I've asked you this before, but I still can't get it straight. You mean to say that you used this print as experimental evidence without having the negative in your hands?'

'I've told you so, often enough.'

'It still seems to me a very curious story. I'm sorry, but I can't imagine anyone doing research like that.'

'I thought the print and the measurements were good enough.'

'That is,' I broke in, 'you took them on Palairet's authority.'

Howard nodded.

Nightingale, with a fresh, open look of incomprehension, was shaking his head.

'Let's leave this for a moment, if you don't mind,' said Dawson-Hill. 'I'm an ignoramus, of course, but I believe this particular print was regarded – before it was exposed as a fraud – as the most interesting feature of the thesis?'

'I shouldn't have said that,' said Howard. (I was thinking, why didn't the fool see the truth and tell it?) 'I should have said it was an interesting feature.'

'Very well. Let me be crude. Without that print, and the argument it was supposed to prove, do you believe, Mr. Howard, that the thesis would have won you a Fellowship?'

'I don't know about that.'

'Do you agree it couldn't have stood the slightest chance?'

Howard paused. (Why doesn't he say Yes, I thought?) 'I shouldn't say that.'

Nightingale again intervened: 'There's not a great deal of substance in the first half, is there?'

'There are those experiments –' Howard seized the thesis and began staring at some graphs.

'I shouldn't have thought that was very original work, by Fellowship standards,' said Nightingale.

'It's useful,' said Howard.

'At any rate, you'd be prepared to agree that without this somewhat providential photograph your chances could hardly have been called rosy?' said Dawson-Hill.

This time Howard would not reply.

Dawson-Hill looked surprised, amused, and broke away into his second attack.

'I wonder if you'd mind giving us some illumination on a slightly different matter,' he said. 'This incident has somewhat, shall I say, disarranged your career?'

'What do you think?' Howard replied.

'Not to put too fine a point upon it, it's meant that you have to say good-bye to being a research scientist, and start again? Or is that putting it too high?'

'That's about the size of it.'

'And you must have realised that, as soon as this Court first deprived you of your Fellowship?'

'But of course I did.'

'That was nearly eighteen months ago, seventeen months, to be precise?'

'You must have the date.' Howard's tone was savage.

'So far as my information goes, during that time, that quite appreciable time, you never took any legal action?'

It was the point Nightingale had challenged us to answer at the Master's dinner-party after Christmas. I had no doubt that Nightingale had put Dawson-Hill up to it.

'No.'

'You've never been to see your solicitor?'

'Not as far as I remember.'

'You must remember? Have you been, or not?'

'No.'

'You never contemplated bringing an action for wrongful dismissal?'

'No.'

'I suggest you weren't willing to face a court of law?'

Howard sat, glowering at the table. I looked at Crawford: for an instant I was going to protest; then I believed it would make things worse.

'I always thought,' Howard replied at long last, 'that the college would give me a square deal.'

'You thought they might give you much more of the benefit of the doubt?'

'I tell you,' Howard said, his voice strained and screeching, 'I didn't want to drag the college through the courts.'

To me this came out of the blue. When Martin and I had pressed him, he had never said so much. Could it be true, or part of the truth? It did not ring true, even to me.

206

A rash of frauds, brought to light in the 1970s and 1980s, were widely publicised and were commonly thought to have done much damage to the cause of science in the USA. William Broad and Nicholas Wade published a study of these episodes and the moral questions that they raised. Here, from Betrayers of the Truth, *is the story of one of the most shameless of deceptions and one that brought scarcely deserved discredit on an ageing and distinguished scientist.*

It is usually a year's work for a graduate student to purify a single enzyme, especially if it is a minor one. But by mid-1980, six months after his arrival in Racker's lab, Spector had purified the ATPase and four kinases. The kinase cascade was a wonderfully interesting mechanism, suggestive to biochemists of all kinds of signal amplification and control systems, but better yet was to come. Spector managed to tie the cascade in with an extremely important new development that had just emerged from the study of viruses that cause tumors in animals.

The tumor-causing gene of these viruses, known as the *src* ('sarc') gene, is one that specifies a protein kinase enzyme. The viruses are thought to have pirated this gene, early in their evolution, from the cells of the species they infect. Cancer researchers had scrutinized animal cells for the present-day versions of these pirated genes, the so-called endogenous *src* genes, but no one had managed to isolate the genes' protein kinase products. Enter Mark Spector with the astonishing news that certain of his cascade kinases were the products of the elusive endogenous *src* genes.

Everything at last seemed to be dropping into place for a unified theory of cancer causation. A tumor virus infects a cell. Its *src* gene makes unmanageably large quantities of a kinase which trips off the cell's otherwise inactive cascade of kinases. The last kinase in the cascade phosphorylates the ATPase enzyme, making it inefficient and thereby setting off the further changes characteristic of malignant cells.

'Seductive' is a word biologists used to describe the compelling intellectual attraction of the Racker-Spector theory. The two of them had picked the most exciting new developments in cancer research and demonstrated, by a sequence of beautifully executed experiments, how each fitted into the overall theory. Before the details had even been published in the scientific literature, Racker started mentioning the theory in lectures given around the country.

A biochemist with a background in psychiatry, the sixty-eight-year-old Racker was an eminent figure in his field, a winner of the U.S. National Medal of Science, and his authority lent to the then unpublished theory a credence it would not otherwise have had. Soon, under Racker's aegis, Spector was collaborating with leading researchers in cancer biology, such as David Baltimore of MIT, and George Todaro and Robert Gallo of the National Cancer Institute.

When Racker gave a lecture about the theory on the campus of the National Institutes of Health in the spring of 1981, some 2,000 people attended. National Cancer Institute director Vincent

DeVita, then in trouble with Congress for not cracking down on cases of scientific fraud, was urged to spread the good news. DeVita held off, but within the biomedical research community enthusiasm was running high. 'We have witnessed in this field a merger of biochemistry and molecular biology which was long overdue,' Racker and Spector declared in a portentous and self-promoting paper published in the July 17, 1981, issue of *Science*.

Leading researchers began to move into the field, but rather than go through the laborious task of replicating Spector's work by purifying their own kinase systems, they would send their reagents to him for testing. 'The striking thing was that when you went there there were samples from all over the world waiting to be tested by this kid. If you looked at the labels on the shelves it was almost a Who's Who of cancer research,' noted Todaro. Some researchers invited the young graduate student to their labs. One by one they became aware of a pattern familiar to Spector's colleagues at Cornell, that often the experiments would work only in Spector's hands and could not be repeated without him. But like Spector's colleagues, they found a simple explanation: Mark was just so good at making experiments go. 'He is technically very gifted,' says Todaro. 'When he came to NIH, people would ask him for practical suggestions in doing their experiments and would get intelligent advice. People didn't talk to him the way one talks to a graduate student, but as to a colleague.'

Among those intrigued by Spector's theory was Volker Vogt, a tumor virologist who worked in the department of biochemistry at Cornell on the floor above Racker's laboratory. In April 1980 Spector performed some experiments on his ATPase enzyme with a student of Vogt's, Blake Pepinsky. 'These results were so clean and beautiful and convincing that I was seduced into putting my time in on this project,' says Vogt.

There was one problem. Sometimes the experiments worked and sometimes they didn't. Vogt was worried that results so beautiful should also be so erratic. He devoted the summer of 1980 to trying to understand why the negative experiments didn't work. Whatever the reason, whether the wrong phase of the moon or having impurities in the distilled water, it was too elusive for Vogt to put his finger on. Eventually he gave up. He also decided he could not publish the experiments, however exciting they might be.

Pepinsky continued to help Spector, however, the two of them often working seventeen-hour days and sometimes through the night. 'The work had nothing to do with my thesis but I would come and do the precipitations when Mark wanted to do it. They only worked when he was around to do them,' notes Pepinsky. A year later, in early 1981, Vogt too was drawn back into the maelstrom when Spector started to find his kinases in cells infected with tumor-causing viruses. The particular experiment that caught Vogt's interest was one that showed that an antiserum to one of Spector's kinases also had an affinity for an important but so far undetected protein, the product of the *src* gene of a widely studied mouse tumor virus.

Pepinsky repeated the experiment twice but it didn't work. Vogt despaired that it seemed to be the same frustrating story as with the ATPase the year before. But this time, he told himself, he was really going to get to the bottom of things and find out why Spector's work was hard to reproduce. In an intensive two-day effort, Pepinsky and Spector redid the experiment.

It was another spectacular success. 'There were fat radioactive bands of protein on the autoradiogram, everything looked as clean as could be,' notes Vogt. 'So I said, "Here at least is something I can work with."' He decided his first step would be to analyze the gels from which the autoradiogram was made. 'Mark was very upset that I had got my hands on the gels. Previously he had done all these analyses himself,' says Vogt. Pepinsky, also uneasy about the lack of reproducibility, had squirreled away the original gels.

The gels were the key pieces of data from Spector's experiments. Cell proteins picked out by antisera and tagged with radioactive phosphorus 32 would be placed on the gel and subjected to an electric field. Each protein would migrate through the gel a particular distance, determined by its size, and mark its presence by darkening a radiosensitive film placed next to the gel.

Hitherto only these films, called the autoradiograms, had been shown by Spector to his colleagues. When Vogt obtained an original gel, his first step was to run a hand Geiger counter over it to locate the radioactive protein bands. He realized instantly from the pattern of clicks that something was terribly wrong. The clicks were not saying phosphorus 32. Judging from the amount of darkening on the autoradiogram, they seemed to be saying iodine 125. A scintillation-counter measurement confirmed this diagnosis. But iodine had no business at all in the experiment.

It was forgery, very cunning but quite simple. The forger was evidently finding proteins of the right molecular weight to reach the desired point in the gel, tagging them with radioactive iodine, and mixing them into the antisera-tagged proteins just before they were put onto the gel.

Vogt was overwhelmed. The day was a Friday, July 24. 'I was pretty shaken up by it. It was something of a nightmare. To begin with, I didn't tell anyone. I knew it was a big event in my career, in everyone's. I went home and brooded over it for one day. Then I went over to see Racker.

'He didn't doubt the actual facts but he was loath to believe that everything was wrong immediately. At that point we thought it might be a recent aberration. The next morning we confronted Mark. We thought he might say "Mea culpa," but he didn't. He didn't dispute the finding that it was iodine, but he said he didn't do it and that he didn't know how it happened.'

With the discovery of the forged experiment, the castle in the clouds, which Racker had announced in *Science* just ten days earlier, began to evaporate.

Racker gave Spector four weeks in which to purify the ATPase enzyme and the four kinases from scratch and to put them in Racker's hands for testing. Spector agreed, saying the task would take him two weeks, not four. But progress was not so fast. It took three attempts before he could provide Racker with an ATPase that worked, but he did provide one. He also produced a kinase that appeared to phosphorylate the ATPase, but there was only enough material for Racker to do the experiment twice. The other kinases Spector supplied were not of the right molecular weight and did not work in the way they should have done. At the end of four weeks Racker told Spector not to come back to his laboratory anymore.

Spector's attempt to vindicate himself was on balance a failure, yet an ambiguous failure. He showed Racker that he could reproduce some, but not all, of his claimed results, thus creating uncertainty as to whether some or none of his previous work was reliable. But he declined to share whatever he knew with his colleagues. 'Mark says that in five years he will be vindicated. But he won't help us find out what is good and what is not,' commented Pepinsky, who in scientific terms knew Spector best.

Did Mark Spector fake all of his results, or none of them, or just some? There may never be a clear answer to the question. All that could be said with certainty was that data in some of his experiments had been deliberately and cunningly contrived by someone. 'If it turns out that the whole thing was made up,' remarked George Todaro, 'it is an incredible tour de force, it really boggles the mind, and in any case something like this really might exist.' 'If we are talking about faking, it is very clever and careful and on a great scale – it's not like painting patches on a mouse,' said another biologist familiar with the work.

As the kinase cascade theory started to fall apart, so did significant parts of Mark Spector's background. On September 9, 1981, he withdrew the thesis that was about to win him a Ph. D. degree earned in one and a half years instead of the usual five. Belated checks that should have been done when Spector entered the Cornell graduate school showed that he possessed neither an M.A. nor a B.A. degree from the University of Cincinnati, as he had claimed. 'A check with law enforcement agencies in Cincinnati,' reported the *Ithaca Journal*, 'shows that Mark B. Spector pleaded guilty on June 12, 1980, to two forgery charges in connection with writing two checks for $4,843.49 to himself from his employer . . . He was sentenced to prison and the sentence was suspended and he was put on three years' probation.'

207

Plagiarism is an offence, seldom forgiven. It embittered the life of Ronald Ross, the discoverer (in the end, of course, vindicated) of the cause of malaria, as this passage from his Memoirs *shows. Ross's expression of his rancour in verse is to be found above (no. 163).*

It will be observed:

1. That these people do not mention where exactly my work was published so that their hearers or readers could not verify their statements regarding it.

2. Their statement that 'it [is] also possible that his two mosquitoes, before having bitten men, had already bitten another animal' was directly contrary to the second sentence in my paper in the *British Medical Journal*, 18 December 1897, in which I said: 'For the last two years I have been endeavouring to cultivate the parasite of malaria in the mosquito. The method adopted has been to feed mosquitoes, *bred in bottles from the larva*, on patients having crescents in the blood and then to examine their tissues for parasites similar to the haemamoeba in man.' The rest of my paper reported

the finding of such parasites; and the whole case was fully proved by my subsequent work on human and avian parasites.

3. Messrs Bastianelli, Bignami, and Grassi must have had this paper of mine before them when they wrote their note (see Charles, page 217). Their statement was therefore a deliberate and intentional lie told in order to discredit my work and so to obtain the priority for themselves and constantly reiterated even after I had exposed it.

4. It is amusing to observe that they themselves say nothing about their own mosquitoes having been bred from the larva. Also they had succeeded only with two insects! Yet they demand 'all certainty' for their observations, which they refuse for mine.

5. The falsification is so obvious that one may doubt the truth of the whole *nota preliminare* given above. It is quite *possible* that these gentlemen were at that time merely showing my preparations which Manson had sent them.

Their next paper was dated 22 December 1898 and was published by the same academy. It describes the finding of 'pigmented cells' in some more *Anopheles claviger* but does not state exactly in how many or whether the insects used had been bred from the larvae. In fact the article might have been transcribed bodily from my proteosoma report and Manson's Edinburgh address. My name is mentioned only once, on the fifth page, where the three authors say (again in Charles's translation): 'The observations reported up till now permit us to reconstruct the life cycle of the human haemosporidia in the body of *Anopheles claviger*: it finds for the most part confirmation in that observed by Ross regarding the proteosoma of birds in the *grey mosquito*.' In other words, my work confirmed that of these rascals! Again, no exact references to my writings are given.

From the intrinsic evidence of these two papers I believe with Koch that it is quite *possible* that the alleged observations were never really made at all but that the papers were written entirely from my specimens and descriptions in order to obtain priority for the discovery of the human parasites in anophelines – which the writers then thought I was making in India. More *probably*, however, they did find the pigmented cells but only in some old *Anopheles claviger* which had been *caught in rooms* where malaria cases were sleeping – as, in fact, they actually say in their second *nota preliminare* – that is, just as I found the cells some months later in *Anopheles costalis* caught in Wilberforce Barracks in Sierra Leone. Some of these old *A claviger* may have been refed subsequently on the malaria cases in hospital; but in neither *nota preliminare* is there any statement whatever that the insects employed had been bred from the larvae in captivity and then fed on the malaria cases in hospital for the first time – no such statement, in fact, as I had clearly made in my article of 18 December 1897. Thus the very ambiguity which the authors falsely asserted against my work existed in their own papers when they made the assertion. Of course the plasmodial nature of the pigmented cells in their mosquitoes was guaranteed 'with all certainty' not by any work of theirs but by my preceding proofs of the whole life cycle of proteosoma in mosquitoes. Can any more impudent scientific frauds than these two preliminary notes of Messrs Grassi, Bignami, and Bastianelli be imagined?

This sort of thing is not uncommon in science, which it impedes and disgraces. All cases of it ought to be publicly exposed when detected and I therefore do not mince my words regarding the present one.

208

Broad and Wade relate a tale to stretch credulity, from which the following extract, of a career, however transient, built on undiluted plagiarism.

How did Alsabti, with a long trail of deceit behind him, manage to climb into ever higher reaches of the academic establishment? The hesitancy on the part of those in the know to make public their knowledge of Alsabti's deceit was certainly a factor. But there is also Alsabti's flair for the art of persuasion, and his subtle knowledge of things human. 'The guy knows the system well,' says Giora Mavligit, a professor of medicine at the M. D. Anderson Hospital in Houston who for five months was Alsabti's boss. 'He went right to the top – right to the president.' In this case, the president was Lee Clark, head of M. D. Anderson. Alsabti showed Clark letters of introduction from Major General David Hanania, the Surgeon General of the Jordanian Armed Forces, letters saying Alsabti was in the United States for post-graduate medical education. In September 1978, Alsabti was assigned to work as a nonpaid volunteer in the lab of Mavligit.

By this time Alsabti was a veritable factory for the production of papers. Each passing month saw another group of Alsabti articles appear in various journals around the world. His method was simplicity itself. He would retype an already published paper, remove the author's name, substitute his own, and send the manuscript off to an obscure journal for publication. His tactics deceived the editors of dozens of scientific journals around the world. Alsabti papers were published in the *Journal of Cancer Research and Clinical Oncology* (U.S.), *Japanese Journal of Experimental Medicine, Neoplasma* (Czechoslovakia), *European Surgical Research* (Switzerland), *Oncology* (Switzerland), *Urologia Internationalis* (Switzerland), *Journal of Clinical Hematology and Oncology* (U.S.), *Tumor Research* (Japan), *Journal of Surgical Oncology* (U.S.), *Gynecologic Oncology* (U.S.), *British Journal of Urology*, and *Japanese Journal of Medical Science and Biology*.

The obscurity of most of the journals ensured that the cases of plagiarism would not be tracked down. The authors whose work was stolen would never read the pirated Alsabti version, and there the matter would rest. However, at M. D. Anderson another demonstrable case of plagiarism took place, an observant author in a distant lab having deduced some of the steps in Alsabti's deceit. In this case, the incident involved a paper that had been sent to a researcher at M. D. Anderson for review prior to publication. What the editor of the journal had not realized was that the researcher, Jeffrey Gottlieb, could not possibly review the paper because he had been dead since July 1975. The manuscript from the *European Journal of Cancer* lay in a mailbox until one day Alsabti picked it up, made a few cosmetic changes, added his name and the names of two fictitious coauthors, Omar Naser Ghalib and Mohammed Hamid Salem, and then proceeded to mail it off to a small Japanese journal for publication. The *Japanese Journal of Medical Science and Biology* published Alsabti's article before the original article from which it was lifted got into print.

'When I first saw the Japanese paper I went into a depression for about a week,' says Daniel Wierda, the actual author of the paper who at the time was a Ph. D. candidate at the University of Kansas. 'I didn't know what to do.' Since Alsabti's paper had appeared first, Wierda feared that colleagues would think he had pinched his paper from Alsabti.

Wierda wrote to the Japanese journal, explaining the circumstances and asking for a retraction. Again, it was only after the Alsabti affair emerged in the international press that a retraction was forthcoming.

Intellectual lapses can come from laziness, ignorance, stupidity or reluctance to face the consequences of the truth. A common feature of laboratory (and for all I know, theoretical) science is the authority that established procedures acquire simply through the passage of time. Some method is found to work, it passes into currency and its basis is never thereafter questioned. And thus many a futile ritual is enacted daily in every laboratory. A fine example follows, taken from Primo Levi's masterly collection of stories, The Periodic Table.

So, returning to boiled linseed oil, I told my companions at table that in a prescription book published about 1942 I had found the advice to introduce into the oil, toward the end of the boiling, two slices of onion, without any comment on the purpose of this curious additive. I had spoken about it in 1949 with Signor Giacomasso Olindo, my predecessor and teacher, who was then more than seventy and had been making varnishes for fifty years, and he, smiling benevolently behind his thick white mustache, had explained to me that in actual fact, when he was young and boiled the oil personally, thermometers had not yet come into use: one judged the temperature of the batch by observing the smoke, or spitting into it, or, more efficiently, immersing a slice of onion in the oil on the point of a skewer; when the onion began to fry, the boiling was finished. Evidently, with the passing of the years, what had been a crude measuring operation had lost its significance and was transformed into a mysterious and magical practice.

Old Cometto told of an analogous episode. Not without nostalgia he recalled his good old time, the times of copal gum: he told how once boiled linseed oil was combined with these legendary resins to make fabulously durable and gleaming varnishes. Their fame and name survive now only in the locution 'copal shoes,' which alludes precisely to a varnish for leather at one time very widespread that has been out of fashion for at least the last half century. Today the locution itself is almost extinct. Copals were imported by the British from the most distant and savage countries, and bore their names, which in fact distinguished one kind from another: copal of Madagascar or Sierra Leone or Kauri (whose deposits, let it be said parenthetically, were exhausted around about 1967), and the very well known and noble Congo copal. They are fossil resins of vegetable origin, with a rather high melting point, and in the state in which they are found and sold in commerce are insoluble in oil: to render them soluble and compatible they were subjected to a violent, semi-destructive boiling, in the course of which their acidity diminished (they decarboxylated) and also the melting point was lowered. The operation was carried out in a semi-industrial manner by direct fire in modest, mobile kettles of four or six hundred pounds; during the boiling they were weighed at intervals, and when the resin had lost 16 percent of its weight in smoke, water vapor, and carbon dioxide, the solubility in oil was judged to have been reached. Around about 1940, the archaic copals, expensive and difficult to supply during the war, were supplanted by phenolic and maleic resins, both suitably modified, which, besides costing less, were directly compatible with the oils. Very well: Cometto told us how, in a factory whose name shall not be uttered, until 1953 a phenolic resin, which took the place of the Congo copal in a formula, was treated exactly like copal itself – that is, by consuming 16 percent of it on the fire, amid pestilenial phenolic exhalations – until it had reached that solubility in oil which the resin already possessed.

Here at this point I remembered that all languages are full of images and metaphors whose origin is being lost, together with the art from which they were drawn: horsemanship having declined to the level of an expensive sport, such expressions as 'belly to the ground' and 'taking the bit in one's teeth' are unintelligible and sound odd; since mills with superimposed stones have disappeared, which were also called millstones, and in which for centuries wheat (and varnishes) were ground, such a phrase as 'to eat like four millstones' sounds odd and even mysterious today. In the same way, since Nature too is conservative, we carry in our coccyx what remains of a vanished tail.

210

In Bernard Shaw's The Doctor's Dilemma *Ridgeon's great discovery (see above, no. 113) is misapplied by the hopelessly incompetent but unshakably confident B. B.*

RIDGEON. What has happened?
SIR PATRICK. Do you remember Jane Marsh's arm?
RIDGEON. Is that whats happened?
SIR PATRICK. Thats whats happened. His lung has gone like

Jane's arm. I never saw such a case. He has got through three months galloping consumption in three days.

RIDGEON. B. B. got in on the negative phase.

SIR PATRICK. Negative or positive, the lad's done for. He wont last out the afternoon. He'll go suddenly: Ive often seen it.

RIDGEON. So long as he goes before his wife finds him out, *I* dont care. I fully expected this.

SIR PATRICK [*drily*] It's a little hard on a lad to be killed because his wife has too high an opinion of him. Fortunately few of us are in any danger of that.

Sir Ralph comes from the inner room and hastens between them, humanely concerned, but professionally elate and communicative.

B. B. Ah, here you are, Ridgeon. Paddy's told you, of course.

RIDGEON. Yes.

B. B. It's an enormously interesting case. You know, Colly, by Jupiter, if I didn't know as a matter of scientific fact that I'd been stimulating the phagocytes, I should say I'd been stimulating the other things. What is the explanation of it, Sir Patrick? How do you account for it, Ridgeon? Have we over-stimulated the phagocytes? Have they not only eaten up the bacilli, but attacked and destroyed the red corpuscles as well? a possibility suggested by the patient's pallor. Nay, have they finally begun to prey on the lungs themselves? Or on one another? I shall write a paper about this case.

Walpole comes back, very serious, even shocked. He comes between B. B. and Ridgeon.

WALPOLE. Whew! B. B.: youve done it this time.

B. B. What do you mean?

WALPOLE. Killed him. The worst case of neglected blood-poisoning I ever saw. It's too late now to do anything. He'd die under the anaesthetic.

B. B. [*offended*] Killed! Really, Walpole, if your monomania were not well known, I should take such an expression very seriously.

SIR PATRICK. Come come! When youve both killed as many people as I have in my time youll feel humble enough about it. Come and look at him, Colly.

Ridgeon and Sir Patrick go into the inner room.

WALPOLE. I apologize, B. B. But it's blood-poisoning.

B. B. [*recovering his irresistible good nature*] My dear Walpole, everything is blood-poisoning. But upon my soul, I shall not use any of that stuff of Ridgeon's again. What made me so sensitive about what you said just now is that, strictly between ourselves, Ridgeon has cooked our young friend's goose.

211

Arthur Hailey's tale of pharmaceutical research, Strong Medicine, *treats of what may well be common occurrences and the ethical difficulties that they bring in their wake: the directors of a large pharmaceutical company are able to convince themselves that their product is sufficiently important for the company, and therefore (it follows) for the community, to justify the cutting of some legal corners and even some downright skulduggery. Such reasoning probably lies behind many an industrial and social catastrophe.*

It was sheer bad luck that the new drug application for Montayne had drawn Dr. Gideon Mace as the reviewer.

Sam Hawthorne had not met Mace, and didn't want to. He had heard more than enough about the man from Vince Lord and others, and about the trouble Mace caused Felding-Roth, first with the unreasonable delay two years ago over Staidpace, and now with Montayne. Why should people like Mace possess the power they had, Sam fumed, and have to be endured by honest business men who sought, from the Maces of this world, no more than equal honesty and fairness?

Fortunately, people like Mace were a minority – at FDA a small minority; Sam was certain of that. Just the same, Mace existed. He was currently sitting on the Montayne NDA, using regulations, procedural tactics, to delay it. Therefore, a way to circumvent Gideon Mace had had to be found.

Well, they had a way. At least, Felding-Roth had, in the person of Vince Lord.

Originally, when Vince had collected – no, make that *bought* – evidence of criminality by Dr. Mace, purchased it with two thousand dollars of Felding-Roth cash, the voucher for that cash now buried deep in the travel expense account where auditors or the Internal Revenue Service would never find it . . . at that time Sam had been angry, critical of Vince, and shocked at the thought that the material might ever be used in the way which Vince envisaged.

But not now. The existing situation affecting Montayne was too critical, too important, for that kind of scruples anymore.

And *that* was another cause for anger. Anger because criminals like Mace begat criminality in others – in this case, in Sam and Vincent Lord – who had to use those same low-grade tactics for reasonable self defence. *Damn Mace!*

Still soliloquizing silently, in the quietness of his office, Sam told himself: A penalty you paid for appointment to the top job in any large company was having to make unpalatable decisions authorizing actions which, if they happened elsewhere or in a vacuum, you would consider unethical and disapprove of. But when you shouldered responsibilities involving so many people, all of them dependent on you – shareholders, directors, executive colleagues, employees, distributors, retailers, customers – it was necessary at times to swallow hard and do what was needed, however tough, unpleasant or repugnant it might seem.

Sam had just done that, an hour ago, in okaying a proposal by Vincent Lord to threaten Dr. Gideon Mace with exposure and therefore criminal charges if he failed to expedite the approval of Montayne.

Blackmail. No point in mincing words or hiding behind euphemisms. It would be blackmail, which was criminal too.

Vince had laid his plan bluntly in front of Sam. Equally bluntly, Vince declared, 'If we don't make use of what we have, putting pressure on Mace, you can forget any idea of marketing Montayne in February, and maybe for another year.'

Sam had asked, 'Could it really be that long – a year?'

'Easily, and more. Mace has only to ask for a repeat of –'

Lord stopped as Sam waved him to silence, cancelling an unnecessary question, remembering how Mace had delayed Staid-pace for longer than a year.

'There was a time,' Sam reminded the research director, 'when you talked of doing what you're proposing without involving me.'

'I know I did,' Lord said, 'but then you insisted on knowing where that two thousand dollars went, and after that I changed my mind. I'll be taking a risk and I don't see why I should take it alone. I'll still handle the frontline attack, the confrontations with Mace. But I want you to know about it, and approve.'

Back in the days of the discovery of the DNA structure, and following the award of the Nobel Prize to Watson, Crick and Wilkins, Erwin Chargaff observed that science had become a spectator sport. Since then it has embraced the techniques of the advertising industry and has sold itself by diligent promotion and by running down rival products. Here is a description from Gary Taubes's Nobel Dreams *of how a politically necessary Nobel Prize was secured for the European consortium, CERN. The snark that they hunted was a boson, or rather the W and Z bosons – elusive kinds of elementary particles.*

For the W and the Z Schopper was not about to make the same mistake. He had begun, with Rubbia's help, to sell van der Meer and Rubbia to the Nobel Prize Committee. The Nobel Prize could be given to at most three people. One of the faults with the gluon discovery was that too many principals had been involved: four experiments, four group leaders, and four sets of results, all nearly identical. Schopper had tried to back Ting's priority claim in various physics publications, but it was a hopeless task. The same went for the neutral-current experiment at CERN a decade earlier. Nobody could decide who should get credit, and the press releases announcing neutral currents mentioned no names at all. It was a faceless discovery.

This time around, every press release came with a supplement that explained how this was all made possible by stochastic cooling, courtesy of van der Meer: and no press release, even the one announcing UA2's confirmation of the Z, was without Rubbia's name.

Behind Schopper's urgency to get the Nobel for CERN was the Large Electron Positron machine, which was Schopper's baby. He had already procured the money from the participating European governments to build it. Although he had had to accept a scaled-down version of the machine – only half as powerful as originally designed. He would still have to fight every year to keep the money coming in, and do it against a grim economic background in Europe of declining outputs, inflation, and unemployment. LEP was a $500 million project according to CERN's bookkeeping, which didn't include such things as shop costs and labor provided by the lab, and nearly a billion dollars according to U. S. bookkeeping, which did include these things. Schopper dearly needed some international prestige, and the Nobel would provide it.

At Rubbia's press conference announcing the Z discovery, a month before the collider run ended, Schopper told reporters that the discoveries were the most important in physics since the invention of the transistor twenty-five years earlier; they put CERN at least six years ahead of all its competitors, and they would surely merit a Nobel Prize in physics. It was one of the few occasions in the history of science that a lab director had told the press to inform the Nobel Foundation where his vote lay – before the experiment was even finished.

213

In point of authenticity The Search *is C. P. Snow's best novel, for it seems beyond doubt that it is deeply rooted in his own experience. The narrator is a rising physicist, and one of the movers behind a plan to set up a national biophysical research institute. He expects to be named as director, after much jockeying among the representatives of different universities. (The Institute, it is resolved, is to be at King's College, London.) And then it transpires that the research on which his professional standing is based is wrong, the result of a mistake by an inadequately supervised technician. Miles decides against trying to rebuild his career and reputation; science is finished for him. C. P. Snow, working with his friend, P. V. Bowden (Francis Getliffe in his* roman fleuve, Strangers and Brothers*), published a paper on the chemistry of vitamin A and its derivatives, which was received with considerable acclaim. The work was seriously flawed and the conclusions wholly wrong. No retraction was ever published and the author's evasions were held by many to have been highly unsatisfactory. Snow never published another paper, and devoted himself instead to his new career as a novelist, civil servant, sage and chronicler of shenanigans in the Corridors of Power (a phrase that he coined). Austin in this extract is a portrait (more loosely based, as it appears, than most) of Rutherford, and Constantine is J. D. Bernal.*

I had been asked to attend the beginning of the final meeting, to help discuss some small matter of Constantine's. It was assumed that I should then withdraw for the last time. As I turned out of Piccadilly I was slightly nervous with the luxurious nervousness that one knows is soon to be substituted by delight; I should listen for a few minutes, I thought, go away for the afternoon, and hear the news at tea-time.

The moment I got inside the room I knew it was all wrong. Someone stopped speaking as I entered. I made some remark to Desmond, who was nearest to me. There was an instant of silence before he replied, in a forced staccato. I felt empty. Constantine was sitting alone, pale and miserable.

Austin coughed. 'This is extremely unfortunate, Miles.'

'I'm afraid I don't know –' I said.

'This latest work of yours,' Austin was heavily troubled, 'someone has told Pritt it won't hold water.'

I turned angrily on Pritt: 'Who?'

Pritt said:

'Archer. I was in his place yesterday. He's doing the same work. He said your stuff couldn't be right' – Pritt smiled – 'you must have forgotten something rather elementary. Either that or your results are different from his. Under the same conditions –'

I thought feverishly, trying to hope, inventing reasons, rejecting them. I tried to keep anything but mild concern from my face.

'This is rather sad,' I said. 'If it's true.'

'Can it be true?' asked Austin loudly.

'I'm not certain,' I said. 'I'll look into it later today.'

I saw Desmond catch Pritt's eye.

'It won't take long,' I added.

I sat through the first piece of business. It was one of the hardest things I had ever done. Constantine was speaking in a dulled voice. I could feel, I could not help but feel, the doubts, the pleasure, the regret, that came between the others and me: I knew what had been said, what would be said when I was gone. I longed to go, either to know the worst or prove myself right. In an hour or two I could look up the results, come back and say, very quietly: 'Professor Pritt's friend is wrong.' I stayed there, with my face as impassive as I could make it, keeping my hands from wandering.

Constantine stopped short. After they had voted, I got up and asked permission to withdraw. There was no friendly resonance in Austin's 'Yes.'

I took a taxi to the laboratory, went to my room, pulled out the records of the experiments, my notes, Jepp's films. I made false starts looking through the books. Thoughts kept branching jaggedly from one another, leading me on with trembling fingers. I spread out the films with quick, uncertain movements. I had to wipe my forehead to stop drops of sweat spoiling the films. Once I thought all was well. That's right – and then that – and, thank God, it all comes out perfectly. I knew I could not have made a mistake.

Then, quite suddenly, quite definitely, I realised I was working on a fact given me by Jepp. Which was wrong. Which he could not know was wrong, because there was a small technical point involved. Which I had looked over twenty times, but passed because of his assurance. Which if I had inspected it with a moment's care would have shouted itself as wrong. Upon that flaw, the whole structure rested.

I felt sick and giddy. But through despair a coolness of mind returned. I was able to work the whole thing out, how the mistake arose, how I had been led astray; on the other hand, I drew out the real inference from the experiments. I even wrote down an account of it. It was a queer exercise. For I knew I was broken, and I could not realise it.

Constantine burst into the room a few minutes afterwards.

'Who have you appointed?' I asked.

'We are meeting again at five,' he said.

He looked at me, appealingly.

'What about this –?'

'I was wrong,' I said. 'Quite wrong.'

'Oh God,' he sighed.

I explained it to him, showed him the description I had just written. Disconsolate on my account as he was, he still became interested in the new possibilities. 'You see that must mean –' he said. Then he remembered:

'Oh, why *couldn't* you be careful?' he cried. He looked helpless, strangely forlorn.

'Perhaps I can persuade them still,' he added. 'After all, one's allowed a mistake or two –'

'Not in circumstances like this,' I said.

'I must go back,' he said. He wanted to stay.

They appointed Tremlin. After Constantine returned to the meeting, I walked the streets until the time we had arranged for dinner. Slowly my numbness passed. I began to feel what had happened to me. All the time I was hoping, though there was no hope.

Constantine told me how the meeting went. He was dejected; we ate rapidly at dinner to hinder a conversation that was hurting both of us. For me, I could not tell whether it was more painful to hear the details of my rejection, or leave them unknown to be guessed at in moments of reproach. I stopped him brusquely in one part of the story; at another I pressed him to tell all. I did not know what I wanted. More than anything I thought I wanted to escape. So that I was never sure of all that happened.

Constantine went back, I gathered, with my news; he had to tell them that Pritt's friend was right, but he made the best show he could of explaining it away. It was a mistake, but a natural and venial mistake. He gave a lavish and complicated description, which was the best way of defending me. He did everything he could; it was a task he loathed, for which he had no gifts, but he did as well as anyone could have done. Finally, he proposed that they should give me the Directorship. 'This triviality,' he said, 'has not affected the position, which is simply that we have the best man we could hope for.' I fancy he was more eloquent than that.

Fane suavely intervened. He had always had doubts, he said, about the prudence of appointing as Director so young and – if he might suggest – so unpedestrian a man. (I could imagine that double-edged sneer.) Now, of course, it was quite impossible. It would cripple the Institute at its inception to appoint a Director whose work had an element of – airiness.

Desmond thought Fane was right, said how sorry he was for everyone, but felt how lucky it was that this came out before they appointed me. Though he himself had always thought they would be wise to make a safer choice. He told them an anecdote of how I had talked lightly at Munich of research being easy. He looked earnestly at the committee. 'That's not the right spirit,' he said. 'I felt it then. I didn't like it. And I must say I feel justified by results.'

That casual remark which I could dimly remember, thrown off when Desmond and I were drinking cheerfully together, seemed to incense them. More perhaps, than anything else I had said or done.

Pritt said that I might be able to talk, but I couldn't get down to good hard spade work. That a man who couldn't do honest spade work himself would simply turn out flashy stuff at the Institute that everyone would laugh at. That I was not a scientist at all; spent my time having holidays on bathing beaches; that I was unbalanced on other things besides science, and should alienate all the future benefactors of the Institute; that I was a charlatan, and the sooner I was got rid of the better.

(Even from Pritt those remarks hurt, hurt too much to hear in full, and I cut Constantine short.)

Constantine said angrily that for Pritt to give an opinion on my character and habits was an impertinence. He lost his temper, and Austin and Desmond tried to soothe him down.

Austin disapproved of the tone of Pritt's remarks, but was disappointed in me, felt that my future was not as certain to be brilliant as he once hoped, and had to withdraw the support he tentatively thought of giving. Austin was angry with me, of course, for his own sake; it was a rebuff to himself, he would no longer have a young man of his in charge, his patriarchate was shattered. And also he was genuinely fond of me, which made him more indignant still; I had disturbed his arrangements, hurt his scientific conventions, repudiated his affections all in one blow.

Constantine had no support, and my name was rejected. Pritt then proposed Tremlin – 'a sound man who won't let us down,' and he was elected after a perfunctory discussion, Constantine and Austin not voting.

'That's all,' Constantine said to me at the end of dinner, 'that's all.'

'It seems enough,' I said.

Constantine was frowning. He was puzzled as well as distressed.

'Does it make any real difference?' he said. 'Oh, I know it's disgusting, I know how you feel. But in practice, now, what difference does it make?'

I had a savage amusement at this practicalness.

He persisted: 'You'll still be able to do your research. Your present job – '

'It isn't much to the point,' I said. 'But, as a matter of fact, they think I'm going. It wouldn't be so easy to stay on –'

It would be impossible, I thought.

Constantine burst in:

'But we can soon settle that. They'll give you the assistant directorship at the Institute. It was mentioned – this evening.'

'Very gracious of them,' I said. 'A good assistant to Tremlin I should make.'

Constantine was hurt.

'Of course, it's annoying, but that job would keep you here and let you do your work in comfort. We could still get on to those ideas –'

'In comfort?' I said tiredly. I could not explain. I searched round for an excuse. 'I want money –'

Constantine smiled eagerly:

'You'd get about as much as assistant director as you do now, and if you want more – well, I can let you have any amount. What can I do with £1,200 a year? You can borrow £300, £400 a year for as long as you like.'

'Thanks,' I said. 'I'm grateful.' I tried to mean it.

'You see,' said Constantine, 'this affair won't be serious at all. It can't affect you in any way except make you angry, and that's only temporary. You won't be angry for very long. In a couple of years you'll have all the new ideas carried through, and everything will be completely unaffected as though this had never happened.'

He was talking fast, in protection from my bitterness. I tried to respond to him, but it was too difficult. He insisted:

'And so everything will adjust itself?'

'No doubt,' I said. 'Oh yes, Leo. Only it's been rather – sudden.'

Then Desmond came up to us; and I felt both maddened and relieved.

'I've just seen you,' he said affably. 'I've been eating alone.'

Constantine muttered something, looked embarrassed, and, after staying a moment whilst Desmond chatted, said that he must go. We watched him leave. Desmond gave me a glance, half-cheerful, half-furtive.

'I'll have to run away soon,' he said. 'But let's get down a drink or two first.'

We walked out to one of the bars near by, and stood in the middle of a crowd so noisy that Desmond's voice became a tone more strident:

'Wonderful days,' he shouted over his glass.

He saw my face set harshly. He said:

'I'm sorry I couldn't do anything for you this afternoon, Miles. Believe me, I would have done anything I could. But it was no use. You'd dropped too big a brick, you know.'

'Yes,' I said.

'Of course I should have overlooked it for myself. We're all human. I've dropped bricks myself. But the others wouldn't. You can't blame them altogether. They have to think of their duty to the Institute.'

'They've done their duty,' I said. 'With their present choice.'

'Tremlin?' Desmond laughed confidentially. 'Oh, Tremlin's a

dull dog. Most of them are dull dogs. But it doesn't matter. We can't have everything we want. We rub along somehow.' He looked at the clock. 'I must run for my train. These drinks are on me.' He gave me some money and rushed off. I remember that he did not give me quite enough.

The night was a long one. For hours I sat in my rooms trying to read. New forms of my humiliation kept rousing themselves. I should have to write to Sheriff; there was not much chance for him now; he and Audrey would talk of my disgrace together. I went to the desk to write the letter. I could not get beyond the first words, and I looked at them so long that they grew faint and I wondered if the light was failing. At last I left it. There were the notes of a new paper on the desk. It had been designed to give the Institute a good send off.

I slept in exhausted snatches, waking each time, it seemed, to an emptier world. When finally I woke up late none of the pain had gone; it was not an imaginary misery that vanished in the morning. With a sick nervousness, I rustled through the newspapers as soon as I was out of bed, but they were merciful. There were neat little paragraphs in one or two: 'Biophysical Institute Director appointed – Honour for young scientist. Yesterday the Executive Committee of the Institute for Biophysical Research elected a billiant young scientist, Dr. R. P. Tremlin, to be the Institute's first Director. Dr. Tremlin, who is now a Senior Lecturer at Birmingham University, has had a most distinguished career at Cambridge and London . . . he is thirty-seven years of age.'

214

Miles behaves chivalrously towards his assistant, Jepp; not so apparently the leader of the huge team of particle physicists in Nobel Dreams *(who was to win his Nobel Prize later, notwithstanding) towards his subordinate.*

Rubbia had asked Sulak to help him with the collider at CERN and, for personal reasons. Sulak had refused. When Sulak went to Rubbia with the idea of working together on a water detector for proton decay, Rubbia claimed he'd been thinking of the same thing. ('Carlo insists to this day,' Cline told me, 'that Sulak had nothing to do with it; the idea all was his.') And instead of working with Sulak, Rubbia chose to compete with him.

At this time Sulak was refused tenure at Harvard and turned to the University of Michigan for both tenure and support on the proton-decay experiment. Rubbia wrote a letter of 'dis-recommendation' for Sulak that has become renowned among Harvard physics alums. Sulak was shown the letter by one of the Michigan physicists. As he described it. 'Essentially everything that Carlo had done wrong in the previous eight years was attributed to me. Many of these things I didn't have anything to do with. It was explicit as to how I had screwed this up or screwed that up.' Sulak received tenure in spite of the letter, however, being helped considerably by recommendations from Glashow and Weinberg, and also by a telegram from Glashow suggesting that Michigan ignore one of the letters from Harvard since one of his colleagues 'might be mad.'

In December, 1978, Dave Cline organized a proton-decay workshop at the University of Wisconsin. After the interested parties had presented their papers, Cline published a volume of conference proceedings that included every talk except Sulak's. Sulak told me he would never forgive Cline for this. However, Cline later volunteered that it was Rubbia who had insisted that he withhold Sulak's paper. When I asked why, Cline said, 'Carlo has a very good smell for competition, and he probably felt that Sulak would be our competition.'

215

Eleazar Lipsky's novel, The Scientists, *appears to have been based on the discovery of the antibiotic, actinomycin, for which Selman Waksman received the Nobel Prize. A junior member of his research group then claimed that it was a perception of his own that led to the discovery, and that Waksman had contributed little to the work. The dispute culminated in a lawsuit (several times repeated since in relation to other Nobel Prizes). At that time such actions were still widely felt to reflect poorly on the dignity of the institutions in which they arose and on the academic calling generally.*

Mackenzie remained suspended in thought. 'Davey, I seem to recall your original paper, *Biocin, a New Bacterial-Resistance Inhibitory Substance.* Wasn't that it?'

'Oh, yes. That was eight years ago.'

'Now wasn't that signed with more than one name?'

'Yes.'

'Whose name appeared first?'

Luzzatto said slowly, 'Dr. Ullman's.'

'How did that happen?'

'I felt it was the fair thing. After all he'd been my chief at the time. I got a good deal of knowledge and guidance from him.'

Mackenzie's deep voice dropped a tone. 'One might take it from the order of names that Dr. Ullman was in fact the senior author.'

'Oh, no!' cried Luzzatto. 'I was.'

'Yet his name precedes yours?' Mackenzie demanded. 'Was that at his request?'

A tempting door had been opened. Luzzatto frowned and thought back and shook his head. 'No. It was at my insistence.'

'You were senior author –?'

'Yes.'

'And yet you put Ullman's name first?' Mackenzie's brows shot up. 'Isn't that surprising?'

'Not at all,' said Luzzatto warmly. 'As a matter of fact, I distinctly recall that we had a stiff argument. I might even be able to give you the date. It was a few weeks before Christmas recess. We met in front of Mackenzie Hall. Jo was there. Jo Ullman,' he added unnecessarily. 'She might remember because she'd brought her Chevvie to pick her father up and to get me to the barber. I wouldn't have paid attention except that Jo said I looked like a wet owl with feathers sticking out around my ears and she was so bossy about the whole thing that –' He paused. 'Now that I think, Jo made an entry in my diary, so it's still there, I guess, in her handwriting. In any case, we were all in the car. I told Victor, I mean, that is, Dr. Ullman that I'd got a call about the paper. It was a question of the credit line. Jo would remember.'

'A call? What about?'

'Oh? Well, you see, Frank Higby, who was editing the *Journal of Cytology* had gotten a note from Dr. Ullman to take his name off the paper. Didn't think he deserved the credit or something like that, and Dr. Higby wanted to know if the note was on the level. I told him I had put down Dr. Ullman's name and to let it ride. Well, Dr Ullman definitely stated to me that he didn't want his name to go on the paper.'

'Why not?'

Luzzatto hesitated for a moment – barely the skip of a heartbeat – and went on. 'At the time I thought it was modesty. Or something. Later I got to wonder whether or not he was offended with me.'

'Why?'

Luzzatto drew a breath. 'I simply don't know. I just assumed it was one of those unpredictable things and that he had his reasons. I told him, Nonsense! I owed him a lot. The fact was he helped to draft the paper. I was so damn fagged out at the time, I was practically neurasthenic. I think the records might show I'd been working steadily around the clock for months. I finally reached the point where I couldn't even take readings. I had double vision, spots, oh, the works. I had to rely on Will Tewksbury for that and a lot of things.

'I mean, when the time came I had to have help. I was mentally blocked and we were anxious to publish before we could be forestalled. Victor had a shrewd idea that something like biocin was being cooked up at McGurk and we couldn't stall.

'Anyhow, I had my notes but I couldn't seem to get off the ground. After ten versions, I gave up and went to Dr. Ullman. At that time I had no experience in writing for publication and his style was first-class. He was very nice about it and told me how to arrange for publication, with his permission. Anyway when he found I had put his name on the paper, he ordered me to get it off.'

'But you didn't?' said Mackenzie.

'Well, I flatly refused and I suppose I got stubborn and angry. I wouldn't vouch for my emotional state at that time. Jo thought we were both getting recriminatory and ridiculous, and finally he threw up his hands and told me I could do whatever I pleased. That's about it except that I was conscious all the time that – I don't want to sound precocious but, well – I felt that something pretty important was happening. I was in a state of exaltation. Not too responsible. Maybe I went overboard.'

'But the fact remains,' said Mackenzie caustically, 'that Ullman's name came first?'

Luzzatto paused with a sense of intolerable dryness in his throat. He had been rushing pell-mell against the old man's disbelieving frown and in the gush of tuumbling words he found control impossible. 'Yes. Dr. Ullman's name came first.'

'Yet you want us to understand that you alone were making these important decisions?'

'I didn't mean to give that impression.'

'Hump! Why would he give you all that authority? You've got to admit it's hard to swallow.'

'It was because he knew I was entirely and solely responsible for the discovery. You can ask anyone in the old crowd.'

'You give him no credit at all?'

Luzzatto hesitated and glanced about. 'Only in the most general way. Only for background support and the kind of help any teacher gives a student. As for the discovery *qua* discovery? He had absolutely nothing –' He paused.

'Nothing to do with it at all?' Mackenzie supplied.

'Exactly.'

'Can you possibly explain that state of affairs to the satisfaction of this group of intelligent men?'

In the long watchful silence Dean Polk's whistling breath was loud. Luzzatto closed his eyes and a vision of the old laboratory, mysterious and promethean, quickening in the dark of night, came unbidden. He was conscious of a deepening scowl, almost of dislike.

Mackenzie demanded harshly. 'Is the patent itself here?'

Seixas glanced up. 'Just a moment,' he said quietly. 'This is exactly what I didn't want, Dr. Foxx. I don't think this line of questioning is advisable.'

Foxx cut this short. 'Mr. Seixas,' he said, 'let Dean Mackenzie see the patent.'

The closely printed document was handed over to Mackenzie who held the paper to the light. Almost four years had passed, he observed, between date of application and final grant of letters patent. No drawings and six tables of data of which only the first four attracted his serious attention.

Four species of bacteria tested by gradient plating in the presence and absence of biocin had been classified in the order of resistance to a score of antibiotic drugs ranging from penicillin to the barbarously named nicotinaldehyde thiosemicarbazone. In the presence of biocin the antibiotic drugs were uniformly effective; in its absence an ominous phenomenon was noted:

'. . . the rapid development of resistant and possibly virulent populations of cells negates a measure of safety in any program of therapy where the rise of exceptional mutants must be taken into account. Dispersion of resistant and virulent mutants of pathogenic species among the human population must therefore eventually impair the usefulness of antibiotic drugs in current use. It is against this danger that biocin may be prescribed.'

He came next to countervailing language:

'The novelty of the present invention includes use of biocin and the process of preparing the same . . .'

The concluding claim was sublime:

'Claim 17. Biocin.'

Mackenzie meditated citations of prior patents and the literature which included, he noted with interest, an article of his own published in 1912 in the *Proceedings of the Society of Experimental Genetics and Medicine*. It baffled and disconcerted him that he had forgotten the paper entirely.

Seixas put a question. 'Professor Luzzatto, is biocin an antibiotic?'

Luzzatto turned. 'No, sir.'

'Why not?'

Luzzatto hesitated. 'It's not bacteriostatic.'

'Is it bacteriocidic?'

'Only in concentration,' said Luzzatto. 'But that's true of almost anything. It's a matter of definition.'

'What do you call it?'

'I've called it, um, a biochemical additive.'

'*Add*-itive?' Mackenzie broke in, looking sick. 'What the hell are ye? A gasoline chemist?' He snorted with disgust. 'Add-i-tive! Why?'

Luzzatto laughed suddenly. 'Because it's something you add, Dean,' he grinned impulsively, 'but I'll relieve your mind. At one time I intended to call the stuff a co-antibiotic.'

Mackenzie looked his horror. 'Ye didn't! Well, anything's better than that, Davey,' he conceded grudgingly. 'Additive, hey?' He tapped the paper. 'But on the clinical side, I gather it's being used like an antibiotic?'

'Not at all. It's got no lethal or inhibitive biological activity of its own. At least none that we know of yet.' Luzzatto turned to the others. 'But in conjunction with most of the known antibiotics, it precludes the emergence of organisms resistant to the major drug with which it's given. It has tremendous public health significance. It's pretty well unique because it creates no resistance to itself so far as we can detect. Its clinical value is to increase the safety of the antibiotic with which it's mixed. However, if you don't like the word additive,' he concluded restively, coming back to Mackenzie, 'you can call it anything you please. Really, though, Dean, I don't see where this is leading. I'm not adding to your knowledge, I'm sure.'

'Oh, but you are.' Mackenzie turned back to the letters patent and examined legend and signatures. 'Davey, you've explained how you came – I'm accepting your version for the moment – to put Ullman's name on your first publication of discovery. But, if he was not a discoverer with you, will you please now account for the fact that, one year later, the man's name appears with yours on the patent?'

Luzzatto looked about the table. 'Well, I don't know why you adopt that tone,' he said impatiently. 'I did that under legal advice.'

216

And, to end, Ogden Nash's vision of the universally incompetent man of science.

THE MIRACULOUS COUNTDOWN

Let me tell you of Dr. Faustus Foster.
Chloe was lost, but he was loster.
He was what the world for so long has missed,
A truly incompetent scientist.
His morals were good and his person cleanly,
He had skied at Peckett's and rowed at Henley.
The only liquor that touched his lips
He drew through pipettes with filter tips.
He could also recite, in his modest manner,
The second verse of The Star-Spangled Banner.
Yet, to his faults we must not be blinded;
He was ineluctably woolly-minded.
When his further deficiencies up are summed,
He was butter fingered and margarine thumbed.
You'd revoke the license of any rhymer
Who ranked him with Teller and Oppenheimer.

It took him, and here your belief I beg,
Twenty minutes to boil a three-minute egg,
Which will give you a hint as to what went on
Whenever he touched a cyclotron.
There wasn't a problem he feared to face,
From smashing atoms to conquering space,
And, should one of his theories expire,
He had other ions in the fire,
Even walking to work to save his carfare
For tackling bacteriological warfare.
For years he went to no end of bother
To explode this planet or reach another.
A more ambitious, industrious savant
You may have encountered; I know I haven't.
One Christmas Eve he was tired and irked,
He had shot the works and nothing worked.
'I'd sell my soul,' he cried to the night,
'To have one experiment come out right.'
No sooner said than his startled eyes
Saw a ghostly stranger materialize,
Who, refraining from legalistic jargon,
Announced, 'You have got yourself a bargain.
Here's a pact with iron-clad guarantees;
Sign here, in the usual fluid, please.'
Faustus disdained to quibble or linger,
He merely remarked, as he pricked his finger,
'It had better be good, you *quid pro quo*;
My blood is especially fine type O.'
(Always in character, come what may
He was down in his doctor's records as A.)
A snicker was heard from the stranger weird,
Then he snatched the parchment and disappeared.
Faustus was filled with wild surmise.
And roseate dreams of the Nobel Prize,
Now certain to drop in his lap with awful ease,
He thought, with the aid of Mephistopheles.
Behold him now in his laboratory,
A modern Merlin, hell-bent for glory.
With a flourish worthy of the Lunts
He triggered every project at once.

Intercontinental ballistic missiles
Blasted the air with roars and whistles,
Rockets punctured the midnight clear,
And the atmosphere and the stratosphere.
Before the human eye could absorb it
A giant satellite entered orbit.
With the germ's equivalent of a howl
The bacteria issued forth to prowl.
Faustus shouted with joy hysterical,
And was then struck dumb as he watched a miracle.
He gazed aghast at his handiwork
As every experiment went berserk.
The bacteria, freed from their mother mold,
Settled down to cure the common cold.
Distant islanders sang Hosanna
As nuclear fall-out turned to manna.
Rockets, missiles and satellite
Formed a flaming legend across the night.
From Cape Canaveral clear to the Isthmus
The monsters spelled out Merry Christmas,
Penitent monsters whose fiery breath
Was rich with hope instead of death.
Faustus, the clumsiest of men,
Had butter-fingered a job again.
I've told you his head was far from level;
He thought he had sold his soul to the devil,
When he'd really sold it, for heaven's sake,
To his guardian angel by mistake.
When geniuses all in every nation
Hasten us towards obliteration,
Perhaps it will take the dolts and geese
To drag us backward into peace.

References

1. William Cooper, *The Struggles of Albert Woods*. London: Jonathan Cape, 1952.

2. Stephen Hall, *Invisible Frontiers: the race to synthesize a human gene*. New York: Atlantic Monthly Press, 1987; London: Sidgwick & Jackson, 1988.

3. Ibid.

4. Gregory Benford, *Timescape*. London: Gollancz; New York: Simon and Schuster, 1980.

5. Stephen Hall, *Invisible Frontiers: the race to synthesize a human gene*. New York: Atlantic Monthly Press, 1987; London: Sidgwick & Jackson, 1988.

6. Robert Frost, 'A Wish to Comply', from *The Poetry of Robert Frost*, edited by Edward Connery Latham. New York: Holt, Rinehart & Winston, 1969; London: Jonathan Cape, 1971.

7. James Thurber, *My Life and Hard Times*. London and New York: Harper & Bros., 1933.

8. Gary Taubes, *Nobel Dreams: power, deceit and the ultimate experiment*. New York: Random House, 1986.

9. James Clerk Maxwell, 'Rigid Body Sings', from Part III of *The Life of James Clerk Maxwell*, edited by Lewis Campbell and William Garnett. London: Macmillan and Co., 1882.

10. Nicholas Wade, *The Nobel Prize Duel*. Garden City, NY: Anchor Press/Doubleday, 1981.

11. A. M. Sullivan, 'Atomic Architecture', from *Stars and Atoms Have No Size*. New York: Dutton, 1946.

12. François Jacob, *La statue intérieure*. Paris: Editions Odile Jacob, 1987. This translation is by Walter Gratzer. The book has now been translated as *The Statue Within*, translated by Franklin Philip. London: Unwin Hyman; New York: Basic Books, 1988.

13. Jeremy Bernstein, *The Life it Brings*. New York: Ticknor & Fields, 1987.

14. Mitchell Wilson, *Live with Lightning*. Boston: Little, Brown, 1949; London: W. H. Allen, 1950.

15. Nigel Balchin, *A Sort of Traitors*. London: Collins, 1949.

16. Louis MacNeice, from *The Kingdom*, from *Collected Poems*. London: Faber & Faber, 1966.

17. Arthur Koestler, *The Sleepwalkers: a history of man's changing vision of the universe*. London: Hutchinson; New York: Macmillan, 1959.

18. Julian Huxley, 'Cosmic Death', from *The Captive Shrew and other poems of a biologist*. Oxford: Basil Blackwell, 1932; New York: Harper & Bros., 1933.

19. Michael Frayn, *The Tin Men*. London: Collins; Boston: Little, Brown, 1965.

20. Jonathan Swift, *Gulliver's Travels*. 1726 and modern editions.

21. Samuel Butler, *Erewhon*. 1872 and modern editions.

22. Gustave Flaubert, *Bouvard and Pécuchet*, translated by A. J. Krailsheimer (from *Bouvard et Pécuchet*, published posthumously, 1881). Harmondsworth: Penguin Books, 1976, 1978.

23. H. G. Wells, *Love and Mr Lewisham*. London and New York: Harper & Bros., 1900.

24. Paul de Kruif, *The Sweeping Wind*. London: Rupert Hart-Davis, 1962.

25. Sinclair Lewis, *Arrowsmith*. London: Jonathan Cape; New York: Harcourt Brace, 1925.

26. Paul de Kruif, *The Sweeping Wind*. London: Rupert Hart-Davis, 1962.

27. Miroslav Holub, 'Suffering', translated by George Theiner, from *Selected Poems*, translated by Ian Milner & George Theiner, with an introduction by A. Alvarez. Harmondsworth: Penguin Books, 1967.

28. Stanislas Ulam, *Adventures of a Mathematician*. New York: Charles Scribner's Sons, 1983.

29. E. C. Large, *Sugar in the Air. A Romance*. London: Jonathan Cape; New York: Charles Scribner's Sons, 1937.

30. John Updike, *Couples*. London: André Deutsch; New York: Alfred Knopf, 1968.

31. Alex Comfort, *Come out to Play*. London: Eyre & Spottiswoode, 1961.

32. Vicki Baum, *The Weeping Wood*. New York: Doubleday, 1943; London: Michael Joseph, 1945.

33. Roald Hoffmann, 'Next Slide Please' from *The Metamict State*. Orlando: University of Central Florida Press, 1987.

34. V. N. Ipatieff, *The Life of a Chemist*, edited by Xenia Jonkoff Endin, Helen Dwight Fisher, Harold H. Fisher; translated by Vladimir Haensel & Mrs Ralph H. Lusher. London: Oxford University Press; Stanford: Stanford University Press, 1946.

35. J. J. Thomson, *Recollections and Reflections*. London: G. Bell & Sons, 1936; New York: Macmillan, 1937.

36. Arthur Hailey, *Strong Medicine*. London: Michael Joseph in association with Souvenir Press; New York: Doubleday, 1984.

37. Ralph A. Lewin, 'Elks, Whelks and Their Ilk', from *The Biology of Algae and diverse other verses*. Pacific Grove, California: Boxwood Press, 1987.

38. E. C. Large, *Asleep in the Afternoon. A Romance*. London, Jonathan Cape, 1938; New York: H. Holt, 1939.

39. R. V. Jones, *Most Secret War*. London: Hamish Hamilton, 1978.

40. G. Y. Craig & E. J. Jones, editors, *A Geological Miscellany*. Oxford: Orbital Press, 1982. Princeton: Princeton University Press, 1985.

41. Anon., 'Epitaph on a Geologist', from *The Penguin Book of Comic and Curious Verse*, collected by J. M. Cohen. Harmondsworth: Penguin Books, 1952.

42. Marianne Moore, 'Four Quartz Crystal Clocks', from *Complete Poems*. New York: Macmillan, Viking, 1981; London: Faber & Faber, 1984.

43. Ronald Clark, *J. B. S.: the life and work of J. B. S. Haldane*. London: Hodder & Stoughton, 1968; New York: Oxford University Press, 1984.

44. Aldous Huxley, *Antic Hay*. London: Chatto & Windus; New York: Harper & Bros., 1923.

45. Ibid.

46. Ibid.

47. P. B. Medawar, *Memoirs of a Thinking Radish*. Oxford and New York: Oxford University Press, 1986.

48. Aldous Huxley, *Point Counter Point*. London: Chatto & Windus; New York: Harper & Bros., 1928.

49. Nicholas Wade, *The Nobel Prize Duel*. Garden City, NY: Anchor Press/Doubleday, 1981.

50. Daniil Granin, *The Geneticist: the life of Nikolai Timofeyev-Ressovsky, known as Ur*. Published in *Novy Mir* (USSR), 1987. Translated here by Walter Gratzer.

51. Olga Metchnikoff, *Life of Elie Metchnikoff 1845–1916*, with a preface by Sir Ray Lankester. London: Constable; Boston: Houghton Mifflin, 1921.

52. Paul de Kruif, *The Sweeping Wind*. London: Rupert Hart-Davis, 1962.

53. W. H. Mallock, *The New Republic: or culture, faith and philosophy in an English country house*. 1877.

54. Ibid.

55. Charles Darwin, *The Voyage of the Beagle*. 1839 and modern editions.

56. H. G. Wells, *Ann Veronica: a modern love story*. London: T. Fisher Unwin; New York: Harper & Bros., 1909.

57. T. H. Huxley, 'Autobiography', in *From Handel to Hallé: biographical sketches with autobiographies of Professor Huxley and Professor Herkomer*, edited by Louis Engel. 1890. See also, Charles Darwin and Thomas Henry Huxley, *Autobiographies*, edited with an introduction by Gavin de Beer. London and New York: Oxford University Press, 1974.

58. H. G. Wells, *Experiment in Autobiography*. London: Gollancz; New York: Macmillan, 1934.

59. Samuel Butler, from *Hudibras*. 1663. A longer extract may be found in *The Penguin Book of Comic and Curious Verse*, collected by J. M. Cohen. Harmondsworth: Penguin Books, 1952.

60. Arthur Conan Doyle, *The Poison Belt: being an account of another amazing adventure of Professor Challenger*. London & New York: Hodder & Stoughton, 1913.

61. Hans Krebs, *Otto Warburg: cell physiologist, biochemist and eccentric*, in collaboration with Roswitha Schmid, translated by Hans Krebs & Anne Martin. Oxford and New York: Clarendon Press, 1981.

62. Kurt Mendelssohn, *The World of Walther Nernst: the rise and fall of German science*. London: Macmillan; Pittsburgh: University of Pittsburgh Press, 1973.

63. Lionel Davidson, *The Sun Chemist*. London: Jonathan Cape; New York: Alfred A. Knopf, 1976.

64. Chaim Weizmann, *Trial and Error*. London: Hamish Hamilton; New York: Harper & Bros., 1949.

65. William Wordsworth, from *The Prelude*, Book III, line 61 ff. 1850 and modern editions.

66. Alexander Pope, 'Epitaph, Intended for Sir Isaac Newton in Westminster Abbey.' 1730.

J. C. Squire, from *Poems in One Volume*. London: Heinemann, 1926.

67. George Bernard Shaw, *In Good King Charles's Golden Days*. London: Constable, 1939.

68. Thomas Moore, *Memoirs, Journal and Correspondence 1853–56*. See also, *Thomas Moore's Journal*, edited by Peter Quennell. London: Batsford; New York: Macmillan, 1964.

George Gordon, Lord Byron, from *Don Juan*, Canto X. 1822 and modern editions.

69. James Clerk Maxwell, 'To the Chief Musician Upon Nabla', from Part III of *The Life of James Clerk Maxwell*, edited by Lewis Campbell and William Garnett. London: Macmillan and Co., 1882.

70. M. L. Oliphant, from *Notes and Records of the Royal Society*, eds R. V. Jones & W. D. M. Paton. Vol. 27, no. 1., August 1972.

71. C. P. Snow, *Variety of Men*. London: Macmillan; New York: Charles Scribner's Sons, 1967.

72. C. P. Snow, *Science and Government*. Cambridge (Mass): Harvard University Press; London: Oxford University Press, 1961.

73. Roy Harrod, *The Prof. A personal memoir of Lord Cherwell*. London: Macmillan, 1959.

74. Ronald Clark, *Einstein: the life and times*. London: Hodder & Stoughton, 1973; New York: Avon, 1979.

75. John Updike, 'Cosmic Gall', from *Telephone Poles and other poems*. New York: Alfred A. Knopf, 1963; London: André Deutsch, 1964.

76. Jeremy Bernstein, *Experiencing Science*. New York: Basic Books, 1978; London: André Deutsch, 1979.

77. Jeremy Bernstein, *The Life it Brings: one physicist's beginnings*. New York: Ticknor & Fields, 1987.

78. Ibid.

79. Richard Feynman, *Surely You're Joking, Mr Feynman! Adventures of a curious character*, as told to Ralph Leighton, edited by Edward Hutchings. New York: W. W. Norton, 1985; London: Unwin paperbacks, 1986.

80. Jeremy Bernstein, *The Life it Brings: one physicist's beginnings*. New York: Ticknor & Fields, 1987.

81. Ibid.

82. W. N. P. Barbellion, *The Journal of a Disappointed Man*. London: Chatto & Windus, 1919.

83. Jeremy Bernstein, *The Life it Brings: one physicist's beginnings*. New York: Ticknor & Fields, 1987.

84. H. G. B. Casimir, *Haphazard Reality: half a century of science*. London and New York: Harper & Row, 1983.

85. Gary Taubes, *Nobel Dreams: power, deceit and the ultimate experiment*. New York: Random House, 1986.

86. Ibid.

87. Nuel Pharr Davis, *Lawrence and Oppenheimer*. London: Jonathan Cape, 1969; New York: Da Capo Press, 1986.

88. Richard Blackmore, from *The Creation: a philosophical poem*. 1712.

89. C. P. Snow, *Variety of Men*. London, Macmillan; New York: Charles Scribner's Sons, 1967.

90. J. E. Littlewood, *A Mathematician's Miscellany*. London: Methuen, 1953; 1957.

Stanislas Ulam, *Adventures of a Mathematician*. New York: Charles Scribner's Sons, 1983.

91. Marc Kac, *Enigmas of Chance: an autobiography*. New York: Harper & Row, 1985.

92. Leopold Infeld, *Quest: the evolution of a scientist*. London: Gollancz; New York: Doubleday, Doran, 1941.

93. Rudolf Peierls, *Bird of Passage: recollections of a physicist*. Princeton: Princeton University Press, 1985.

Jeremy Bernstein, *Experiencing Science*. New York: Basic Books, 1978; London, André Deutsch, 1979.

94. Charles Darwin, *Autobiography*, in *Life and Letters of Charles Darwin*, edited by Francis Darwin. 1887. See also, Charles Darwin & Thomas Henry Huxley, *Autobiographies*, edited with an introduction by Gavin de Beer. London & New York: Oxford University Press, 1974.

Joel Shurkin, *Engines of the Mind: a history of the computer*. New York: W. W. Norton, 1984.

95. Charles Babbage, *Passages from the Life of a Philosopher*. 1864.

96. Logan Pearsall Smith, *Unforgotten Years. Reminiscences*. London, Constable, 1938; Boston: Little, Brown, 1939.

97. Robert Frost, 'Why Wait for Science', from *The Poetry of Robert Frost*, edited by Edward Connery Latham. New York: Holt, Rinehart & Winston, 1969; London: Jonathan Cape, 1971.

98. Charles Babbage, *Passages from the Life of a Philosopher*. 1864.

99. François Jacob, *La Statue intérieure*. Paris: Editions Odile Jacob, 1987. This translation is by Walter Gratzer, but the book has now been translated as *The Statue Within: an autobiography* by Franklin Philip. London: Unwin Hyman; New York: Basic Books, 1988.

100. Sinclair Lewis, *Arrowsmith*. London: Jonathan Cape; New York: Harcourt Brace, 1925.

101. Richard Feynman, *Surely You're Joking, Mr Feynman! adventures of a curious character*, as told to Ralph Leighton, edited by Edward Hutchings. New York: W. W. Norton, 1984; London: Unwin paperbacks, 1986.

102. Richard Feynman, *What do you care what other people think?: further adventures of a curious character*, with Ralph Leighton. New York: W. W. Norton, 1988; London: Unwin Hyman, 1989.

103. A. D. Godley, 'The Megalopsychiad', from *Fifty Poems by A. D. Godley*, edited by C. L. Graves & C. R. L. Fletcher. London: Humphrey Milford, 1927.

104. Isaac Asimov, 'Pâté de foie gras', from *The Edge of Tomorrow*. New York: Tor Books, 1985; London: Harrap, 1986.

105. Arthur Conan Doyle, 'A Physiologist's Wife', from *Tales of Adventure and Medical Life*. London: John Murray, 1924.

106. Edgar Allan Poe, 'Von Kempelen and His Discovery', in his *Works* (1850), vol. I, and modern editions.

107. Edgar Allan Poe, 'To Science', from *Al Aaraaf, Tamerlane and Minor Poems*. 1829 and modern editions.

108. J. B. S. Haldane, 'The Gold-Makers', from *The Inequality of Man*. London: Chatto & Windus, 1932.

109. Edgar Allan Poe, 'The Thousand-and-Second Tale of Scheherazade', in his *Works* (1850), vol. I, and modern editions.

110. Julian Huxley, 'The Tissue-Culture King', in *The Cornhill Magazine*, vol. 60, April 1926; *The Yale Review*, n.s. 15, April 1926. See also *Great Science Fiction by Scientists*, edited by Groff Conklin. New York: Collier Books, 1962.

111. Gwyn MacFarlane, *Alexander Fleming: the man and the myth*. London: Chatto & Windus; Cambridge (Mass): Harvard University Press, 1984.

112. George Bernard Shaw, *The Doctor's Dilemma*. London: Constable, 1906; New York: Brentano's, 1911.

113. Ibid.

114. Leo Szilard, 'Report on Grand Central Terminal', from *The Voice of the Dolphins and other stories*. London: Gollancz; New York: Simon & Schuster, 1961.

115. Robert Jungk, *Brighter than a Thousand Suns: the moral and political history of the atomic scientists*, translated by James Cleugh. London: Gollancz, Rupert Hart-Davis; New York: Harcourt Brace, 1958.

116. John Updike, 'To Crystallization', from *Facing Nature. Poems*. London: André Deutsch; New York: Alfred A. Knopf, 1986.

117. Dan Greenberg, *The Politics of Pure Science*. Harmondsworth: Penguin Books, 1969. As *The Politics of American Science*. New York: New American Library, 1968.

118. William Cooper, *Memoirs of a New Man: a novel*. London: Macmillan, 1966.

119. Aldous Huxley, *After Many a Summer*. London: Chatto & Windus; New York: Harper & Bros., 1939.

120. Arthur Shipley, 'Ere You Were Queen of Sheba', from *Life. A book for elementary students*. Cambridge: Cambridge University Press, 1923.

121. H. G. Wells, *The Food of the Gods and how it came to earth*. London: Macmillan; New York: Charles Scribner's Sons, 1904.

122. Karel Capek, *War with the Newts*, translated by M. & R. Weatherall. London: George Allen & Unwin; New York: G. P. Putnam's Sons, 1937.

123. Leo Szilard, 'The Voice of the Dolphins', from *The Voice of the Dolphins and other stories*. London: Gollancz; New York: Simon & Schuster, 1961.

124. Vicki Baum, *Helene*. London: G. Bles, 1932; New York: Doubleday, 1933.

125. Arthur Hailey, *Strong Medicine*. London: Michael Joseph in association with Souvenir Press; New York: Doubleday, 1984.

126. Gregory Benford, *Timescape*. London: Gollancz; New York: Simon & Schuster, 1980.

127. Flann O'Brien, *The Third Policeman*. London: MacGibbon & Kee, 1967; New York: Walker, 1968.

128. P. B. Medawar, *Memoirs of a Thinking Radish*. Oxford and New York: Oxford University Press, 1986.

129. Ronald Clark, *Einstein: the life and times*. London: Hodder & Stoughton, 1973; New York: Oxford University Press, 1984.

130. J. J. Thomson, *Recollections and Reflections*. London: G. Bell & Sons, 1936; New York: Macmillan, 1937.

131. H. G. Wells, *Experiment in Autobiography*. London: Gollancz; New York: Macmillan, 1934.

132. John Updike, *Roger's Version*. London: André Deutsch; New York: Alfred A. Knopf, 1986.

133. William Blake, 'Mock on, Mock on . . .' from Notebook Poems c. 1800–1806.

134. Leopold Infeld, *Quest: the evolution of a scientist*. London: Gollancz; New York: Doubleday, Doran, 1941.

135. Jeremy Bernstein, *The Life it Brings: one physicist's beginnings*. New York: Ticknor & Fields, 1987.

136. Ronald Clark, *Einstein: the life and times*. London: Hodder & Stoughton, 1973; New York: Oxford University Press, 1984.

137. Stephen Hall, *Invisible Frontiers: the race to synthesize a human*

gene. New York: Atlantic Monthly Press, 1987; London: Sidgwick & Jackson, 1988.

138. Kurt Vonnegut, *Cat's Cradle*. London: Gollancz; New York: Holt, Rinehard & Winston, 1963.

139. Kurt Vonnegut, 'Address to the American Physical Society', from *Wampeters, Foma & Granfalloons (Opinions)*. New York: Delacorte Press/Seymour Lawrence, 1974; London: Jonathan Cape, 1975.

140. Fred Hoyle, *The Black Cloud*. London: Heinemann; New York: Harper, 1957.

141. Edgar Allan Poe, *Eureka: an essay on the material and spiritual universe,* in *Works* (1850), vol. II.

142. Michael Frayn, *The Tin Men*. London: Collins; Boston: Little, Brown, 1965.

143. Ben Jonson, *The Alchemist*. 1610 and modern editions.

144. E. C. Large, *Sugar in the Air. A Romance*. London: Jonathan Cape; New York; Charles Scribner's Sons, 1937.

145. Primo Levi, *The Periodic Table*, translated by Raymond Rosenthal, (from *Il sistema periodico*, 1975). New York: Schocken Books, 1984; London: Michael Joseph, 1985.

146. Robert Graves, stanza 33 from *The Marmosite's Miscellany*. London: the Hogarth Press, 1924.

147. Mitchell Wilson, *Live with Lightning*. Boston: Little, Brown, 1949; London: W. H. Allen, 1950.

148. Ibid.

149. Nuel Pharr Davis, *Lawrence and Oppenheimer*. London: Jonathan Cape, 1969; New York: Da Capo Press, 1986.

150. Michael Roberts, 'Note on θ, ϕ and ψ' from *Selected Poems and Prose*, edited by Frederick Grubb. Manchester: Carcanet Press, 1979.

151. Gregory Benford, *Timescape*. London: Gollancz; New York: Simon & Schuster, 1980.

152. Paul Preuss, *Broken Symmetries*. New York: Timescape, 1983; Harmondsworth: Penguin Books, 1984.

153. Gary Taubes, *Nobel Dreams: power, deceit and the ultimate experiment*. New York: Random House, 1986.

154. John Banville, *Kepler: a novel*. London: Secker & Warburg, 1981; Boston (Mass): David Godine, 1983.

155. Benjamin Franklin, *Autobiography*. London: 1793; USA: 1818 and modern editions.

156. Stephen Hall, *Invisible Frontiers: the race to synthesize a human gene*. New York: Atlantic Monthly Press, 1987; London: Sidgwick & Jackson, 1988.

157. Miroslav Holub, 'Evening in a Lab', translated by Dana Hábová & Stuart Friebert, from *Sagittal Section*. Oberlin, OH: Oberlin College Press, 1980.

158. Olga Metchnikoff, *Life of Elie Metchnikoff 1845–1916*, with a

preface by Sir Ray Lankester. London: Constable; Boston: Houghton Mifflin, 1921.

159. Georges Duhamel, *St John's Eve*, translated by Béatrice de Holthoir (from *La Nuit de la Saint-Jean*, 1935; number four of the ten *Chronique des Pasquier*). London: Howard Baker, 1970.

160. H. G. Wells, 'The Moth', from *Tales of the Unexpected.* London: Collins, 1922.

161. John Masefield, *Multitude and Solitude.* London: Grant Richards, 1907; New York: Mitchell Kennerly, 1911.

162. Ibid.

163. Ronald Ross, 'The Anniversary', from *Poems.* London: E. Mathews & Marrot, 1928.

164. Alex Comfort, *Come out to Play*. London: Eyre & Spottiswoode, 1961.

165. H. G. Wells, *The Invisible Man. A grotesque romance.* 1897 and modern editions.

166. J. D. Watson, *The Double Helix: a personal account of the discovery of the structure of DNA.* London: Weidenfeld & Nicolson; New York: Atheneum, 1968.

167. Michael Bliss, *The Discovery of Insulin.* Chicago: University of Chicago Press, 1982; Edinburgh: Harris, 1983.

168. Stephen Hall, *Invisible Frontiers: the race to synthesize a human gene.* New York: Atlantic Monthly Press, 1987; London: Sidgwick & Jackson, 1988.

169. Alexander Todd, *A Time to Remember: the autobiography of a chemist.* Cambridge: Cambridge University Press, 1983; New York: Cambridge University Press, 1984.

170. John Updike, 'Ode to Growth', from *Facing Nature. Poems.* London: André Deutsch; New York: Alfred A. Knopf, 1986.

171. Olga Metchnikoff, *Life of Elie Metchnikoff 1845–1916*, with a preface by Sir Ray Lankester. London: Constable; Boston: Houghton Mifflin, 1921.

172. Nicholas Wade, *The Nobel Prize Duel.* Garden City, NY: Anchor Press/Doubleday, 1981.

173. André Malraux, *The Walnut Trees of Altenburg*, translated by A. W. Fielding (from *Les Noyers de l'Altenburg*, 1948). London: John Lehmann, 1952.

174. Chaim Weizmann, *Trial and Error.* London: Hamish Hamilton; New York: Harper & Bros., 1949.

175. Vicki Baum, *The Weeping Wood.* New York: Doubleday, 1943; London: Michael Joseph, 1945.

176. Daniel Kevles, *The Physicists: the history of a scientific community in modern America.* New York: Alfred A. Knopf, 1977.

177. Otto Frisch, *What Little I Remember.* Cambridge: Cambridge University Press, 1979. New York: Cambridge University Press, 1980.

178. C. P. Snow, *Science and Government.* Cambridge (Mass): Harvard University Press; London: Oxford University Press, 1961.

179. Nigel Balchin, *The Small Back Room.* London: Collins, 1943; Boston: Houghton Mifflin, 1945.

180. C. P. Snow, *Science and Government.* Cambridge (Mass): Harvard University Press; London: Oxford University Press, 1961.

181. R. V. Jones, *Most Secret War.* London: Hamish Hamilton, 1978.

182. Otto Frisch, *What Little I Remember.* Cambridge: Cambridge University Press, 1979. New York: Cambridge University Press, 1980.

183. Sam Goudsmit, *Alsos.* London: Sigma Books; New York: H. Schuman, 1947.

184. Friedrich Dürrenmatt, *Die Physiker.* Zurich, 1962. The translation here is by Walter Gratzer. See Dürrenmatt, *Four Plays*, translated by Gerhard Nelhaus et al. London: Jonathan Cape, 1964; New York: Grove Press, 1965.

185. Ibid.

186. Nuel Pharr Davis, *Lawrence and Oppenheimer.* London: Jonathan Cape, 1969; New York: Da Capo Press, 1986.

187. Emile Zola, *Paris.* 1898. Translated by E. A. Vizetelly, 1898.

188. George Gamow, *My World Line.* New York: Viking, 1970.

189. Marc Kac, *Enigmas of Chance: an autobiography.* New York: Harper & Row, 1985.

190. Primo Levi, *If this is a man*, translated by Stuart Woolf (from *Se questo e un uomo*, 1958). Harmondsworth: Penguin Books, 1979.

191. June Goodfield, *Courier to Peking.* London: Hart-Davis, MacGibbon; New York: Dutton, 1973.

192. Martin Gardner, *Fads and Fallacies in the Name of Science.* 2nd edition. New York: Dover, 1957.

193. Denis Hamilton, *The Monkey Gland Affair.* London: Chatto & Windus, 1986.

194. Martin Gardner, *Fads and Fallacies in the Name of Science.* 2nd edition. New York: Dover, 1957.

195. Edmund Gosse, *Father and Son: a study of two temperaments.* London: Heinemann; New York: Charles Scribner's Sons, 1907.

196. Gustave Flaubert, *Bouvard and Pécuchet*, translated by A. J. Krailshamer (from *Bouvard et Pécuchet*, published posthumously, 1881). Harmondsworth: Penguin Books, 1976, 1978.

197. Thomas Hardy, 'Epitaph for G. K. Chesterton', from *Complete Poems*, edited by James Gibson. London: Macmillan, 1976.

198. Mark Twain, *Life on the Mississippi.* 1883 and modern editions.

199. Thomas McMahon, 'Bell and Langley', in *Granta*, issue 16, *Science.* Cambridge: Granta Publications; New York: Viking Penguin, 1985.

200. V. Dudintsev, *Not by Bread Alone*, translated by Dr Edith Bone. London: Hutchinson; New York: Dutton, 1957.

201. Zhores Medvedev, *The Rise and Fall of T. D. Lysenko*, translated by I. Michael Lerner. New York & London: Columbia University Press, 1969.

202. Daniil Granin, *The Geneticist: the life of Timofeyev-Ressovsky, known as Ur*. Published in *Novy Mir* (USSR), 1987. This translation by Walter Gratzer.

203. Mark Azbel, *Refusenik: trapped in the Soviet Union*. Boston (Mass): Houghton Mifflin, 1981; London: Hamish Hamilton, 1982.

204. Martin Gardner, *Science: Good, Bad and Bogus*. Buffalo, NY: Prometheus Books, 1981; Oxford: Oxford University Press, 1983.

205. C. P. Snow, *The Affair*. London: Macmillan; New York: Charles Scribner's Sons, 1960.

206. William Broad & Nicholas Wade, *Betrayers of the Truth*. London: Century; New York: Simon & Schuster, 1983.

207. Ronald Ross, *Memoirs: with a full account of the great malaria problem and its solution*. London: John Murray, 1923.

208. William Broad & Nicholas Wade, *Betrayers of the Truth*. London: Century; New York: Simon & Schuster, 1983.

209. Primo Levi, *The Periodic Table*. translated by Raymond Rosenthal (from *Il sistema periodico*, 1975). New York: Schocken Books, 1984; London: Michael Joseph, 1985.

210. George Bernard Shaw, *The Doctor's Dilemma*. London: Constable, 1906; New York: Brentano's, 1911.

211. Arthur Hailey, *Strong Medicine*. London: Michael Joseph in association with Souvenir Press; New York: Doubleday, 1984.

212. Gary Taubes, *Nobel Dreams: power, deceit and the ultimate experiment*. New York: Random House, 1986.

213. C. P. Snow, *The Search*. London: Gollancz, 1934; Indianapolis: Bobbs-Merrill, 1935.

214. Gary Taubes, *Nobel Dreams: power, deceit and the ultimate experiment*. New York: Random House, 1986.

215. Eleazar Lipsky, *The Scientists*. London: Longmans; New York: Appleton-Century-Crofts, 1959.

216. Ogden Nash, 'The Miraculous Countdown', from *Everyone but Thee, Me and Thee*. Boston: Little, Brown, 1962.

Index of Names and Institutions

References are to extract numbers. Bold type indicates author entries.

Adler, Friedrich 136
Alsabti, E. A. 208
Andrade, E. N. da C. 68
Anon **41**
Appleton, E. V. 180
Arimura, Akira 172
Asimov, Isaac **104**
Auden, W. H. p.xvi
Azbel, Mark **203**

Babbage, Charles 94, **95, 98**
Balchin, Nigel **15, 179**
Balfour, A. J. 63
Banach, Stefan 28, 91
Banting, Frederick 167
Banville, John **154**
Barbellion, W. N. P. **82**
Baum, Vicki **32, 124, 175**
Beerbohm, Max p.xviii
Behring, Emil 124
Bell, Alexander Graham 199
Bell Telephone Laboratories 42
Belloc, Hilaire **p.xvi**
Benford, Gregory **4, 126, 151**
Bernstein, Jeremy **13, 76, 77, 78, 80, 81, 83, 93, 135**
Berthelot, Marcellin 187
Black, Joseph p.xvii
Blackett, Patrick 178, 180
Blackmore, Richard **88**
Blake, William 97, **133**
Bliss, Michael **167**
Block, R. 78
Bodenstein, Max 62
Bohr, Niels 74, 83, 84, 87, 115, 177
Bonhoeffer, Karl 62
Bordet, Jules 52
Bosch, Carl 39
Boswell, James pp.xvi–xvii
Bowers, C. Y. 10
Bowra, C. M. 181
Bragg, Lawrence 166
Brahe, Tycho 17, 88, 154
Broad, William **206, 208**
Brod, Max p.xix, 154
Brougham, Henry **p.xvii**
Budaev, Professor 34
Butler, Samuel (*Erewhon*) **21**
Butler, Samuel (*Hudibras*) **59**
Byron, George Gordon, Lord **68**

Californian Institute of Technology (Caltech) 78, 87
Capek, Karel **122**
Carlyle, Thomas 94
Carrel, Alexis 26
Casimir, H. G. B. **84**
Cavendish laboratory 71, 72, 166
CERN 8, 85, 153, 212
Charles II 67
Chaucer, Geoffrey **p.xvii**
Chesterton, G. K. 197
Churchill, Winston 63, 64, 180
Clark, Ronald **43, 74, 129, 136**
Cockcroft, John 70, 71
Coleridge, Samuel Taylor pp.xviii–xix

Collip, J. B. 167
Comfort, Alex **31,** 47, 113, **164**
Connolly, Cyril p.xix
Cooper, William p.xviii, **1, 118**
Copernicus, Nicolas 88, 154
Craig, G. Y. **40**
Crick, Francis 166

Daghlian, Harry 182
Darwin, Charles 22, **55,** 58, **94,** 195
Davidson, Lionel **63**
Davidson, Randall 130
Davis, Nuel Pharr **87, 149, 186**
Davis, Robert 43
Davy, Humphry p.xix, 106
De Gaulle, Charles 99
De Kleine, William 100
De Kruif, Paul **24, 26, 52**
De Moivre, Abraham 90
Denegri, Daniel 8
Dickens, Charles 22
Dirac, P. A. M. 78, 92
Donald, Kenneth 43
Doyle, Arthur Conan **60, 105**
Dudintsev, V. **200**
Dürrenmatt, Friedrich **184, 185**
Duhamel, Georges **159**

Edgeworth, David 40
Efstratiadis, Argiris 3, 5, 168
Ehrenfest, Paul 83
Einstein, Albert p.xix, 26, 28, 66, 71, 73, 74, 76, 130, 131, 136, 192

Faraday, Michael 71, 98
Feynman, Richard 78, **79, 101, 102,** 126, 151
Flaubert, Gustave **22, 196**
Fleming, Alexander 111
Folkers, Karl 10
Forster, E. M. 93
Franklin, Benjamin **155**
Franklin, Rosalind 166
Frayn, Michael **19, 142**
Frisch, Otto 115, **177, 182**
Frost, Robert **6, 97**
Fuller, Forrest 5
Funk, Casimir 119
Furry, Wendell 135

Gamow, George **188**
Gardner, Martin **192, 194, 204**
Garfield, James 199
Gawehn, Karlfried 61
Gell-Mann, Murray 78, 151
Geller, Uri 204
Gerlach, Walther 183
Gilbert, Walter 2, 3, 5, 137, 156, 168
Godley, A. D. **103**
Goeddel, David 156
Goodfield, June **191**
Goodman, Howard 2
Gosse, Edmund **195**
Gosse, Phillip 195
Goudsmit, Sam **183**
Granin, Daniil **50, 202**
Graves, Robert **146**
Greenaway, Peter 168
Greenberg, Dan **117**
Guillemin, R. 10, 49, 172

Haber, Fritz 173, 174
Hahn, Otto 115, 183
Hailey, Arthur **36, 125, 211**
Haldane, J. B. S. p.xviii, 43, 44, 47, **108**
Haldane, J. S. 43, 48
Haldane, R. B. 130, 131
Hall, Stephen **2, 3, 5, 137, 156, 168**
Hamilton, Denis **193**
Hardy, G. H. 71, 89
Hardy, Thomas **197**
Harrod, Roy **73**
Harvard University 3, 5, 135, 137
Heisenberg, Werner 83, 92
Hetper, Wladek 189
Hill, Archibald 180
Hill, Leonard 43

Hinshelwood, C. N. 93
Hitler, Adolf 115, 181, 183
Hoffmann, Roald **33**
Holub, Miroslav **27, 157**
Howes, T. G. B. 56, 58
Hoyle, Fred **140**
Huxley, Aldous p.xviii, **44, 45, 46, 48, 119**
Huxley, Julian p.xviii, **18, 110**
Huxley, T. H. 53, 55, **57,** 58

Imperial Institute (Kensington) 23, 56
Infeld, Leopold **93, 134**
Institut Pasteur 24, 51, 52, 62, 99
Ipatieff, V. N. **34**

Jacob, François **12, 99**
Jeffery, Francis 33
Johnson, Samuel pp.xvi–xvii, 71
Joliot-Curie, Irène 115
Jones, E. J. **40**
Jones, R. V. **39, 181**
Jonson, Ben **143**
Joseph, H. W. B. 73
Jowett, Benjamin 96
Jungk, Robert **115**

Kac, Marc **91, 189**
Kaiser Wilheim Institute 115, 174
Kekulé, August 146
Kepler, Johann p.xix, 88, 141, 154
Kevles, Daniel **176**
Kingsley, Charles 195
Kipling, Rudyard **p.xv**
Kleid, Denis 156
Koestler, Arthur **17**
Krebs, Hans **61**

Lamarck, Jean Baptiste 55
Landau, Lev Davidovich 83
Lane, William Arbuthnot 113
Lang, Cosmo Gordon 71
Langmuir, Irving 139
Lankester, Ray 60
Laputa 20
Large, E. C. **29, 38, 144**
Lawrence, E. O. 149
Lederman, Leon 5
Le Prince Ringuet, Louis 13, 78
Levi, Primo p.xviii, 30, **145, 190, 209**
Lévy, Maurice 13, 78
Lewin, Ralph **37**
Lewis, G. N. 149
Lewis, Sinclair p.xviii, **25, 100**
Lindemann, Frederick (Lord Cherwell) 72, 73, 178, 189
Lipsky, Eleazar **215**
Littlewood, J. E. 89, 90
Locci, Elizabeth 8
Loeb, Jacques 24, 26
Los Alamos 182, 186
Lwoff, André 12
Lyell, Charles 195
Lysenki, Trofim Denisovitch 50, 200, 201

McCarthy, Joseph 135, 186
Macfarlane, Gwyn **111**
Macleod, J. R. R. 167
McMahon, Thomas **199**
MacNeice, Louis **16**
Mallock, W. H. **53, 54**
Malraux, André **173**
Markov, A. A. 90
Masefield, John **161, 162**
Matsuo, Keizo 49, 172
Matveyev, Boris Stepanovich 50
Mawson, Douglas 40
Max-Planck Institutes 61
Maxwell, James Clerk **9, 69**
Medawar, P. B. **47, 128,** 191
Medvedev, Zhores **201**
Meitner, Lise 115
Mencken, H. L. p.xix, 199
Mendelssohn, Kurt **62**
Mencken, H. L. p.xix
Metchnikoff, Elie 51, 52, 119, 158, 159, 171
Metchnikoff, Olga **51, 158, 171**

Michel, Louis 13, 78
Mikhail Military Academy 34
Millikan, R. A. 6, 87, 88
Moore, Marianne **42**
Moore, Thomas **68**

NASA 102
Nash, Ogden **216**
Nernst, Walther 62
Newton, Isaac 65, 66, 67, 68, 92, 109 n.4, 192, 203
Nightingale, Florence 112
Noguchi, Hideyo 24
Nollet, Abbé 155
Novy, Frederick G. 24

O'Brien, Flann **127**
Oliphant, M. L. **70,** 71, 176
Oppenheimer, Robert 80, 81, 87, 135, 184, 186

Pais, Abraham 74
Pannwitz, Doctor 190
Pauli, Wolfgang 83, 84
Pauling, Linus 166
Peierls, Rudolf 93, 178
Pepinsky, Blake 206
Perl, Martin 85, 153
Planck, Max 115
Poe, Edgar Allan **106, 107, 109, 141**
Pope, Alexander **66,** 68
Porton Down 168, 169
Povzner, Professor 203
Preuss, Paul **152**
Price, Melvin 117
Ptashne, Mark 137

Racker, Efraim 206
Ramanujan, S. 89, 90
Reich, Wilhelm 194
Reynolds, Osborne 35
Richardson, Lewis Fry **155**
Rideal, E. K. 1
Roberts, Michael **150**
Robinson, Robert 1
Rockefeller Institute 24, 26, 52
Rosenfeld, L. 84
Ross, Ronald **163, 207**
Rous, Peyton 24, 26
Roux, Emile 24, 51
Royal Society 1, 67, 121
Rubbia, Carlo 8, 85, 86, 153, 212, 214
Rutherford, Ernest 70, 71, 72, 213
Rutter, William 2

Schally, A. 10, 49, 172
Schopper, Herwig 212
Schwinger, Julian 76, 77
Senkevich, Lev Alexandrovich 50
Shaw, George Bernard **67,** 111, **112, 113,** 192, **210**
Shelley, Percy Bysshe p.xviii
Shipley, Arthur **120**
Shurkin, Joel **94**
Singer, Maxine 137
Slotin, Louis 182
Smith, Logan Pearsall **96**
Smith, Sydney 33
Snow, C. P. p.xvii, p.xviii, 1, **71, 72, 89,** 118, **178, 180, 205, 213**
Spector, Mark 206
Squire, J. C. **66**
Steinhaus, Hugo 91
Strassmann, Fritz 115
Sulak, Lawrence 214
Sullivan, A. M. **11**
Swift, Jonathan **20**
Szilard, Leo 39, **114,** 115, **123**

Tamm, Igor 188
Taubes, Gary **8, 85, 86, 153, 212, 214**
Taylor, J. G. 204
Teller, Edward 186
Tennyson, Alfred 94
Thomson, J. J. **35, 130**
Thurber, James **7**

Timofeyev-Ressovsky, Nikolai 50, 202
Tizard, Henry 72, 178
Todd, Alexander (Lord) 1, **169**
Touch, Gerald 39
Trilling, Lionel p.xvi
Twain, Mark 155, **198**
Tyndall, John 53, 57, 69

UCSF 2
Ulam, Stanislas **28, 90,** 186
Ullrich, Axel 2
University College, London 43, 47
Updike, John p.xviii, **30, 75, 116, 132, 170**

Vellucci, Alfred 137
Vignal, Jean 13
Villa-Komaroff, Lydia 168
Vogt, Volker 206
Von Laue, Max 183
Vonnegut, Kurt **138, 139**
Voronoff, Serge 26, 119, 193

Wade, Nicholas **10, 49, 172, 206, 208**
Waksmann, Selman 215
Warburg, Otto 61
Watson, J. D. **166**
Weizmann, Chaim 63, **64, 174**
Wells, H. G. p.xviii, **23, 56, 58,** 113, **121, 131,** 139, **160, 165**
Wheeler, John 126, 151
Whewell, William **68**
Wiener, Norbert 90
Wilkins, Maurice 166
Wilson, Mitchell **14, 147, 148**
Wolfe, Tom p.xx
Wordsworth, William **65**
Wright, Almroth 111, 113

Zola, Emile **187**

Index of Titles

'Address to the American Physical Society' 139
Adventures of a Mathematician 28, 90
The Affair 205
After Many a Summer 119
The Alchemist 143
Alsos 183
Ann Veronica 56
'The Anniversary' 163
Antic Hay 44, 45, 46
Arrowsmith p.xviii, 25, 100
Asleep in the Afternoon 38
'Atomic Architecture' 11
Autobiography (Darwin) 55
Autobiography (Franklin) 155
'Autobiography' (Huxley) 57

'Bell and Langley' 199
Betrayers of the Truth 206, 208
'Big whirls have little whirls . . .' 155
Bird of Passage 93
The Black Cloud 140
Bouvard and Pécuchet 22, 196
Brighter than a Thousand Suns 115
Broken Symmetries 152

The Canon's Yeoman's Tale p.xvii
Cat's Cradle 138
Come out to Play 31, 164
'Cosmic Death' 18
'Cosmic Gall' 75
Couples 30
Courier to Peking 191
The Creation 88

The Discovery of Insulin 167
The Doctor's Dilemma 112, 113, 210
Don Juan 68
The Double Helix 166

Einstein 74, 129, 136
'Elks, Whelks and Their Ilk' 37
Engines of the Mind 94
Enigmas of Chance 91, 189
'Epitaph for G. K. Chesterton' 197
'Epitaph. Intended for Sir Isaac Newton . . .' 66
'Epitaph on a Geologist' 41
'Ere You Were Queen of Sheba' 120
Erewhon 21
Eureka 106
'Evening in a Lab' 157
Experiencing Science 76, 93
Experiment in Autobiography 58, 131

Fads and Fallacies in the Name of Science 192, 194
Father and Son 195
Alexander Fleming 111
The Food of the Gods 121
'Four Quartz Crystal Clocks' 42

The Geneticist 50, 202
A Geological Miscellany 40
'The Gold-Makers' 108
Gulliver's Travels 20

Haphazard Reality 84
Helene 124

Hudibras 59

If this is a man 190
In Good King Charles's Golden Days 67
Invisible Frontiers 2, 3, 5, 137, 156, 168
The Invisible Man 165
'It did not last . . .' 66

J.B.S. 43
The Journal of a Disappointed Man 82
Journal of Thomas Moore 68

Kepler 154
The Kingdom 16

Lawrence and Oppenheimer 87, 149, 186
The Life it Brings 13, 77, 78, 80, 81, 83, 135
The Life of a Chemist 34
Life of Elie Metchnikoff 1845–1916 51, 158, 171
Life on the Mississippi 198
Littlewood's Miscellany 90
Live with Lightning 14, 147, 148
Love and Mr Lewisham 23

The Marmosite's Miscellany 146
A Mathematician's Miscellany 90
'The Megalopsychiad' 103
Memoirs (Ross) 163
Memoirs of a New Man 118
Memoirs of a Thinking Radish 47, 128
'The Miraculous Countdown' 216
'The Microbe' p.xvi
'Mock on, mock on . . .' 133
The Monkey Gland Affair 193
Most Secret War 39, 181
'The Moth' 160
Multitude and Solitude 161, 162
My Life and Hard Times 7
My World Line 188

The New Republic 53, 54
Newton and the Science of his Age 68
'Next Slide Please' 33
Nobel Dreams 8, 85, 86, 153, 212, 214
The Nobel Prize Duel 10, 49, 172
Not By Bread Alone 200
'Note on θ, φ and ψ' 150
Notes and Records of the Royal Society 70

'Ode to Growth' 170

Paris 187
Passages from the Life of a Philosopher 95, 98
'Pâté de foie gras' 104
The Periodic Table p.xviii, 145, 209
The Physicist 184, 185
The Physicists 176
'A Physiologist's Wife' 105
Point Counter Point 48
The Poison Belt 60
The Politics of Pure Science 117
The Prelude 65
The Prof. 73

Quest 92, 134

Recollections and Reflections 35, 130
The Redemption of Tycho Brahe p.xix
Refusenik 203
'Report on Grand Central Terminal' 114
'Rigid Body Sings' 9
The Rise and Fall of T. D. Lysenko 201
Roger's Version 132

St John's Eve 159
Science and Government 72, 178, 180
Science: Good, Bad and Bogus 204
The Scientists 215
The Search 213
The Sleepwalkers 17
The Small Back Room 179

A Sort of Traitors 15
La Statue intérieure 12, 99
Strong Medicine 36, 125, 211
The Struggles of Albert Woods 1
'Suffering' 27
Sugar in the Air 29, 144
The Sun Chemist 63
Surely You're Joking, Mr Feynman! 79, 101
The Sweeping Wind 24, 26, 52

The Third Policeman 127
'The Thousand-and-Second Tale of Scheherezade' 109
A Time to Remember 169
Timescape 4, 126, 151
The Tin Men 19, 142
'The Tissue-Culture King' 110
'To Crystallization' 166
'To the Chief Musician upon Nabla' 69
'To Science' 107
Trial and Error 64, 174

Unforgotten Years 96

Variety of Men 71, 89
'The Voice of the Dolphins' 123
'Von Kempelen and His Discovery' 106
The Voyage of the Beagle 55

The Walnut Trees of Altenburg 173
War with the Newts 122
Otto Warburg 61
The Weeping Wood 32, 175
What do you care what other people think? 102
What Little I Remember 177, 182
'Why Wait for Science' 97
'A Wish to Comply' 6
The World of Walther Nernst 62